WHITE DWARFS AS PROBES OF FUNDAMENTAL PHYSICS:
TRACERS OF PLANETARY, STELLAR AND GALACTIC EVOLUTION

IAU SYMPOSIUM 357

COVER ILLUSTRATION:

Rainbow over Hilo bay, Big Island, Hawai'i, viewed from the Grand Naniloa hotel (credit: Rachel Barstow).

IAU SYMPOSIUM PROCEEDINGS SERIES

INTERNATIONAL ASTRONOMICAL UNION

UNION ASTRONOMIQUE INTERNATIONALE

International Astronomical Union

IAU

WHITE DWARFS AS PROBES OF FUNDAMENTAL PHYSICS: TRACERS OF PLANETARY, STELLAR AND GALACTIC EVOLUTION

PROCEEDINGS OF THE 357th SYMPOSIUM OF THE INTERNATIONAL ASTRONOMICAL UNION HELD IN HILO, HAWAII, USA 21–25 OCTOBER, 2019

Edited by

MARTIN A. BARSTOW
University of Leicester, United Kingdom

SCOT J. KLEINMAN
Gemini Observatory, USA

JUDITH L. PROVENCAL
University of Delaware, USA

and

LILIA FERRARIO
The Australian National University, Australia

CAMBRIDGE
UNIVERSITY PRESS

CAMBRIDGE UNIVERSITY PRESS
University Printing House, Cambridge CB2 8BS, United Kingdom
1 Liberty Plaza, Floor 20, New York, NY 10006, USA
10 Stamford Road, Oakleigh, Melbourne 3166, Australia

First published 2020

Printed in the UK by Bell & Bain, Glasgow, UK

Typeset in System LaTeX 2ε

A catalogue record for this book is available from the British Library Library of Congress Cataloguing in Publication data

This journal issue has been printed on FSC™-certified paper and cover board. FSC is an independent, non-governmental, not-for-profit organization established to promote the responsible management of the world's forests. Please see www.fsc.org for information.

ISBN 9781108492027 hardback
ISSN 1743-9213

Table of Contents

Foreword

IAU Symposium 357 - White Dwarfs as probes of fundamental physics and tracers of planetary, stellar and galactic evolution – was held in October 21st to 25th 2019, at the Grand Naniloa hotel in Hilo on the Big Island of Hawai'i, USA. At the time of writing this foreword, approximately 5 months later, the World is in the middle of the COVID-19 crisis. Many communities are confined to their homes, with many of us, working remotely. It is sobering to realise how our usual privilege and freedom to travel the World to meet fellow scientists can be brought to a halt in such a short space of time. It also brings into focus discussions we had in Hawai'i about how to organise remote meetings to reduce costs and the travel-related carbon footprint of our community.

Hawai'i is among the most remote and beautiful places on the planet. This makes it a wonderful location for a scientific meeting. Participants feel apart from the rest of the world and the effort of travel encourages long-duration visits. I am very grateful to the Local Organising Committee for an exceedingly well-organised, fruitful and enjoyable meeting. Thanks also go to my co-chairs and other members of the Scientific Organising Committee for helping with the original symposium application and devising a vigorous scientific programme. All the organisers and participants are grateful to the IAU for selecting the symposium and supporting the attendance of early career scientists. We would also like to thank the Association of Universities for Research in Astronomy (AURA), the Royal Astronomical Society (RAS) and the University of Leicester for financial support for the meeting organisation and travel. The image chosen for the cover was fortuitously recorded by my wife Rachel at the opening of the meeting, when the whole audience was completely distracted from my opening remarks by that stunning rainbow – impossible to compete with... but also a fitting and encouraging start to the symposium.

White dwarfs are the most numerous members of the stellar graveyard. Over 90 percent of all stars currently on the main sequence will end their lives as white dwarfs. As such, they are important laboratories for fundamental studies of the evolution of stars, the formation and history of the Milky Way Galaxy and of planetary systems. Furthermore, white dwarfs give us crucial insights on the behavior of matter at extreme temperatures and densities. Surveys such as SDSS, SPY and ELM have given us access to an unprecedented wealth of information on the white dwarf population. Recent studies incorporating these databases have initiated a revolution in our understanding of its global properties that will continue to grow with the *Gaia* data releases and upcoming LSST results.

Once a white dwarf is formed, its evolution is only dominated by cooling. As white dwarfs cool over billions of years, determinations of the age of the oldest and therefore coolest white dwarfs place limits on the ages of the components of the Galaxy, such as the thin disk, and the thick disks, the halo, and the system of open and globular clusters. The characteristics, such as temperature and mass, of the white dwarf population contain invaluable information on the star formation history of the Galaxy.

White dwarfs are also extremely important indicators for cosmology. Type Ia supernovae are the standard candles that allow us to study the acceleration history of cosmic expansion. However, although it is crucial to identify the progenitor systems, the evolutionary paths leading to these explosions are still poorly understood. Recent surveys have begun to reveal the properties of single and double degenerate progenitors, but the picture is still very unclear and more work is needed.

In the past few years, white dwarfs have also begun to influence our understanding of the evolution of planetary systems. We have strong evidence that some white dwarfs harbour planets. We now know that white dwarfs can disrupt terrestrial planets, asteroids and other minor bodies and the resulting debris is accreted onto the white dwarf. White dwarfs have a unique atmospheric characteristic. The high surface gravity ($\log g \sim 8$) naturally leads to chemically pure hydrogen or helium photospheres. This means that the spectral features produced by the accreted material are not contaminated by original abundances. The observed features provide a unique opportunity to study the bulk composition of extrasolar planetary material. A subset of accreting white dwarfs contains spectral features of highly ionized heavy elements. Furthermore, accurate measurements of observed wavelengths can be compared with laboratory measurements to probe the possible variation of the fine structure constant in a strong gravitational field.

White dwarf research is fascinating in its own right, since it requires developments in atomic data and the study of properties of matter under extreme conditions. However, the impact that these studies have on other areas of astrophysics is also enormous. Thus, the time was ripe to bring together experts from different branches of science so that they could share their knowledge and provide feedback to each other.

The Symposium was highly interdisciplinary, bringing together not just astronomers working on white dwarfs, but also astronomers with expertise in a wide range of relevant disciplines. Such a gathering presented an opportunity to formulate the direction of white dwarf studies for the next decade.

The programme consisted of sessions organized around a number of key themes: SN Ia progenitors, debris from extrasolar planetary systems, fundamental physics, precision studies of white dwarf structure, stellar physics and galactic evolution. Each session included one or two invited keynote talks plus a number of contributed papers. Time was set aside for extensive discussion following the sessions associated with each them. These were moderated by members of the SOC, posing a number of questions of the audience to stimulate the discussion. The nature of such discussions makes them hard to record in detail, but a number of key points have been extracted and incorporated into a short concluding paper in these proceedings.

Martin Barstow – 31st March 2020

Editors

Martin A. Barstow
University of Leicester, United Kingdom

Scot J. Kleinman
Gemini Observatory, USA

Judith L. Provencal
University of Delaware, USA

Lilia Ferrario
The Australian National University, Australia

Organising Committee

Scientific Organising Committee

SOC Chairs

Martin Barstow	University of Leicester, United Kingdom
Barbara Castanheira-Endl	Baylor University/UT Austin, USA
Lilia Ferrario	Australian National University, Australia
S.O. Kepler	Universidade Federale do Rio Grande do Sul, Brazil

SOC Members

Pierre Bergeron	University of Montreal, Canada
Zhanwen Han	Yunnan Observatory, China
Daniel Maoz	Tel Aviv University, Israel
Jayant Murthy	Indian Institute of Astrophysics, India
Judi Provencal	University of Delaware, USA
Lydia Tchang-Brillet	Observatoire de Paris/Sorbonne Université, France
Siyi Xu	Gemini Observatory, USA
G. C. Anupama	Indian Institute of Astrophysics, India
Shazreen Mohamed	South African Astronomical Observatory, South Africa

Local Organising Committee

LOC Chairs

Siyi Xu	Gemini Observatory, USA
Atsuko Nitta	Gemini Observatory, USA
Scot Kleinman	Gemini Observatory, USA
Sandy Leggett	Gemini Observatory, USA

LOC Members

Sarah Casewell	University of Leicester, United Kingdom
Chris Stark	Gemini Observatory, USA
Terry Lee	Gemini Observatory, USA
Erik Dennihy	Gemini Observatory, USA
Trent Dupuy	Gemini Observatory, USA
Peter Michaud	Gemini Observatory, USA
Ben Shappee	Institute for Astronomy, Manoa, USA
Andre-Nicolas Chene	Gemini Observatory, USA

Participants

First Name	Last Name	Affiliations
Carlos	Badenes	University of Pittsburgh
Andrzej	Baran	Cracow Pedagogical University
Joseph	Barnett	Texas A&M University-Commerce
Martin	Barstow	University of Leicester
Evan	Bauer	UC Santa Barbara
Antoine	Bédard	Universite de Montreal
Keaton	Bell	University of Texas, Austin
Presanta	Bera	University of Southampton
Simon	Blouin	Universite de Montreal
Kevin	Burdge	California Institute of Technology
Barbara	Castanheira	Baylor College
Stephane	Charpinet	Observatoire Midi-Pyrenees
André-Nicholas	Chené	Gemini Observatory
Sihao	Cheng	Johns Hopkins University
Patricia	Cho	University of Texas, Austin
Aleksandar	Cikota	Lawrence Berkeley National Laboratory
Matthew	Coleman	Institute for Advanced Study
Alejandro Hugo	Córsico	Universidad Nacional de La Plata
Elena	Cukanovaite	University of Warwick
Tim	Cunningham	University of Warwick
Francisco	De Geronimo	Instituto de Astrofisica de La Plata-CONICET
Jacqueline	den Hartogh	Konkoly Observatory
Erik	Dennihy	Gemini Observatory
Alexandra	Doyle	UC Los Angeles
Patrick	Dufour	Universite de Montreal
Bart	Dunlap	University of Texas, Austin
Trent	Dupuy	Gemini Observatory
Nick	Fantin	University of Victoria
Lilia	Ferrario	Australian National University
Nicolle	Finch	University of Leicester
Nicola	Gentile Fusillo	University of Warwick
James	Green	University of Colorado
Marcin	Hajduk	University of Warmia and Mazury
Na'ama	Hallakoun	Weizmann Institute of Science
Gerald	Handler	Nicolas Copernicus Astronomical Centre
François	Hardy	Universite de Montreal
Tyler	Heintz	Boston University
JJ	Hermes	Boston University
Kenneth	Hinkle	National Optical Astronomy Observatory
Mark	Hollands	University of Warwick
Jordi	Isern	Institut de Ciencies de l'Espai
Elizabeth	Jeffery	Cal Poly
Surajit	Kalita	Indian Institute of Science
Steven	Kawaler	Iowa State University
Adela	Kawka	Curtin University
Bhusan	Kayastha	NAOC
S. O.	Kepler	Universidade Federal do Rio Grande do Sul
Agnes	Kim	Pennsylvania State University
Markus	Kissler-Patig	European Space Agency
Scot	Kleinman	Gemini Observatory
Jerzy	Krzesinski	Astronomical Observatory of the Jagiellonian University
Nadège	Lagarde	Institut UTINAM
Susana	Landau	IFIBA-CONICET-UBA
Chien-Hsiu	Lee	National Optical Astronomy Observatory

First Name	Last Name	Affiliations
Sandy	Leggett	Gemini Observatory
Lisa	Löbling	Universitat Tubingen
Christopher	Manser	University of Warwick
Nadine	Manset	Canada France Hawaii Telescope
Paola	Marigo	University of Padova Vicolo
Kento	Masuda	Princeton University
Lucy	McNeill	Monash University
Pavana	Muralimohan	Indian Institute of Astrophysics
Atsuko	Nitta	Gemini Observatory
Terry	Oswalt	Embry-Riddle Aeronautical University
Thomas	Prince	California Institute of Technology
Judith	Provencal	Unversity of Delaware
Thomas	Rauch	Universitat Tubingen
Harvey	Richer	University of British Columbia
Laura	Rogers	Cambridge University
Ashley	Ruiter	University of New South Wales
Didier	Saumon	Los Alamos National Laboratory
Ben	Shappee	University of Hawaii
Henry	Shipman	Unversity of Delaware
Paulina	Sowicka	Nicolaus Copernicus Astronomical Centre
Valery	Suleymanov	Universitat Tubingen
Paula	Szkody	University of Washington
Lydia	Tchang-Brillet	Observatiore de Paris, Meudon
Patrick	Tremblay	Universite de Montreal
Michael	Tucker	University of Hawaii
Zach	Vanderbosch	University of Texas, Austin
Eva	Villaver	Universidad Autónoma de Madrid
Olivier	Vincent	Universite de Montreal
Ted	von Hippel	Embry-Riddle Aeronautical University
Kurtis	Williams	Texas A&M University-Commerce
Emily	Wilson	Rochester Institute of Technology
David	Wilson	McDonald Observatory
Don	Winget	University of Texas, Austin
Matt	Wood	Texas A&M University-Commerce
Siyi	Xu	Gemini Observatory
Yossef	Zenati	Technion Israel

White Dwarfs as Probes of Fundamental Physics: Tracers of Planetary, Stellar and Galactic Evolution
Proceedings IAU Symposium No. 357, 2019
M. A. Barstow, S. J. Kleinman, J. L. Provencal & L. Ferrario, eds.
doi:10.1017/S1743921320000587

Type Ia supernova sub-classes and progenitor origin

Ashley J. Ruiter

School of Science, University of New South Wales Canberra
The Australian Defence Force Academy, 2600 ACT, Canberra, Australia
email: ashley.ruiter@adfa.edu.au

Abstract. This paper presents a short review on the current state of SN Ia progenitor origin. Type Ia supernova explosions (meaning thermonuclear disruption of a white dwarf) are observed to be widely diverse in peak luminosity, lightcurve width and shape, spectral features, and host stellar population environment. In the last decade alone, theoretical simulations and observational data have come together to seriously challenge the long-standing paradigm that all SNe Ia arise from explosions of Chandrasekhar mass white dwarfs. In this review I highlight some of the major developments (and changing views) of our understanding of the nature of SN Ia progenitor systems. I give a brief overview of binary star configurations and their plausible explosion mechanisms, and infer links between some of the various (observationally-categorized) SN Ia sub-classes and their progenitor origins from a theoretical standpoint.

Keywords. supernovae, white dwarfs, binaries: close, stars: evolution

---

1. Introduction

It is accepted that Type Ia supernovae (SNe Ia) – or thermonuclear supernovae – are explosions of white dwarf (WD) stars in interacting binaries. When we talk about supernovae, SNe Ia are sometimes left out of the conversation since their progenitor population is poorly understood, unlike the case of core-collapse supernovae which are clearly understood to arise from the gravitational collapse of the cores in massive stars. For SNe Ia it is still unclear **(1)** what type of companion star is donating matter to the WD and how this mass transfer proceeds (red giant or WD etc.; thermal timescale or dynamical timescale etc.), and **(2)** what the WD mass (density) is when it explodes. This second point is most important for understanding what the explosion mechanism(s) is (are) in SNe Ia (e.g. Fisher *et al.* 2019 and references therein). Understanding point (1) is necessary for knowing what ages of stellar populations give rise to SNe Ia and, characteristically, what observational properties the progenitor will exhibit. For progenitors with non-degenerate companions it might be possible to observe the binary system through pre-explosion observations (McCully *et al.* 2014, see also Liu *et al.* 2015), and it has been argued that in some cases, an excess flux would be produced soon after explosion due to SN ejecta-companion interaction (Bianco *et al.* 2011). On the other hand, with double degenerate progenitors it is challenging to detect such binaries without space-based gravitational wave observatories such as the future Laser Interferometer Space Antenna (LISA). With binary star evolution modelling it is possible to make predictions about which types of binary star systems are more likely to produce WDs that could explode as SNe Ia (birthrates), and what their characteristic ages (or 'delay times') will be. Point (2), which needs to be investigated with sophisticated burning and explosion simulations, is crucial for informing which scenarios from part (1) are realistic, and for predicting observables such as ejecta velocities, abundance stratification, and nucleosynthetic yields (Thielemann *et al.* 1986; Hillebrandt *et al.* 2013).

Despite a lack of a cohesive explanation for their progenitor origin, the role SNe Ia play in iron-group element nucleosynthesis, in driving a detailed understanding of interacting stars, and in determining cosmological parameters, is unequivocal. It has long been observed that the rate at which SN Ia lightcurves rose and declined was connected to the peak luminosity of the explosion (Arnett 1979; Phillips 1993), which itself was linked to the amount of ^{56}Ni synthesized in the SN Ia (Pankey 1962). Though their explosion physics remains unclear, this information has nonetheless empowered SNe Ia to become our most useful objects for studying the expansion history (and future) of our Universe, and by extension, the nature of dark energy, which makes up most of the energy density of our Universe (Nobel prize in physics, 2011). SNe Ia are also responsible for more than half of the iron in our own solar neighbourhood (Maoz & Graur 2017), mostly from the nickel synthesized in the explosion through the radioactive decay chain ^{56}Ni $\rightarrow$ ^{56}Co $\rightarrow$ ^{56}Fe. While very heavy elements (e.g. heavier than atomic number $\sim$40) are primarily synthesized in low-mass stars during the late stages of evolution or compact object mergers, elements around the iron-peak (atomic numbers $\sim$22 – 30, titanium to zinc) are created in supernovae; SNe Ia being dominant per event over core-collapse SNe.

Since it is difficult to disentangle which progenitor scenario led to any hitherto observed SN Ia explosion,† and delay time distributions are intrinsically difficult to derive observationally especially for complex star formation histories, another useful constraint on SN Ia progenitors is nucleosynthetic yields. Manganese is an interesting element in this context because it is synthesized at high (white dwarf) densities, thus requiring at least some Chandrasekhar (or close to Chandrasekhar, which is $\sim$1.4 $M_\odot$) mass WDs to have exploded as SNe Ia in the history of our Galaxy (Seitenzahl *et al.* 2013). Otherwise, the solar abundance of manganese cannot easily be explained. Using different analysis methods, various authors have argued there is a requirement for a combination of Chandrasekhar mass and sub-Chandrasekhar mass progenitors, though the relative fractions do not usually agree (e.g. Scalzo *et al.* 2014, Flörs *et al.* 2019, see also Maguire *et al.* 2018).

1.1. *Diversity: What are the SN Ia progenitors?*

Unlike core-collapse supernovae which are confined to exploding among young stellar populations, SNe Ia may explode as early as $\sim$40 Myr up to (and beyond) a Hubble time after starburst. Figure 1 illustrates that while a large fraction of events thought to be of thermonuclear origin follow the Phillips relation (black curve; the empirical relationship between SN lightcurve width and peak luminosity that makes SNe Ia standardizable candles), a growing number (over 30%) of sources do not (see also recent review by Jha *et al.* 2019). The term 'normal' SN Ia is usually adopted to refer to a SN that tends to obey the Phillips relation and is therefore useful for cosmological studies; the SNe that do not follow this relationship are generally not used in such calculations (see e.g. D'Andrea *et al.* 2018).

It has become very apparent that SNe Ia are a rather diverse group of objects, contrary to what was assumed over a decade ago. While it is becoming an accepted view that more than one progenitor scenario probably contributes (perhaps even to explain 'normal' SNe Ia), the 'progenitor problem' remains one of the longest-standing unsolved problems in stellar astrophysics (Maoz & Mannucci 2012). However, through a combination of numerical modelling (explosion simulations, radiative transfer models, binary

† This has been attempted extensively for the nearby SN 2011fe which exploded in M101 (Röpke *et al.* 2012).

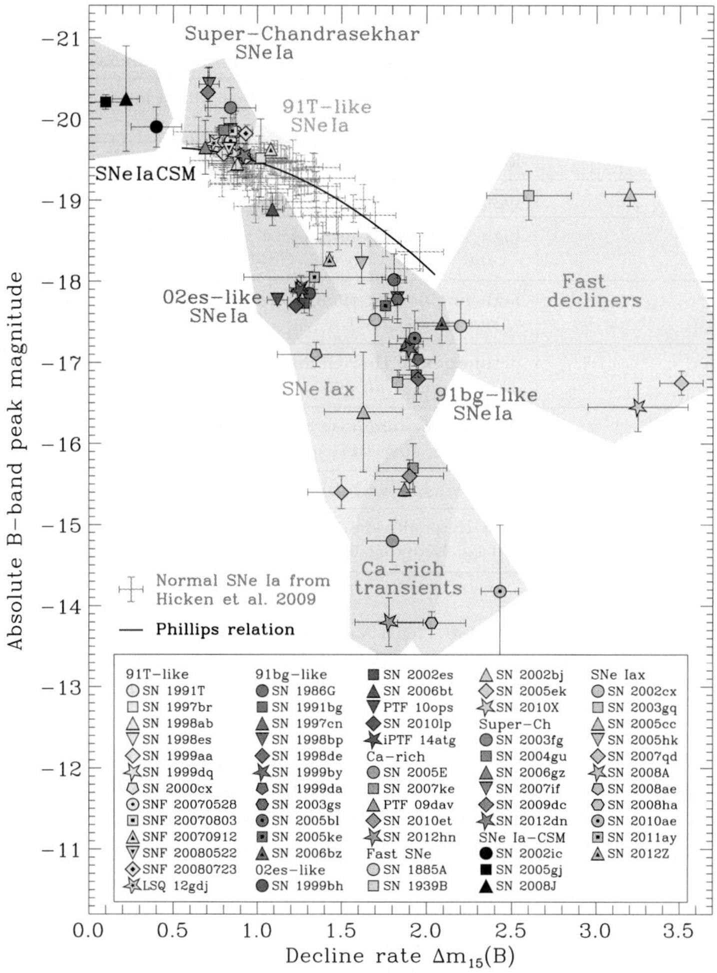

Figure 1. Figure by S. Taubenberger showing various sub-classes of observed thermonuclear supernovae (see Taubenberger 2017).

star population synthesis) and observational data, it has been possible to make some progress toward solving this problem. It seems likely that the explosion originates in a WD of carbon-oxygen composition either through *delayed detonation* of a Chandrasekhar mass WD (deflagration-to-detonation transition, Khokhlov 1991), or a *prompt detonation* (e.g. via double-detonation) in a sub-Chandrasekhar mass WD (Woosley & Weaver 1994). A few studies have shown that some thermonuclear events could plausibly arise from heavier, ONe WDs (Marquardt *et al.* 2015; Jones *et al.* 2016), and there is a probable contribution from WDs that contain some non-negligible fraction of helium in the case of WD mergers (Perets *et al.* 2010; Pakmor *et al.* 2013; Crocker *et al.* 2017; Zenati *et al.* 2019). Table 1 shows a summary of the different labels attributed to sub-classes of the more peculiar SNe Ia in the literature and their main features from an observational standpoint (see Figure 1, which is adapted from Taubenberger 2017). The very luminous sub-class of 'super-Chandrasekhar mass' explosions were proposed to originate from WDs that explode above the Chandrasekhar mass limit, though recent simulations have indicated that such systems, while indeed luminous, do not exhibit properties that are similar to this, or any other, sub-class (see Fink *et al.* 2018). It is still not clear which

Table 1. Synopsis of major SN Ia sub-classes (based on Figure 1 from Taubenberger 2017)

SN Ia sub-class	**observed characteristics**
Ca-strong (previously 'Ca-rich'):	high Ca to O nebular line ratios. Often far from potential galaxy host, see e.g. Shen *et al.* 2019.
1991bg-likes:	lower luminosity; faster than normal (e.g. narrow) light curves.
1991T-likes:	higher luminosity; slower than normal (e.g. broad) light curves.
super-Chandra:	high luminosity; low ejecta velocities. Extra energy source unknown and possibly diverse. CSM?
SNe Ia-CSM:	high luminosity; thought to be due to interaction with circumstellar material (CSM).
.Ia ("dot" one A):	detonation of He-rich material on WD, preceded by weaker He-flashes; Bildsten *et al.* 2007.
Iax:	diverse; low-luminosity, slow ejecta velocities, often young stellar pops; see e.g. Jha 2017.
2002es-likes:	sub-luminous with broad light curves (Ganeshalingam *et al.* 2012).
fast decliners	near-IR peaks similar to 'normals'; possibly not thermonuclear origin (e.g. Drout *et al.* 2013).

explosions in Fig. 1 & Table 1 arise from Chandrasekhar mass WDs (through delayed detonations) or sub-Chandrasekhar mass WDs (through prompt detonations).† However, in a Chandrasekhar mass WD explosion, the WD may not always unbind because the (subsonic) deflagration fails to develop into a (supersonic) detonation. It is mostly well-accepted that this weak explosion can leave behind an 'injured' WD as a remnant (Jordan *et al.* 2012; Kromer *et al.* 2013). Such an explosion of a CO or CONe WD is currently the leading explanation for the origin of SNe Ia of the "Iax" sub-class. Such left-over bound remnants are expected to exhibit unusual abundance signatures (Vennes *et al.* 2017). Numerous searches by different groups for donors of (presumably Chandrasekhar mass) supernovae have been mostly inconclusive so far (e.g. see Li *et al.* 2019), though possibly, donor stars that survived a SN Ia explosion from a sub-Chandrasekhar mass WD have already been found. Such donor stars in double degenerate systems that produce SNe Ia through a double-detonation – if they survive – will be flung out of orbit maintaining their pre-explosion velocity, which could be quite significant (on the order of 10^3 km/s). Shen *et al.* 2018 found three such plausible donor stars in recent Gaia DR2 data, with one object exhibiting a likely path of origin that traces back to the Galactic supernova remnant G70.0-21.5.

Currently, the three leading binary star configurations, and possible matching explosion scenarios (see Section 4 for details), thought to lead to SNe Ia (including sub-classes) are:

(*a*) A Chandrasekhar mass white dwarf that is accreting matter in a single degenerate system (Whelan & Iben 1973), where the white dwarf explodes successfully via delayed detonation (OR, the WD may not completely unbind in explosion if the initial deflagration does not transition into a detonation, a so-called 'failed deflagration', or failed detonation). Multiple donor types are possible.

(*b*) A sub-Chandrasekhar mass WD that is accreting helium-rich material through RLOF and explodes below the Chandrasekhar mass limit via double-detonation (Livne & Arnett 1995). Donors are hydrogen-poor, helium burning stars, or white dwarfs that contain a non-negligible fraction of helium in their outer layers, either helium or HeCO 'hybrid' WDs (Ruiter *et al.* 2014).

† Sometimes prompt detonations are referred to as 'violent', e.g. violent WD mergers.

(*c*) A double white dwarf merger, where mass transfer is unstable (dynamical timescale), and the explosion could be the result of either a Chandrasekhar mass white dwarf (Webbink 1984) through delayed detonation similar to (a), or a sub-Chandrasekhar mass WD likely via double-detonation similar to (b) (Pakmor *et al.* 2012), but likley less helium is required. The latter is currently the favoured scenario in the double degenerate case.

2. The path to simulating 'Normal' SNe Ia

Until recently, theoretical predictions found CO+CO WD mergers were more likely to produce a neutron star via merger-induced collapse (MIC) rather than a thermonuclear explosion (see Section 4.1). Pakmor *et al.* 2010 was the first work demonstrating that the merger of two carbon-oxygen WDs, with roughly equal mass (~0.9 Msun each) could result robustly in a thermonuclear runaway in one WD that would rapidly unbind the star, burning to iron-group elements. The other WD would also be burned but would produce primarily lighter, intermediate-mass elements with little or no contribution to production of ^{56}Ni, which also meant that the *peak brightness of the SN directly depended on the mass of the primary* (not the total binary mass). This study found however that these particular mergers did not produce enough ^{56}Ni to look like normal SNe Ia, but their spectra and lightcurves matched fairly well to the sub-luminous 1991bg sub-class. It's important to note that in this pioneering study, the primary (exploding) WD underwent a single detonation in a sub-Chandrasekhar (0.9 solar mass) WD, not a double-detonation (see Section 4.2).

The realization in the last decade that SNe Ia could very plausibly arise from merging WDs in hydrodynamical simulations, and that the exploding WD did not have to approach the Chandrasekhar mass limit before detonating, has expanded the allowed parameter space for the SN Ia progenitor population. It has also called into question the viability of SNe Ia as standardizable cosmological candles.

3. Evolutionary channels, SN Ia scenarios, and population synthesis

While detailed explosion model simulations and radiative transfer calculations of a single explosion configuration are critical for predicting synthetic observables, we also need to know how often said configuration would occur in Nature to explain the population that we see. It has long been assumed that all SNe Ia originated in a binary star system. It has however been hypothesized that a thermonuclear explosion may be possible in a single star within a specific mass range under the right conditions (a Type I 1/2 SN, Iben & Renzini 1983, see also Chiosi *et al.* 2015). More recently, the idea that a non-negligible number of SNe Ia may arise in triple systems has gained some well-deserved attention (Di Stefano 2019). Collisions of WDs (distinct from WD mergers) have also been proposed as a possible origin for SNe Ia (Kushnir *et al.* 2013), but WD collisions are unlikely to account for the delay time distribution and rates of the SN Ia population (Toonen *et al.* 2018; Hallakoun & Maoz 2019). The core-degenerate scenario (see early work by Sparks & Stecher 1974; Livio & Riess 2003), in which a WD merges with the core of a giant-like star during a common envelope phase, has also been proposed as a possible scenario and has gained some recent attention (see review by Soker 2019). So far such SNe Ia are mostly predicted to have short delay times (Wang *et al.* 2017) and, contrary to the majority of SNe Ia, would likely show hydrogen in their spectra unless some mechanism prevented these systems from exploding soon after the merger begins (see also Ilkov & Soker 2012). While such mergers undoubtedly do occur in Nature and may produce interesting transients, to date, a lack of numerical simulations of this scenario precludes us from considering the core-degenerate scenario as a main formation channel for SNe Ia. Indeed, a number of formation channels could be contributing to the observed SN Ia population. The best way to quantify which types of stars produce SNe Ia, and in what relative numbers, is through rapid binary star evolution population synthesis (BPS).

In a nutshell: BPS enables the rapid evolution of interacting stars through use of initial probability distribution functions for stellar orbital parameters (a, e, ZAMS masses and mass ratio q) and additionally incorporates treatment of the physical processes that should be taken into account in binary systems (tidal interactions, Roche lobe overflow [RLOF] mass transfer and accretion [including common envelope events], magnetic braking, gravitational radiation emission, supernova natal kicks, etc., see Toonen *et al.* 2014a for a BPS code-comparison study). What BPS lacks in detailed calculations of stellar structure and chemical composition (which can be investigated with detailed stellar evolution codes), it makes up for in rapid calculation of large populations of single and binary stars. While many star systems will not produce potentially explosive events, all 'interesting' systems (e.g. potential SNe Ia) are flagged and can be studied in detail post-processing to uncover their evolutionary history, delay times, birthrates, and binary physical properties†. For BPS, initial distributions for orbital configurations were historically adopted using either a Salpeter or Kroupa IMF, the Abt (1983) initial separation, with a flat (meaning equal probability between 0 to 1) mass ratio distribution q where $M_{\rm b} = qM_{\rm a}$ (with $M_{\rm b} < M_{\rm a}$), and a 'thermal' distribution for initial eccentricities. More recently, especially when concerning stars that will end up as compact objects, it is reasonable to adopt initial distributions for orbital period, mass ratio and eccentricity from Sana *et al.* 2012 (see also Moe & Di Stefano 2017).

The main scientific challenge in describing (numerically) interacting binary stars always stems back to how a star loses and/or gains mass (see e.g. Toonen *et al.* 2014b). This is a complex problem and has been tackled from a number of angles from different people over the years (Ritter 1988; van den Heuvel *et al.* 1992; Hachisu *et al.* 1996; Yaron *et al.* 2005; Nomoto *et al.* 2007; Goliasch & Nelson 2015). Phases of unstable mass transfer involving at least one non-degenerate star with a distinct core-envelope structure are referred to as common envelope events. Such evolutionary phases are integral to SN Ia progenitors and are poorly-understood. Common envelope events are parameterized in BPS studies, with the possibility of adopting different formalisms and efficiency factors (Claeys *et al.* 2014). I will note that changing the efficiency (α) or binding energy parameter (λ) does not appear to affect the overall rates of SN Ia progenitors too drastically; e.g. changing efficiency by an order of magnitude may only change birthrates by a factor of a few. This is because there is a shift, rather than an extreme suppression or enhancement, of progenitor parameter space. There is a noticeable effect on the types of progenitors that give rise to plausible SN Ia events. Detailed discussion of common envelopes was covered elsewhere in this symposium.

4. Type Ia supernova progenitor breakdown

Below I give a brief summary of the two (delineated by explosion mechanism, not donor type) categories: Chandrasekhar mass and sub-Chandrasekhar mass explosions.

4.1. *Chandrasekhar mass white dwarf SNe Ia*

Traditionally, it was thought that SN Ia progenitors originated from either a 'textbook' single degenerate, or (sometimes) a double degenerate, binary star system. It was

† Typically one BPS simulation involves the evolution from the ZAMS to a Hubble time with a single metallicity, but most BPS codes are capable of simulating stars with a wide range of metallicities.

believed that in either the case of a stably accreting WD (single degenerate) or in the case of a merger (double degenerate), the primary star (the more massive WD in the double degenerate case) would increase its mass via accretion, and would explode as it approached the Chandrasekhar mass limit ($\sim$1.4 M_{sun}, e.g. once the central density achieved a high enough value such that carbon burning could take place; Lesaffre *et al.* 2006). In the single degenerate case when mass is transferred via RLOF, the donor can be a wide range of stellar types: a MS star, a sub-giant, a red giant, a hydrogen-stripped, helium-burning star (and possibly an AGB star in the case of efficient wind-accretion, Chiotellis *et al.* 2012). Chandrasekhar mass SNe Ia with hydrogen-poor, helium-burning donors will have very short delay times, and undergo different evolutionary phases, compared to Chandrasekhar mass SNe Ia with H-rich donors (see Section 5), even though the explosion mechanism in the WD is expected to be the same.

It was found quite early on that if the entire 1.4 $M_{\odot}$WD exploded immediately as a detonation then the amount of ^{56}Ni (and thus ^{56}Fe) produced would be too high, being in tension with observations (Arnett *et al.* 1971). The 'deflagration to detonation transition' was hypothesized to ameliorate this, in which the WD undergoes a period of expansion following sub-sonic carbon burning, leading up to the (super-sonic) explosion that unbinds the star. Since many SNe Ia observed before this century appeared to follow the Phillips relation, the notion that all explosions occurred from WDs with approximately the same (Chandrasekhar) mass, thus producing the same mount of ^{56}Ni that would power the lightcurve, seemed to be a natural consequence. Note that in this explosion scenario carbon is ignited once the central density of the WD attains a critical value, but the star only explodes $\sim$centuries after this 'simmering' phase of carbon burning.

Early theoretical studies of merging WDs indicated that rapid accretion on the primary would not result in a SN Ia since off-centre carbon burning was more likely to occur (Yoon *et al.* 2007), thereby enabling the carbon-oxygen WD to escape an explosive fate and rather evolve into an oxygen-neon-rich WD. An ONe WD is more likely to collapse to form a neutron star via accretion induced collapse (Miyaji *et al.* 1980; Saio & Nomoto 1985) when it approaches the Chandrasekhar mass (but see Jones *et al.* 2016).

As mentioned previously, searches for stellar companions have focused on finding possible donors in the Chandrasekhar mass scenario, since until very recently, it was thought that double WD progenitors would leave behind no companion star. While searching for companions from single degenerate scenario explosions has not led to any definitive discovery of a progenitor configuration, observations of ionized gas in the vicinity of SN Ia remnants is another way of constraining the nature of the donor star: The pre-supernova luminosity (and thus stellar type) of the donor can be inferred from the level of hydrogen and/or helium ionization experienced in nebular region of the remnant (Woods *et al.* 2018; Kuuttila *et al.* 2019). Thus far, the main conclusion is that in most remnants studied, the progenitor system could not have contained a hot, luminous donor which we could expect in the presence of a nuclear-burning WD undergoing accretion from a non-degenerate companion. In other words: a 'textbook' Chandrasekhar mass progenitor is disfavoured for these remnants.

4.2. *Sub-Chandrasekhar mass white dwarf SNe Ia*

Studies in the 1990s investigated the possibility of sub-Chandrasekhar mass SNe Ia, in which a carbon-oxygen WD could explode at a mass well below 1.4 $M_{\odot}$ under the right conditions: the WD needed to be accreting matter that was helium-rich (Woosley & Weaver 1994; Livne & Arnett 1995).† At fairly low ($\sim$few $\times 10^{-8}$ $M_{\odot}$/yr) accretion rates

† The sub-Chandrasekhar mass scenario was also explored in the context of symbiotic binary systems (Yungelson *et al.* 1995).

of helium, a WD can build up a thick helium shell; the low accretion rate suppresses the effect of helium nova flashes (e.g. Kato *et al.* 2008). Eventually the base of the layer can become degenerate and ignite. This helium-shell detonation can be successful in driving a second detonation close to the WD core (Shen & Moore 2014), thus the term 'double-detonation'. However, early work showed that the thick $\sim$0.2 $M_\odot$ layer of helium shell needed to initiate the detonation was in severe tension with observations of SN ejecta; the models over-produced iron-peak elements at high velocities. It was shown only over the last decade or so (Fink *et al.* 2010; Woosley & Kasen 2011) that the helium shell did not need to be so massive; only on the order of 5% of a solar mass, for a shell detonation to proceed (the shell mass at detonation is dependent on the WD mass, Piersanti *et al.* 2015). With the decrease in the required shell mass, theoretical lightcurves and spectra showed an improved match to observations (Kromer *et al.* 2010) – on par with what was being found for the WD merger and Chandrasekhar mass scenarios. Thus, the double-detonation model was rejuvinated. Interestingly, recent work has shown that explosions from sub-Chandrasekhar mass WDs accreting helium in this 'classical' double-detonation scenario (e.g. helium accretion rates on the order of $\sim$few$\times 10^{-8}$ Msun/yr) do not exhibit properties of normal SNe Ia but may produce rapid transients, possibly events of the Ca-strong sub-class (see Neunteufel *et al.* 2017, which includes effects of magnetic fields and WD rotation).

Historically, WD mergers were not expected to produce SNe Ia very readily. However, since the 1980s, the idea that WD mergers could make thermonuclear supernovae started to gain ground (Webbink 1984; Iben & Tutukov 1984). These SN Ia scenarios of WD mergers proposed decades ago generally presumed that during the merger, the secondary WD would become disrupted, form an accretion torus around the primary, and the primary would accrete matter and eventually (after thousands of years) explode as it approached the Chandrasekhar mass limit. WD mergers that did not lead to a primary WD approaching this mass limit could form R Coronae Borealis stars, sdO stars, and other objects (Ruiter talk HDEF 2018). While as discussed in 4.1, this 'double degenerate merger to Chandrasekhar mass' scenario may indeed be plausible, WDs that explode by unstable mass transfer are now expected to do so early in the merging process, while still at sub-Chandrasekhar mass. After the Pakmor *et al.* 2010 study that demonstrated this was numerically viable in three-dimensional WD merger simulations, other groups began exploring the parameter space more extensively (see Schwab *et al.* 2012; Dan *et al.* 2014; Raskin *et al.* 2014; Zhu *et al.* 2015; Sato *et al.* 2016). Guillochon *et al.* 2010 demonstrated that WD mergers involving a helium-rich secondary could successfully form a detonation in the primary WD's accreted helium-rich envelope, which could possibly lead to a detonation in the primary CO WD. The idea that the presence of helium in a WD merger could facilitate the detonation process in a sub-Chandrasekhar mass WD was another important leap forward, enabling sub-Chandrasekhar mass WDs to explode more easily through double-detonations. It was proposed by Pakmor *et al.* (2013) that CO+CO and CO+He-rich WD mergers might form a continuum of progenitors, accounting for both normal and sub-luminous events. It is expected that so-called 'pure' CO WDs would also contain some helium on their surfaces left over from previous stellar evolution, and this small amount of helium could possibly be sufficient to enable the first (surface) detonation in the double-detonation process during unstable mass transfer. It was demonstrated with AREPO (Springel 2010) simulations consisting of CO WDs in Pakmor *et al.* 2013) that a helium detonation in a thin helium shell on the surface of the primary WD could be ignited during a merger. This opened the door to not only facilitating the thermonuclear explosion mechanism for sub-Chandrasekhar WDs, but also meant that it was now easier for WD mergers to produce 'normal' SNe Ia as well as other sub-classes, with the exploding (primary) WD mass being the main determinant for absolute peak luminosity

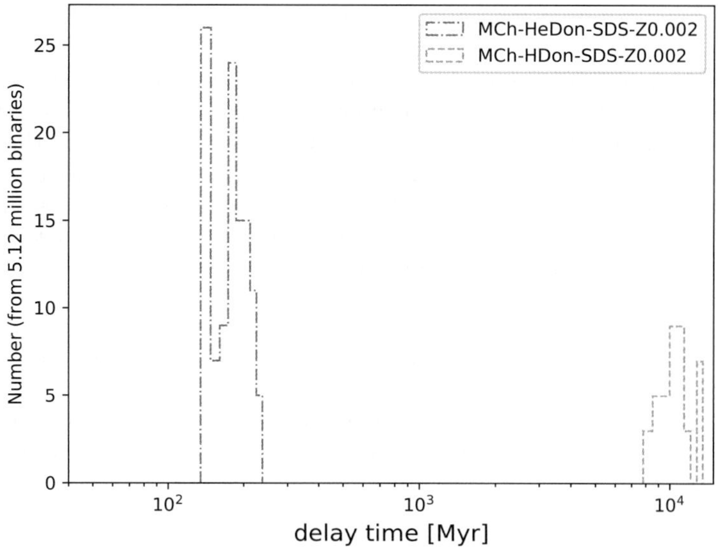

Figure 2. Theoretical delay time distribution for single degenerate, Chandrasekhar WD mass SNe Ia computed with `StarTrack` assuming a metallicity of Z = 0.002 (∼10% solar). Accretion is through RLOF only (no wind accretion assumed). Explosions with H-stripped, He-burning star donors have short delay times while those with H-burning star donors more typically have longer delay times.

through production of ^{56}Ni. In a study that combined binary population synthesis, 1D hydrodynamic explosion models and synthetic observables through radiative transfer calculations, it was shown that the peak luminosity distribution of local (within 80 Mpc) SNe Ia could indeed be reproduced by sub-Chandrasekhar mass CO WDs undergoing a merger. It was revealed that nearly half of the progenitors end up going through an extra phase of RLOF mass transfer long before the formation of the secondary WD (Ruiter *et al.* 2013, fig. 2). It is this extra mass transfer stage that leads to a shift of the average primary WD mass toward higher masses in WD merger progenitors. This primary WD mass shift is responsible for bringing the predicted peak brightness distribution into agreement with observations (Ruiter *et al.* 2013, figs. 5 and 8).

5. Rates and delay times

While for decades astronomers have known about real binary star systems that fit the profile for the textbook single degenerate scenario of SNe Ia (e.g. recurrent novae; see Hillman *et al.* 2016 for a modern review of such systems), many population synthesis studies have found that it is difficult to substantially increase the mass of a CO WD via hydrogen accretion, though some optimistic prescriptions for mass accretion efficiency and retention can bring up the potential rates in this scenario (see e.g. Nelemans *et al.* 2013). The theoretically-predicted rate of SNe Ia from the 'textbook' single degenerate Chandrasekhar mass channel still appears to fall rather short compared to the theoretically-predicted rates of double degenerate mergers. While delay time distributions of WD mergers from BPS models are canonically found to have a mostly smooth (beyond a few 100 Myr) shape, the single degenerate Chandrasekhar mass and sub-Chandrasekhar mass delay time distributions are each bimodal: those with hydrogen-poor, helium-burning donors occur at short delay times (see Hillebrandt *et al.* 2013).

An observational study of the delay time distribution of SNe Ia† in old stellar populations by Totani *et al.* (2008) demonstrated that this distribution followed a power law shape with $\sim t^{-1}$. A distribution of delay times that follows such a power-law is expected if the initial distribution of binary stars even roughly follows an initial distribution where the separation is flat in log (see Ruiter *et al.* 2009). This type of delay time distribution shape is not especially expected for single degenerate progenitors where the delay time is primarily governed by the evolutionary timescale of the donor rather than a decrease in orbital period due to gravitational wave radiation. By its nature however, the Totani *et al.* study could not probe delay times shorter than ~ 0.2 Gyr. Measuring the delay time distribution at short delay times is indeed more challenging, since we have to either be successful in studying high-redshift SNe (Friedmann & Maoz 2018), or re-construct the SN delay time from supernova remnants of Type Ia in local galaxies with active star-formation (Maoz & Badenes 2010). It is still unclear why the rate of SNe Ia in galaxy clusters is measured to be higher than the rate derived from field galaxies (see Maoz & Graur 2017, fig. 2), though metallicity may play some role (see Summary).

5.1. *Metallicity*

Preliminary studies of the impact of ZAMS metallicity on SN Ia progenitors have shown that initial metallicity can play an important role regarding which types of evolutionary channels for SNe Ia may be favoured at lower vs. higher metallicity. It is probably wise to keep in mind that using lower redshift (higher metallicity) SNe Ia to standardize SNe Ia at large redshift (lower metallicity) in order to perform precise calculations of cosmological quantities may then lead to difficult-to-trace uncertainties in the derivation of cosmological parameters.

Figures 2 and 3 show the delay time distribution from `StarTrack` (Belczynski *et al.* 2008; Belczynski *et al.* 2002) for Chandrasekhar mass WD explosions originating from binaries in RLOF where the donor was either H-rich (typically a red giant or sometimes a star in the Hertzsprung gap), or a H-stripped, He-burning star (typically a helium sub-giant). The systems involving H-stripped, He-burning stars arise from more massive progenitors that leave the main sequence rather quickly, and tend to always have delay times $\lesssim$300 Myr. The formation of such a progenitor normally involves two common envelope events and one phase of stable mass transfer. In the case of the 'textbook' H-rich donor Chandrasekhar mass channel, the formation of such a progenitor usually involves only one common envelope event and one stable phase of mass transfer. It should be noted that a trend was found in the `StarTrack` data where at very low metallicities (e.g. $Z = 0.0001$), the rate of SNe Ia from H-rich donors can become suppressed because it is more favourable to produce ONe WDs rather than CO WDs due to the decreased stellar wind mass-loss rates.

Changing the initial metallicity will affect the stellar core mass and thus will influence the birth mass of the WD. The H-poor, He-burning donor systems with delay times above 200 Myr do not occur in the higher Z model because the same binary would undergo a different evolutionary pathway on account of the smaller core (and hence WD) mass. The change in mass (and mass ratio) not only affects the chance of the WD reaching the Chandrasekhar mass limit, but is complicated by the fact that a given binary configuration may or may not reach contact and undergo stable RLOF, and if contact is achieved, that mass transfer may or may not be stable (see Côté *et al.* 2018 for an analagous example involving nova progenitors). For example, a typical system evolved with $Z = 0.002$ is not likely to produce a SN Ia if it is evolved with $Z = 0.03$. Even if the

† The time at which a supernova occurs relative to its ZAMS formation time; applicable to coeval star systems.

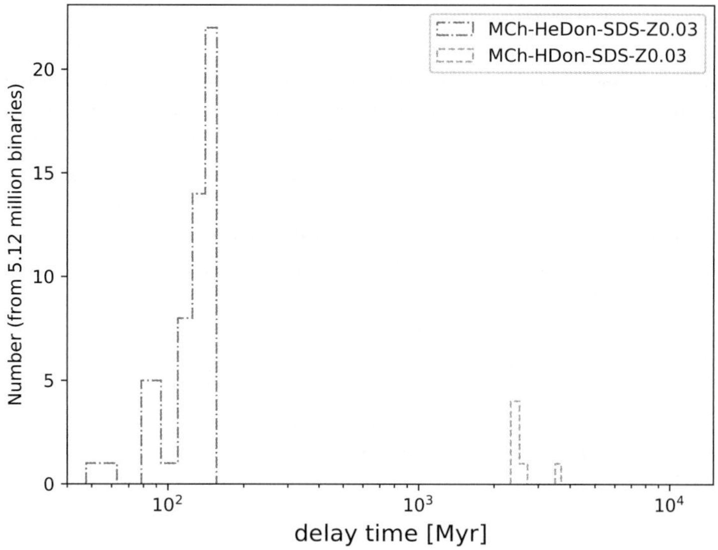

Figure 3. Theoretical delay time distribution for single degenerate, Chandrasekhar WD mass SNe Ia computed with `StarTrack` assuming a metallicity of Z = 0.03 (∼50% above solar). Accretion is through RLOF only (no wind accretion assumed). Explosions with H-stripped, He-burning star donors have short delay times while those with H-burning star donors more typically have longer delay times.

higher-Z system could result in a RLOF binary within a Hubble time, the lower core mass (due to increased wind mass loss rates) will inhibit formation of a Chandrasekhar mass WD. On the other hand, for the same binary evolved from very low-Z (Z = 0.0001), the primary star is more likely to evolve into an ONe WD than a CO WD, possibly resulting in an increased number of accretion-induced collapse neutron stars relative to Chandrasekhar mass SNe Ia (Ruiter *et al.* 2019).

The systems with H-stripped, He-burning donors – if they fail to produce a detonation after ignition – are good candidates for explaining some members of the SN Iax population, such as SN 2008ha (Foley *et al.* 2014; Kromer *et al.* 2015) or possibly SN 2012Z (McCully *et al.* 2014).

6. Summary

In what ways can we constrain the theoretically-predicted scenarios of SNe Ia, and match these to various members of the SN zoo? In terms of progenitor clues, the merger scenario progenitors discussed in Ruiter et al. 2013 predict a phase of binary evolution where the first-formed WD undergoes a short period ($\lesssim$1 Myr) of accretion of helium-rich material long before the double WD is formed. This phase of He-rich accretion would result in a helium nova system. If such progenitors are contributing significantly to the SN Ia population then we would expect on the order of a few to 10 of such He novae in the Galaxy today (V445 Pup is such a system). With the launch of LISA projected to be in 2034, we can look forward to unprecedented data from double WD binaries that will give us further insight into the formation and evolution of these systems. Currently, WD mergers appear to tick many of the boxes in explaining SNe Ia: delay time distribution, lack of hydrogen, peak brightness distribution, explosion mechanism, and the predicted rates are on par with the empirically-derived 'Galactic' SN Ia rate (though usually fall a factor of a few short in explaining rates in galaxy clusters). A subset of He+CO WD mergers are predicted to have rather long delay times, and if these binaries undergo helium detonations, they could explain the Galactic antimatter signal (Timmes *et al.* 1996; Crocker *et al.* 2017) and account for the 1991bg SN sub-class (Panther *et al.* 2019). For SNe Ia of Chandrasekhar mass, the predicted rates remain lower than those of double degenerate mergers, though the 'failed deflagration' (or rather failed detonation) model involving a H-stripped, He-burning star donor seems very promising in explaining the faint thermonuclear SNe of the Iax sub-class. Regarding accretion of helium-rich material – the 'classical' double-detonation scenario in which a sub-Chandrasekhar mass WD accretes stably from a He-rich WD or a H-stripped, He-burning star: it is still not clear what range of He shell masses is acceptable if these systems have a hand in contributing to normal SNe Ia. SN 2018byg was a peculiar, rare transient thought to originate from a double-detonation that is predicted to have contained a rather thick helium shell (De *et al.* 2019). A recent study by Seitenzahl *et al.* (2019) used MUSE integral field spectroscopy data of SN Ia remnants to interpret the nature of these explosions. Using optical coronal lines of [Fe XIV] to constrain their shock models, they found that in one case the progenitor was best-explained by a Chandrasekhar mass WD, while in another case it could be best-explained by a WD of sub-Chandrasekhar mass. This exciting result has given the supernova community a new method for revealing the origin of SNe Ia.

A general trend found with StarTrack BPS data is that SN Ia efficiency decreases with metallicity (more SNe at lower Z). This trend is not yet fully understood in the context of binary star evolution, but it does go the right way considering cosmological SN Ia rates are observed to be higher than local field rates. Separately, recent observations also indicate that binary fraction (of solar-type stars) goes up as Z goes down (Moe *et al.* 2019). To first order this implies that there is more opportunity for stellar interactions at lower Z which may account for an increase in the SN Ia rate at lower Z. In any case, The BPS models investigated here employed the same binary star fraction at all metallicities, so there are likely to be additional metallicity effects that play important roles regarding birthrates (e.g. wind mass loss rates affecting mass transfer; stellar core and hence WD mass). Indeed metallicity seems to play some role in SN Ia evolution and/or progenitor formation, but pinpointing all of the reasons how is still a work in progress.

In the coming years, surveys such as ZTF and LSST will find more supernovae. Detailed observations of a handful of nearby SNe will be more useful in constraining the progenitor nature than having a large (possibly incomplete) sample of faint events. But explaining the progenitor of any one explosion will be difficult unless a pre-explosion progenitor is confirmed, the explosion occurs nearby, or, in the case of SN remnants, is within a certain age range. Despite those unlikelihoods, it is becoming possible to set limits on what types of SNe exploded among ancient stellar populations. With the recent advances in theoretical yield predictions from different SN Ia explosion models, it is within our reach to delineate between different explosion scenario histories in small galaxies with relatively simple star formation history (Venn *et al.* 2016; Cescutti & Kobayashi 2017; Kirby *et al.* 2019).

Acknowledgments

AJR acknowledges financial support from the Australian Research Council under grant FT170100243, and thanks the SOC of the IAUS 357 for the invitation. AJR also thanks Stefan Taubenberger for the updated plot (Figure 1), and thanks Ivo Seitenzahl for some useful references. This research was undertaken with the assistance of resources and services from the National Computational Infrastructure (NCI), which is supported by the Australian Government, through the National Computational Merit Allocation Scheme and the UNSW HPC Resource Allocation Scheme. AJR is thankful for the New South Wales Rural Fire Service, who are mostly volunteers, for their hard and important work. They were constantly in my thoughts while I was finishing this paper in the early days of 2020, while Australia entered its worst-ever bushfire season.

References

Abt, H. A. 1983, *ARA&A*, 21, 343
Arnett, W. D. 1979, *ApJ*, 230, L37
Arnett, W. D., Truran, J. W., & Woosley, S. E. 1971, *ApJ*, 165, 87
Belczynski, K., Kalogera, V., & Bulik, T. 2002, *ApJ*, 572, 407
Belczynski, K., Kalogera, V., Rasio, F. A., *et al.* 2008, *ApJS*, 174, 223
Bianco, F. B., Howell, D. A., Sullivan, M., *et al.* 2011, *ApJ*, 741, 20
Bildsten, L., Shen, K. J., Weinberg, N. N., *et al.* 2007, *ApJ*, 662, L95
Cescutti, G. & Kobayashi, C. 2017, *A&A*, 607, A23
Chiosi, E., Chiosi, C., Trevisan, P., *et al.* 2015, *MNRAS*, 448, 2100
Chiotellis, A., Schure, K. M., & Vink, J. 2012, *A&A*, 537, A139
Claeys, J. S. W., Pols, O. R., Izzard, R. G., *et al.* 2014, *A&A*, 563, A83
Crocker, R. M., Ruiter, A. J., Seitenzahl, I. R., *et al.* 2017, Nature Astronomy, 1, 0135
Côté, B., Denissenkov, P., Herwig, F., *et al.* 2018, *ApJ*, 854, 105
D'Andrea, C. B., Smith, M., Sullivan, M., *et al.* 2018, arXiv e-prints, arXiv:1811.09565
Dan, M., Rosswog, S., Brüggen, M., *et al.* 2014, *MNRAS*, 438, 14
De, K., Kasliwal, M. M., Polin, A., *et al.* 2019, *ApJ*, 873, L18
Di Stefano, R. 2019, AAS/High Energy Astrophysics Division, 112.72
Drout, M. R., Soderberg, A. M., Mazzali, P. A., *et al.* 2013, *ApJ*, 774, 58
Fink, M., Röpke, F. K., Hillebrandt, W., *et al.* 2010, *A&A*, 514, A53
Fink, M., Kromer, M., Hillebrandt, W., *et al.* 2018, *A&A*, 618, A124
Fisher, R., Mozumdar, P., & Casabona, G. 2019, *ApJ*, 876, 64
Flörs, A., Spyromilio, J., Taubenberger, S., *et al.* 2019, *MNRAS*, 2650
Foley, R. J., McCully, C., Jha, S. W., *et al.* 2014, *ApJ*, 792, 29
Friedmann, M. & Maoz, D. 2018, *MNRAS*, 479, 3563
Ganeshalingam, M., Li, W., Filippenko, A. V., *et al.* 2012, *ApJ*, 751, 142
Goliasch, J. & Nelson, L. 2015, *ApJ*, 809, 80
Guillochon, J., Dan, M., Ramirez-Ruiz, E., *et al.* 2010, *ApJ*, 709, L64
Hachisu, I., Kato, M., & Nomoto, K. 1996, *ApJ*, 470, L97
Hallakoun, N. & Maoz, D. 2019, *MNRAS*, 490, 657
Hillebrandt, W., Kromer, M., Röpke, F. K., *et al.* 2013, Frontiers of Physics, 8, 116
Hillman, Y., Prialnik, D., Kovetz, A., *et al.* 2016, *ApJ*, 819, 168
Iben, I. & Renzini, A. 1983, *ARA&A*, 21, 271
Iben, I. & Tutukov, A. V. 1984, *ApJS*, 54, 335
Ilkov, M. & Soker, N. 2012, *MNRAS*, 419, 1695
Jha, S. W. 2017, Handbook of Supernovae, 375
Jha, S. W., Maguire, K., & Sullivan, M. 2019, Nature Astronomy, 3, 706
Jones, S., Röpke, F. K., Pakmor, R., *et al.* 2016, *A&A*, 593, A72
Jordan, G. C., Perets, H. B., Fisher, R. T., *et al.* 2012, *ApJ*, 761, L23
Kato, M., Hachisu, I., Kiyota, S., *et al.* 2008, *ApJ*, 684, 1366
Khokhlov, A. M. 1991, *A&A*, 245, L25
Kirby, E. N., Xie, J. L., Guo, R., *et al.* 2019, *ApJ*, 881, 45
Kromer, M., Sim, S. A., Fink, M., *et al.* 2010, *ApJ*, 719, 1067
Kromer, M., Ohlmann, S. T., Pakmor, R., *et al.* 2015, *MNRAS*, 450, 3045
Kromer, M., Fink, M., Stanishev, V., *et al.* 2013, *MNRAS*, 429, 2287
Kushnir, D., Katz, B., Dong, S., *et al.* 2013, *ApJ*, 778, L37
Kuuttila, J., Gilfanov, M., Seitenzahl, I. R., *et al.* 2019, *MNRAS*, 484, 1317
Lesaffre, P., Han, Z., Tout, C. A., *et al.* 2006, *MNRAS*, 368, 187
Li, C.-J., Kerzendorf, W. E., Chu, Y.-H., *et al.* 2019, *ApJ*, 886, 99
Liu, Z.-W., Stancliffe, R. J., Abate, C., *et al.* 2015, *ApJ*, 808, 138
Livio, M. & Riess, A. G. 2003, *ApJ*, 594, L93

Livne, E. & Arnett, D. 1995, *ApJ*, 452, 62
Maguire, K., Sim, S. A., Shingles, L., *et al.* 2018, *MNRAS*, 477, 3567
Maoz, D. & Mannucci, F. 2012, *PASA*, 29, 447
Maoz, D. & Badenes, C. 2010, *MNRAS*, 407, 1314
Maoz, D. & Graur, O. 2017, *ApJ*, 848, 25
Marquardt, K. S., Sim, S. A., Ruiter, A. J., *et al.* 2015, *A&A*, 580, A118
McCully, C., Jha, S. W., Foley, R. J., *et al.* 2014, *Nature*, 512, 54
Miyaji, S., Nomoto, K., Yokoi, K., *et al.* 1980, *PASJ*, 32, 303
Moe, M., Kratter, K. M., & Badenes, C. 2019, *ApJ*, 875, 61
Moe, M. & Di Stefano, R. 2017, *ApJS*, 230, 15
Nelemans, G., Toonen, S., & Bours, M. 2013, Binary Paths to Type Ia Supernovae Explosions, 225 (IAUS 281)
Neunteufel, P., Yoon, S.-C., & Langer, N. 2017, *A&A*, 602, A55
Nomoto, K., Saio, H., Kato, M., *et al.* 2007, *ApJ*, 663, 1269
Pakmor, R., Kromer, M., Taubenberger, S., *et al.* 2013, *ApJ*, 770, L8
Pakmor, R., Kromer, M., Taubenberger, S., *et al.* 2012, *ApJ*, 747, L10
Pakmor, R., Kromer, M., Röpke, F. K., *et al.* 2010, *Nature*, 463, 61
Pankey, T. 1962, Ph.D. Thesis
Panther, F. H., Seitenzahl, I. R., Ruiter, A. J., *et al.* 2019, *PASA*, 36, e031
Perets, H. B., Gal-Yam, A., Mazzali, P. A., *et al.* 2010, *Nature*, 465, 322
Phillips, M. M. 1993, *ApJ*, 413, L105
Piersanti, L., Yungelson, L. R., & Tornambé, A. 2015, *MNRAS*, 452, 2897
Raskin, C., Kasen, D., Moll, R., *et al.* 2014, *ApJ*, 788, 75
Ritter, H. 1988, *A&A*, 202, 93
Röpke, F. K., Kromer, M., Seitenzahl, I. R., *et al.* 2012, *ApJ*, 750, L19
Ruiter, A. J., Ferrario, L., Belczynski, K., *et al.* 2019, *MNRAS*, 484, 698
Ruiter, A. J., Belczynski, K., Sim, S. A., *et al.* 2014, *MNRAS*, 440, L101
Ruiter, A. J., Sim, S. A., Pakmor, R., *et al.* 2013, *MNRAS*, 429, 1425
Ruiter, A. J., Belczynski, K., & Fryer, C. 2009, *ApJ*, 699, 2026
Ruiter, A. J., Invited talk at Hydrogen Deficient Stars IV in Armagh, UK 2018. doi:10.5281/ZENODO.1436922
Saio, H. & Nomoto, K. 1985, *A&A*, 150, L21
Sana, H., de Mink, S. E., de Koter, A., *et al.* 2012, Science, 337, 444
Sato, Y., Nakasato, N., Tanikawa, A., *et al.* 2016, *ApJ*, 821, 67
Scalzo, R. A., Ruiter, A. J., & Sim, S. A. 2014, *MNRAS*, 445, 2535
Schwab, J., Shen, K. J., Quataert, E., *et al.* 2012, *MNRAS*, 427, 190
Seitenzahl, I. R., Ghavamian, P., Laming, J. M., *et al.* 2019, *Phys. Rev. Lett.*, 123, 041101
Seitenzahl, I. R., Cescutti, G., Röpke, F. K., *et al.* 2013, *A&A*, 559, L5
Shen, K. J., Boubert, D., Gänsicke, B. T., *et al.* 2018, *ApJ*, 865, 15
Shen, K. J., Quataert, E., & Pakmor, R. 2019, *ApJ*, 887, 180
Shen, K. J. & Moore, K. 2014, *ApJ*, 797, 46
Soker, N. 2019, arXiv e-prints, arXiv:1912.01550
Sparks, W. M., & Stecher, T. P. 1974, *ApJ*, 188, 149
Springel, V. 2010, *MNRAS*, 401, 791
Taubenberger, S. 2017, Handbook of Supernovae, 317
Thielemann, F.-K., Nomoto, K., & Yokoi, K. 1986, *A&A*, 158, 17
Timmes, F. X., Woosley, S. E., Hartmann, D. H., *et al.* 1996, *ApJ*, 464, 332
Toonen, S., Voss, R., & Knigge, C. 2014, *MNRAS*, 441, 354 (2014b)
Toonen, S., Perets, H. B., & Hamers, A. S. 2018, *A&A*, 610, A22
Toonen, S., Claeys, J. S. W., Mennekens, N., *et al.* 2014, *A&A*, 562, A14 (2014a)
Totani, T., Morokuma, T., Oda, T., *et al.* 2008, *PASJ*, 60, 1327
van den Heuvel, E. P. J., Bhattacharya, D., Nomoto, K., *et al.* 1992, *A&A*, 262, 97
Venn, K. A., Jablonka, P., Hill, V., *et al.* 2016, The General Assembly of Galaxy Halos: Structure, Origin and Evolution, 159 (IAUS 317)

Vennes, S., Nemeth, P., Kawka, A., *et al.* 2017, Science, 357, 680
Wang, B., Zhou, W.-H., Zuo, Z.-Y., *et al.* 2017, *MNRAS*, 464, 3965
Webbink, R. F. 1984, *ApJ*, 277, 355
Whelan, J. & Iben, I. 1973, *ApJ*, 186, 1007
Woods, T. E., Ghavamian, P., Badenes, C., *et al.* 2018, *ApJ*, 863, 120
Woosley, S. E. & Kasen, D. 2011, *ApJ*, 734, 38
Woosley, S. E. & Weaver, T. A. 1994, *ApJ*, 423, 371
Yaron, O., Prialnik, D., Shara, M. M., *et al.* 2005, *ApJ*, 623, 398
Yoon, S.-C., Podsiadlowski, P., & Rosswog, S. 2007, *MNRAS*, 380, 933
Yungelson, L., Livio, M., Tutukov, A., *et al.* 1995, *ApJ*, 447, 656
Zenati, Y., Toonen, S., & Perets, H. B. 2019, *MNRAS*, 482, 1135
Zhu, C., Pakmor, R., van Kerkwijk, M. H., *et al.* 2015, *ApJ*, 806, L1

White Dwarfs as Probes of Fundamental Physics: Tracers of Planetary, Stellar and Galactic Evolution
Proceedings IAU Symposium No. 357, 2019
M. A. Barstow, S. J. Kleinman, J. L. Provencal & L. Ferrario, eds.
doi:10.1017/S1743921320000411

Convection in common envelopes and the formation of double white dwarfs

E. C. Wilson[1] **and J. Nordhaus**[1,2]

[1]Center for Computational Relativity and Gravitation, Rochester Institute of Technology, NY 14623, USA
email: ecw7497@rit.edu

[2]National Technical Institute for the Deaf, Rochester Institute of Technology, NY 14623, USA
email: nordhaus@astro.rit.edu

Abstract. Formation of close double white dwarfs likely requires the initial binary system to evolve through two successive common envelope (CE) phases. A prominent method for describing CE outcomes involves defining an ejection efficiency, $\bar{\alpha}_{\rm eff}$, which quantifies the fraction of orbital energy available to unbind the envelope. Reproducing observed post-CE orbital parameters has proven difficult for numerical simulations, as the companion's decaying orbit fails to eject the envelope. The ejection failure seen in numerical simulations may be resolved with a proper treatment of convection, whereby the binary orbit shrinks before energy can drive ejection. Where the orbital decay timescale exceeds the convective transport timescale, the energy released during inspiral is carried to the stellar surface and radiated away. By including convection, we produce sub-day post-CE orbital periods, a result consistent with observations. We comment on the effects of convection for the population of double white dwarfs that evolve through two CEs.

Keywords. (stars:) binaries: general, stars: AGB and post-AGB, planet-star interactions – convection, (stars:) white dwarfs, (stars:) binaries: close

---

1. Introduction

The common envelope (CE) phase is believed to be the primary mechanism for producing short-period (sub-day) compact white dwarf (WD) binary systems (Toonen & Nelemans (2013); Canals *et al.* (2018)). As the more massive star in a binary system leaves the main sequence it begins to expand significantly during the Red Giant and Asymptotic Giant Branch phases (RGB and AGB). At this stage, the primary may reach the orbital radius of the companion star, overflow its Roche Lobe, or experience orbital decay through tidal dissipation (Nordhaus & Blackman (2006); Nordhaus *et al.* (2010); Chen *et al.* (2017)). In each of these scenarios, the companion can be engulfed in the outer layers of the primary star such that the primary's core and the companion orbit within a shared common envelope (Paczynski (1976); Ivanova *et al.* (2013); Kochanek *et al.* (2014)).

Orbital decay transfers energy and angular momentum from the orbit to the CE which can be tapped to eject the envelope which, if successful, results in a post-CE tight binary (Iben & Livio (1993)). Quantifying this process via energy arguments involves defining an ejection efficiency which describes what fraction of the companion's lost orbital energy can contribute to the unbinding of the envelope (Iben & Tutukov (1984); Webbink (1984); Livio & Soker (1988); DeMarco *et al.* (2011)). This is often parameterized by an α-prescription in the following way:

$$\bar{\alpha}_{\rm eff} = E_{\rm bind}/\Delta E_{\rm orb} \tag{1.1}$$

If the efficiency and energy of the inspiraling companion exceeds that of the binding energy of the primary star's layers, the envelope is ejected leaving behind a companion star in orbit around the primary's core, i.e. a star in orbit around a WD. Should this companion star now engulf the WD during its subsequent ascent up either the RGB or AGB, a second CE phase would ensue. After the envelope is ejected yet a second time, the system will end as a double white dwarf (DWD) whose orbit continues to decay via gravitational-wave radiation.

A giant star's envelope is almost fully convective, yet models of CEs completely neglect convection which can qualitatively change the outcomes. For example, convective eddies redistribute locally deposited energy globally. In addition, convective eddies can transport orbital energy to the surface of the CE where it can be lost via radiation, thereby lowering $\bar{\alpha}_{\mathrm{eff}}$ and bringing the two orbiting bodies closer together before the envelope is ejected (Wilson & Nordhaus (2019)). Below, we detail how including the effects of convection modify the outcomes of common envelopes.

2. Convective CE Model

During a CE, the companion plunges into a nearly-fully convective envelope of the primary star. In order to determine if the orbital energy of the companion contributes to unbinding the envelope, the convective transport timescale must be compared to the orbital decay timescale. The convective transport timescale is given by:

$$t_{\mathrm{conv}}[r] = \int_{r}^{R_*} \frac{1}{v_{\mathrm{conv}}[r]} dr \tag{2.1}$$

(Grichener, Sabach & Soker (2018)), and the time for the companion's orbit to decay is given by:

$$t_{\mathrm{inspiral}}[r] = \int_{r_{\mathrm{i}}}^{r_{\mathrm{shred}}} \frac{\left(\frac{\mathrm{d}M}{\mathrm{d}r} - \frac{M[r]}{r}\right)\sqrt{v_r^2 + \bar{v}_\phi^2}}{4\xi\pi G m_2 r \rho[r]} \mathrm{d}r, \tag{2.2}$$

where r_{shred} can be estimated as $r_{\mathrm{shred}} \sim R_2 \sqrt[3]{2M_{\mathrm{core}}/m_2}$. If the orbital decay timescale exceeds the convective transport timescale, the convective motion of the envelope's matter redistributes the energy deposited by the inspiraling companion and can carry the liberated energy to the surface of the star where it is radiated away from the system altogether. A representative timescale comparison plot can be seen in Figure 1. For higher-mass companions, the orbit decays more quickly, allowing those companions to feel the effects of convection for less of the duration of the common envelope phase.

The companion's released orbital energy is given by

$$\Delta E_{\mathrm{orb}}[r] = \frac{G m_2}{2}\left(\frac{M_{\mathrm{i}}}{r_{\mathrm{i}}} - \frac{M[r]}{r}\right). \tag{2.3}$$

This energy cannot contribute to unbinding the envelope until it reaches a region of the CE where the convective transport timescale is longer than the orbital decay timescale. This certainly occurs in non-convecting regions but can also occur in convecting layers. This allows the companion to inspiral deep into the star before orbital energy can be tapped to drive ejection. The radius within the primary at which the change in orbital energy exceeds the binding energy of the primary is the radius at which the envelope is unbound and ejected. A representative plot which shows the location where the binding energy is exceeded can be seen in Figure 2. With the envelope expelled from the system, a white dwarf with a short-period companion remains. By including convective effects, CE phases with M-dwarf companions result in sub-day orbital periods around WDs (Wilson & Nordhaus (2019)), consistent with observations (Davis *et al.* (2010)).

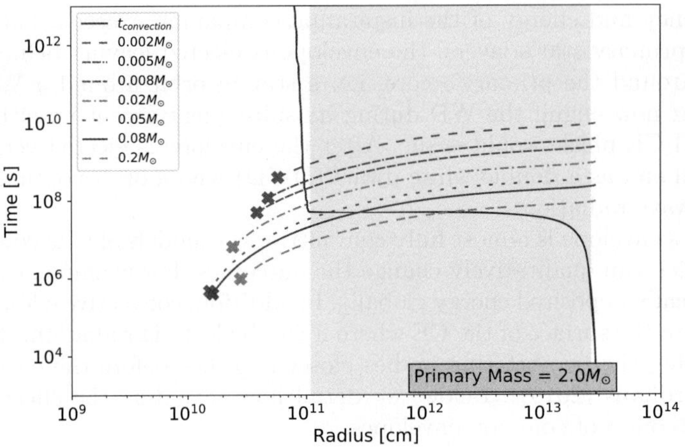

Figure 1. The convective transport timescale (solid curve) is compared to the orbital decay timescales of several companion masses (dashed curves). The shaded region that extends from $\sim 10^{11}$ cm to the surface of the star indicates the convective region of the primary star. Where the orbital decay timescale exceeds the convective transport timescale, the convective motion of matter within the primary's envelope redistributes the orbital energy released during inspiral. This energy can be carried to the surface where it is radiated away.

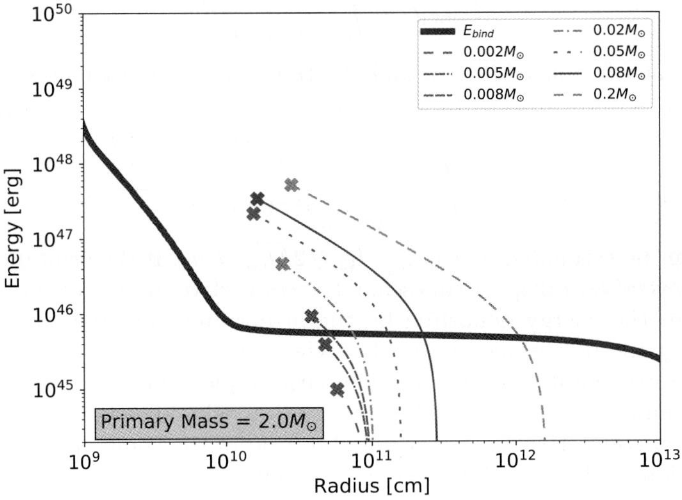

Figure 2. The binding energy (solid curve) of the primary is compared to the change in orbital energy of several companions (dashed curves). Energy released by the inspiraling companion only contributes to unbinding the envelope of the primary where the convective transport time exceeds the orbital decay time. The radius at which the change in orbital energy exceeds that of the binding energy is the radius at which the envelope is ejected. This results in M-dwarf+WD systems with short-period orbits ($\lesssim$ 1 day).

3. Implications for Double White Dwarfs

The cessation of the first CE produces the initial conditions for the second phase which commences when the companion leaves the main-sequence. Since RGB/AGB stars possess rigorous convective envelopes, the effects of convection must also be incorporated in the second common envelope phase. Once again, the inspiraling white dwarf companion travels deep into the primary star before its energy can be tapped to eject the primary's extended envelope. With the employment of the convective $\bar{\alpha}_{\rm eff}$-prescription, DWDs in extremely short-period orbits are a common result following two successive CE phases.

The vast majority of observed DWD systems have orbital separations of less than 3×10^{11} cm, a result which can be quickly matched with a comparison of the depth the companion may travel before depositing its energy into the envelope. When the cores of primary stars have grown to the mass of observed WDs, the radius at the lower limit of the convective zone is also very frequently between 1×10^{11} and 4×10^{11} cm. This correlation has interesting implications for the formation and ultimate population of DWDs.

References

Canals, P., Torres, S., & Soker, N. 2018, *MNRAS*, 489, 4519

Chen, Z., Frank, A., Blackman, E. G., Nordhaus, J., & CarrollNellenback, J. 2017, *MNRAS* 468, 4465

Davis, P. J., Kolb, U., & Willems, B. 2010, *MNRAS* 403, 179

De Marco, O., Passy, J.-C., Moe, M., Herwig, F., Low, M.-M., Mac, & Paxton, B. 2011, *MNRAS*, 411, 2277

Grichener, A., Sabach, E., & Soker, N. 2018, *MNRAS*, 478(2), 1818

Iben, I., J., & Tutukov, A. V. 1984, *ApJ*, 284, 719

Iben, Icko, J., & Livio, M. 1993, *PASP* 105, 1373

Ivanova, N., Justham, S., Chen, X., De Marco, O., Fryer, C. L., Gaburov, E., & Webbink, R. F. 2013, *A&AR*, 21(1), 59

Kochanek, C. S., Adams, S. M. & Belczynski, K. 2014, *MNRAS*, 443, 1319

Livio, M. & Soker, N. 1988, *ApJ*, 329, 764

Nordhaus, J. & Blackman, E. G. 2006, *MNRAS* 370(4), 2004

Nordhaus, J., Spiegel, D. S., Ibgui, L., Goodman, J., & Burrows, A. 2010, *MNRAS* 408, 631

Paczynski, B. 1976, *Structure and Evolution of Close Binary Systems*, Proc. IAU Symposium No. 73, 75

Toonen, S. & Nelemans, G. 2013, *A&A*, 557, 87

Tutukov, A., & Yungelson, L. 1979, *Mass Loss and Evolution of O-Type Stars*, Proc. IAU Symposium No. 83, 401

Webbink, R. F. 1984, *ApJ*, 277, 355

Wilson, E. C. & Nordhaus, J. 2019, *MNRAS*, 485, 4492

White Dwarfs as Probes of Fundamental Physics: Tracers of Planetary, Stellar and Galactic Evolution
Proceedings IAU Symposium No. 357, 2019
M. A. Barstow, S. J. Kleinman, J. L. Provencal & L. Ferrario, eds.
doi:10.1017/S1743921320001040

Double degenerate candidates in the open cluster NGC 6633

Joseph W. Barnett and Kurtis A. Williams

Department of Physics & Astronomy, Texas A & M University-Commerce, P.O. Box 3011, Commerce, TX, 75429, USA
email: jbarnett7@leomail.tamuc.edu

Abstract. The study of white dwarfs, the end stage of stellar evolution for more than 95% of stars, is critical to bettering our understanding of the late stages of the lives of low mass stars. In particular, the post main sequence evolution of binary star systems is complex, and the identification and analysis of double degenerate systems is a crucial step in constraining models of binary star systems. Binary white dwarfs in open star clusters are particularly useful because cluster parameters such as distance, metal content, and total system age are more tightly constrained than for field double degenerates. Here we use the precision astrometry from the Gaia Data Release 2 catalog to study two other white dwarfs which were identified as candidate double degenerates in the field of the open star cluster NGC 6633. One of the two objects, LAWDS 4, is found to have astrometric properties fully consistent with that of the cluster. In such a case, the object is significantly overluminous for a single white dwarf, strongly indicating binarity. The second candidate binary, LAWDS 7, appears to be inconsistent with cluster membership, though a more thorough analysis is necessary to properly quantify the probability. At present we are proceeding to model the photometric and spectroscopic data for both objects as if they were cluster member double degenerates. Results of this latter analysis are forthcoming. Our results will add crucial data to the study of binary star evolution in open star clusters.

Keywords. binaries: general, white dwarfs, open clusters and associations: individual (NGC 6633)

1. Introduction

White dwarfs are the final stage of stellar evolution for low mass stars. Over 95% of stars, including the Sun, will eventually become a white dwarf. This makes white dwarfs critical remnants for the study of stellar evolution. Furthermore, the evolution of binary star systems is more complex than the evolution of single star systems, and since the majority of stars are part of binary star systems, the evolution of such systems is another crucial component to our understanding of stellar evolution. In particular, the evolution of binary star systems in well-studied open clusters provides observational constraints on difficult theoretical questions, such as the binary frequencies and mass distributions resulting from cluster star formation and subsequent evolution of these systems in complex dynamical gravitational systems (e.g., Geller *et al.* 2013).

In this work we further the ongoing study of two candidate binary white dwarf systems in the intermediate-age open cluster NGC 6633. In Williams & Bolte (2007), two white dwarfs in the cluster field were identified as potential binary systems, as they are twice as luminous as would be expected for a single cluster-member white dwarf based on their spectroscopically measured effective temperatures and masses.

Since binary white dwarfs with short orbital periods will eventually merge and may explode as Type Ia supernovae, Williams *et al.* (2015) used high resolution spectra to

Table 1. Gaia DR2 astrometric measurements for NGC 6633 and the candidate binary white dwarfs. *P* is the likelihood that a random identical cluster member would have measured astrometry more discrepant than that of the white dwarf.

Object	μ_α mas yr^{-1}	σ_{μ_α} mas yr^{-1}	μ_δ mas yr^{-1}	σ_{μ_δ} mas yr^{-1}	ϖ mas	$\sigma\varpi$ mas	P
NGC 6633	1.1584	0.0199	−1.7371	0.0200	2.5232	0.0023	
NGC 6633 LAWDS 4	0.9219	0.5546	−1.8654	0.5332	2.7179	0.3116	0.428
NGC 6633 LAWDS 7	−0.7518	0.9580	−4.2236	0.9916	2.6701	0.3981	0.0013

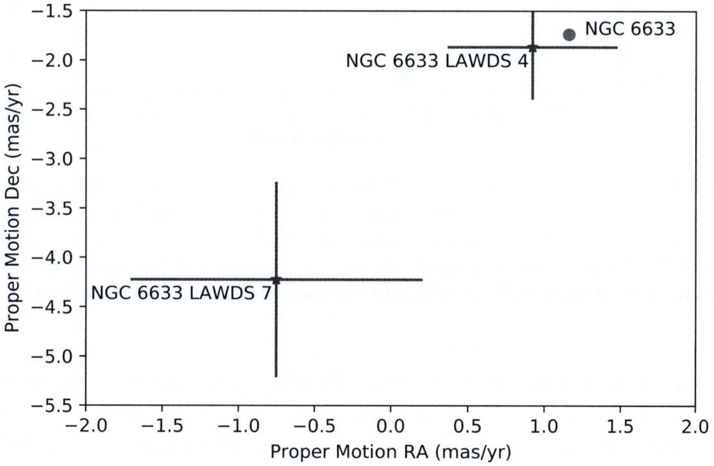

Figure 1. Gaia DR2 proper motion data for NGC 6633 and the two candidate double degenerates, with 1σ error bars. The uncertainties and dispersions for the cluster are smaller than the data point. This plot shows that the proper motions of LAWDS 4 are consistent with cluster membership, while LAWDS 7 appears to be inconsistent.

determine that while both candidates may be binary, their orbital periods are much longer than those needed for a merger within a Hubble time.

Here we present initial results from our most recent study that uses astrometric data from Gaia Data Release 2 to test the cluster membership of both candidate binary systems.

2. Results

The Gaia mission is tasked with making the most precise measurements of stellar parallax and proper motions ever taken for more than 1.3 billion stars across the entire sky, including both white dwarfs. We identified both white dwarfs in the Gaia DR2 catalog (Gaia Collaboration; Brown *et al.* 2018) by searching for coordinate matches within 1 arcsecond resulting in a unique match for each star; the Gaia photometry is consistent with previously published broadband measurements. The position of NGC 6633 LAWDS 4 is RA(J2000) = 18h27m10.40s and Dec(J2000) = + 06d26m15.70s, while NGC 6633 LAWDS 7 has a RA(J2000) = 18h27m49.90s and Dec(J2000)= +06d20m51.80s.

We then extracted the measured proper motions and parallaxes for each star after checking appropriate data quality flags. We used the cluster parallax and proper motions from Gaia observations of open star clusters (Gaia Collaboration; Babusiaux *et al.* 2018). These data are given in Table 1 and illustrated in Figures 1 and 2.

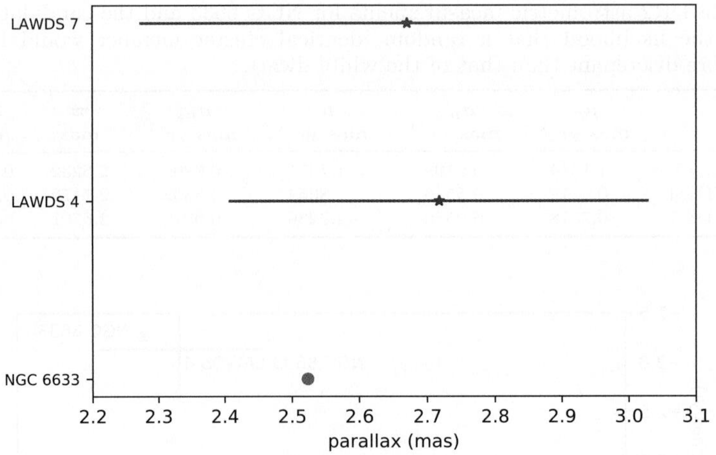

Figure 2. Gaia DR2 parallax data for NGC 6633 compared with that of the double degenerate candidates, with 1σ error bars. Note that the error bars for NGC 6633 are smaller than the point. This plot shows that the parallax of both white dwarfs, while as yet poorly constrained due to the stars' faint fluxes, is nonetheless consistent with cluster membership. Future Gaia data releases should more precisely determine the parallax of each white dwarf.

Quantitative measurements of cluster membership probability require iterative procedures considering the astrometric properties of both cluster and field stars as a function of position and magnitude (e.g., Gaia Collaboration; Babusiaux *et al.* 2018, Dias *et al.* 2006, Sanders 1971). At present we have not finished this analysis. However, we estimate membership probabilities by assuming that the stated Gaia measurement uncertainties have Gaussian distributions and are uncorrelated, and that the field contamination is negligible. These assumptions are not technically correct but are likely reasonable given that the Gaia DR2 errors on the white dwarfs' astrometry are likely primarily due to their faint magnitudes.

In Table 1 we list the probability that a bona fide cluster member with identical measurement uncertainties would exhibit deviations from the cluster astrometry larger than those observed for each white dwarf. As inferred from qualitative inspection of Figures 1 and 2, we find that LAWDS 4 is fully consistent with cluster membership and that LAWDS 7 is likely not consistent with cluster membership. Even so, based on parallax data alone LAWDS 7 is likely to still be overluminous and thus a double degenerate.

3. Ongoing Efforts

Our ongoing work strongly suggests that at least one of our candidate binaries, LAWDS 4, is indeed a member of NGC 6633 and therefore a double degenerate. The next step in our analysis is to fit binary white dwarf models to the existing spectra of both objects under the assumption of cluster membership.

Once we have these parameters, we will be able to consider various evolutionary scenarios for each system, including the potential of past binary interactions and estimates of the progenitor star masses. These data will inform our planned future analyses of candidate double degenerates in numerous additional open clusters. Combined, these data will allow glimpses into the evolution and fate of the numerous multiple star systems observed in open clusters.

References

Dias, W. S., Assafin, M., Flório, V., *et al.* 2006, *A&A*, 446, 949
Gaia Collaboration, Babusiaux, C., van Leeuwen, F., *et al.* 2018, *A&A*, 616, A10
Gaia Collaboration, Brown, A. G. A., Vallenari, A., *et al.* 2018, *A&A*, 616, A1
Geller, A. M., Hurley, J. R., & Mathieu, R. D. 2013, *AJ*, 145, 8
Sanders, W. L. 1971, *A&A*, 14, 226
Williams, K. A. & Bolte, M. 2007, *AJ*, 133, 1490
Williams, K. A., Serna-Grey, D., Chakraborty, S., *et al.* 2015, *AJ*, 150, 194

White Dwarfs as Probes of Fundamental Physics: Tracers of Planetary, Stellar and Galactic Evolution
Proceedings IAU Symposium No. 357, 2019
M. A. Barstow, S. J. Kleinman, J. L. Provencal & L. Ferrario, eds.
doi:10.1017/S1743921320000290

ANTARES: A gateway to ZTF and LSST alerts

Chien-Hsiu Lee[1], Monika Soraisam[2,3], Gautham Narayan[3], Thomas Matheson[1], Abhijit Saha[1], Carl Stubens[1] and Nicholas Wolf[1]

[1]NSF's National Optical-Infrared Astronomy Research Laboratory, AZ 85719, USA
emails: lee@noao.edu, matheson@noao.edu, saha@noao.edu, cstubens@noao.edu, nwolf@noao.edu

[2]National Center for Supercomputing Applications, IL 61801, USA
email: soraisam@illinois.edu

[3] University of Illinois at Urbana-Champaign, IL 61801, USA
email: gsn@illinois.edu

Abstract. Studies of white dwarfs have greatly benefited from time-domain surveys and subsequent follow-up observations. However, with the avalanche of alerts delivered by ZTF and LSST and the limited resources for follow-up, we will need brokers to select intriguing alerts that warrant follow-up in a timely manner. At the University of Arizona and NSF's OIR Lab, we are developing the Arizona-NOAO Temporal Analysis and Response to Events System, to hunt for the rarest of the rare events in the time-domain. In this work, we provide an overview of the ANTARES system, how we use ZTF as a training set, and the way forwards to LSST.

Keywords. Surveys, astrostatistics techniques, transient sources, variable stars, small solar system bodies.

1. Introduction

The advent of wide-field camera, large-area, and high-cadence surveys have revolutionized time-domain astronomy that relates to many aspects of white dwarf studies. For example, the explosion of supernova type Ia (SN Ia) was conventionally pictured as the end product of a white dwarf accreting materials from a non-degenerate companion. However, there is mounting evidence that some SNe Ia are products of two merging white dwarfs: the Kepler satellite has delivered exquisitely cadenced light curves at the rise of several SNe Ia with no sign of ejecta interaction with a stellar companion, thus suggesting double-degenerate progenitors (Olling *et al.* 2015). There are also heavily polluted white dwarfs, likely the product of accreting planetary debris as there are infrared signatures indicating existence of debris discs around those white dwarfs (see, e.g., Jura 2003). In this regard, time-domain surveys have revealed circumstellar discs transiting two white dwarfs, WD1154+017 (Vanderburg *et al.* 2015) and ZTF J0139+5245 (Vanderbosch *et al.* 2019), providing crucial evidence about the circumstellar material around polluted white dwarfs. To test the theories of white dwarf formation and evolution, we also need better estimates of their physical parameters (i.e., mass and radius). In this regard, the Catalina Sky Surveys have also revealed a handful of ultra short period eclipsing binaries (Drake *et al.* 2014), where one of the stars is likely to be a white dwarf, hence providing a unique opportunity to probe the fundamental physical parameters that are otherwise hard to determine. There are also eclipsing binaries with double white dwarfs (Burdge *et al.* 2019) that are not only valuable for studies of white dwarf properties, but can serve as potential gravitational wave sources for LISA as well.

While time-domain surveys have identified numerous promising candidate white dwarfs, it requires dedicated follow-up observations, especially spectroscopically, to reveal their true nature. With the Zwicky Tranisent Facility underway and the Large Synoptic Survey Telescope on the horizon, there are already millions of alerts per night and will be ten millions of alerts per night in the future that require follow up. This is beyond the capability of any single astronomer, and we will need to rely on community brokers to digest and select the most intriguing events worthy of follow-up.

2. ANTARES

In order to cope with the flood of millions of alerts from ZTF and LSST, we will need a software infrastructure to sift through the alerts. At the University of Arizona and NSF's National Optical Infrared Astronomy Research Laboratory (NSF's OIR Lab), we are developing the Arizona-NOAO Temporal Analysis and Response to Events System (ANTARES Saha *et al.* 2014, 2016; Narayan *et al.* 2018). After several years of planning and prototype testing, ANTARES beta† was launched in December 2018 to digest alerts from the Zwicky Transient Facility and disseminate value-added information from multi-wavelength catalogs in real time. ANTARES aggregates alerts at the same location of the sky (dubbed "locus") to form light curves. Each locus is also cross-matched with external catalogues, including, but not limited to, Gaia, 2MASS, WISE, SDSS, GALEX, NED, RC3, Veron Catalog of Quasars & AGN, NYU Value-Added Galaxy Catalog, Catalina Sky Survey Variables, ASAS-SN variable catalog, etc., to provide contextual information. With these pieces of information in hand, users can design and submit their own filters to select targets of interests in real time. Users can interact with ANTARES via different interfaces. We have a web portal that allows users to browse and search alerts (see Fig. 1) and submit user-defined filters (including sophisticated machine learning models). We also have a custom API that allows users to download alerts in batch mode, and can query our alert database in a programmatic manner. Finally, we also have a Slack channel that can notify users right away when an alert of interest is flagged by the user-defined filters.

In addition to user-defined filters, ANTARES is now actively broadcasting alerts in streams with pre-defined filters that are of potential interest and warrant spectroscopic observations. There is a filter flagging alerts associated with external galaxies ("extra-galactic stream") that can be supernovae. We also monitor alerts within 2×2 square degrees centered on M31 that may contain extragalactic novae and output to the "in_m31 stream". There is also a filter that flags alerts with high signal-to-noise ratio detection ("high_snr stream"), which are mostly bright transients in the Milky Way (e.g., novae, dwarf novae) that even small telescopes can provide decent spectra for classifications. As a demonstration, the ANTARES team has been conducting spectroscopic observations and classifications of alerts flagged by our customized filters. We have issued 19 Astronomer's Telegrams so far, and many of those classifications are related to white dwarfs, ranging from SNe Ia, to fast novae in M31, and highly extinguished Galactic novae and dwarf novae. A summary of the Astronomer's Telegrams issued by the ANTARES team and collaborations can be found in Table 1.

3. Prospect

The community alert broker will play a pivotal role in the future time-domain ecosystem, especially enabling rapid responses to the rarest of the rare events. Our broker, ANTARES, has been ingesting and disseminating alerts from ZTF as a step forwards for LSST. On top of that, ANTARES is also well-connected to a suite of tools that are currently under development at NSF's OIR Lab and by the community. ANTARES can

† http://antares.noao.edu

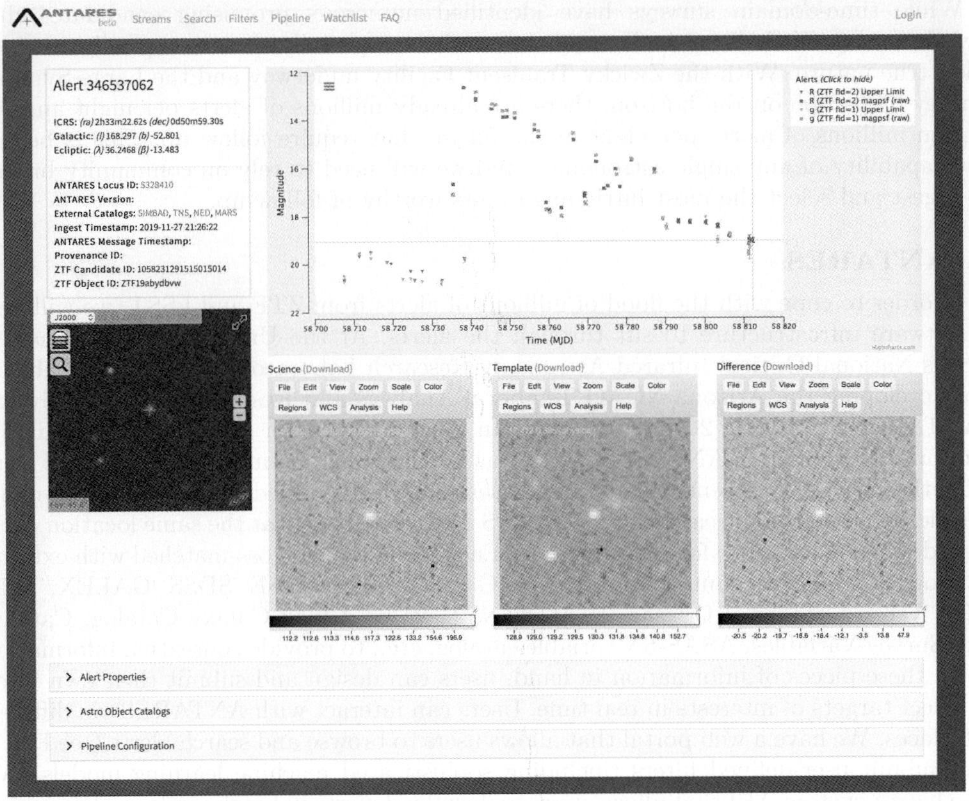

Figure 1. ANTARES alert information page. We aggregated alerts on the same position of the sky and formed light curves. The ANTARES web portal also displays the science, reference, and difference image from ZTF. We also provide contextual information by cross-matching multi-wavelength catalogues. All of the information (including light curves) are downloadable from the web portal. The example shown here is a Galactic dwarf nova HP Cet during superoutburst. Follow-up spectroscopic observations by the ANTARES team and collaborations can be found in Table 1 and in Soraisam *et al.* (2019).

connect to the Las Cumbre Observatory Target and Observation Manager† (TOM) that can handle target information, observation scheduling, and data visualization. ANTARES not only provides a searchable database for TOM to query targets, but also enables TOMs to subscribe to pre-defined or user-specified filters and streams. The ANTARES team also obtained classification spectra from telescopes that are part of the Astronomical Event Observatory Network‡ (AEON), in particular the Las Cumbres Observatory and the Gemini telescope. In the future, we can use TOMs to cross-talk with ANTARES and the facilities in the AEON, and stream-line an end-to-end follow-up infrastructure with minimum human interactions to maximize the science outcome from time-domain surveys.

Acknowledgements

The ANTARES project has been supported by the National Science Foundation through a cooperative agreement with the Association of Universities for Research in Astronomy (AURA) for the operation of NOAO, through an NSF INSPIRE grant to the

† https://lco.global/tomtoolkit/
‡ http://ast.noao.edu/data/aeon

Table 1. Summary of ATels by ANTARES team and collaborations

ATel#	Name	Classification	Report date	Facilities
12935	ZTF19aazcxwk	SN Ia	7/12/2019	LCO 2m telescope
12943	ZTF19abfqlzi	M31 recurrent nova	7/15/2019	Gemini telescope
12946	ZTF19abdooly	dwarf nova	7/18/2019	LCO 2m telescope
12980	ZTF19abgsssu	dwarf nova	8/2/2019	Shane 3m Telescope, Lick Observatory
13053	ZTF19abpmetl	SN Ia	8/30/2019	Shane 3m Telescope, Lick Observatory
13055	ZTF19abraqpf	dwarf nova	8/30/2019	Shane 3m Telescope, Lick Observatory
	ZTF19abqstxq	dwarf nova	8/30/2019	Shane 3m Telescope, Lick Observatory
13115	ZTF19abtuflm	SN II	9/18/2019	Shane 3m Telescope, Lick Observatory
13119	ZTF19abpvysx	SN Ia	9/21/2019	Shane 3m Telescope, Lick Observatory
	ZTF19abrelog	SN Ia-91T	9/21/2019	Shane 3m Telescope, Lick Observatory
	ZTF19abulrfa	SN IIP	9/21/2019	Shane 3m Telescope, Lick Observatory
13141	M31N2019-09b	M31 nova	9/28/2019	Gemini telescope
13149	ZTF19abyukuy	Galactic nova	10/1/2019	Shane 3m Telescope, Lick Observatory
13153	ZTF19abxnerq	M31 nova	10/1/2019	Gemini telescope
13178	ZTF19abzpkss	dwarf nova	10/9/2019	Shane 3m Telescope, Lick Observatory
13183	ZTF19abydbvw	dwarf nova	10/11/2019	Shane 3m Telescope, Lick Observatory
13200	ZTF19acbwmqd	SN IIP	10/18/2019	Shane 3m Telescope, Lick Observatory
13210	ZTF19acbzgog	M31 nova	10/21/2019	Gemini telescope
13231	ZTF19acfsteg	M31 nova	10/28/2019	Gemini telescope
13261	AT2019tsc	M31 nova	11/4/2019	Gemini telescope
13286	ZTF19acmdpyr	SN Ia	11/12/2019	Shane 3m telescope, Lick Observatory
	ZTF19acklbjr	SN Ia	11/12/2019	Shane 3m telescope, Lick Observatory
13317	ZTF19acnfsij	M31 nova	11/29/2019	Shane 3m telescope, Lick Observatory

University of Arizona (CISE AST-1344024, PI: R. Snodgrass), and through a grant from the Heising-Simons Foundation.

References

Burdge, K. B., Coughlin, M. W., Fuller, J., *et al.* 2019, *Nature*, 571, 528
Drake, A. J., Djorgovski, S. G., García-Álvarez, D., *et al.* 2014, *ApJ*, 790, 157
Jura, M. 2003, *ApJL*, 584, L91
Narayan, G., Zaidi, T., Soraisam, M. D., *et al.* 2018, *ApJS*, 236, 9
Olling, R. P., Mushotzky, R., Shaya, E. J., *et al.* 2015, *Nature*, 521, 332
Saha, A., Matheson, T., Snodgrass, R., *et al.* 2014, *Proc. SPIE*, 914908
Saha, A., Wang, Z., Matheson, T., *et al.* 2016, *Proc. SPIE*, 99100F
Soraisam, M., Nunez, R., McKinnon, K., *et al.* 2019, *The Astronomer's Telegram*, 13183, 1
Vanderbosch, Z., Hermes, J. J., Dennihy, E., *et al.* 2019, arXiv e-prints, arXiv:1908.09839
Vanderburg, A., Johnson, J. A., Rappaport, S., *et al.* 2015, *Nature*, 526, 546

White Dwarfs as Probes of Fundamental Physics: Tracers of Planetary, Stellar and Galactic Evolution
Proceedings IAU Symposium No. 357, 2019
M. A. Barstow, S. J. Kleinman, J. L. Provencal & L. Ferrario, eds.
doi:10.1017/S1743921320000939

Exoplanetary oxygen fugacities from polluted white dwarf stars

Alexandra E. Doyle[1], Beth Klein[2], Ben Zuckerman[2], Hilke E. Schlichting[1,2,3] and Edward D. Young[1]

[1]Earth, Planetary, and Space Sciences, University of California, Los Angeles
email: a.doyle@ucla.edu

[2]Physics and Astronomy, University of California, Los Angeles

[3]Earth, Atmospheric and Planetary Sciences, Massachusetts Institute of Technology

Abstract. The intrinsic oxygen fugacity of a planet profoundly influences a variety of its geochemical and geophysical aspects. Most rocky bodies in our solar system formed with oxygen fugacities approximately five orders of magnitude higher than that corresponding to a hydrogen-rich gas of solar composition. Here we derive oxygen fugacities of extrasolar rocky bodies from the elemental abundances in 15 white dwarf (WD) stars polluted by accretion of rocks. We find that the intrinsic oxygen fugacities of rocks accreted by the WDs are similar to those of terrestrial planets and asteroids in our solar system. This result suggests that at least some rocky exoplanets are geophysically and geochemically similar to Earth.

Keywords. white dwarfs, planetary systems

1. Introduction

White dwarf stars are the last observable stage of stellar evolution for stars less than 8–10 solar masses. WDs are approximately $0.6M_\odot$ and $\sim 1R_\oplus$ making them compact objects with immense gravitational fields. As a result, elements in the atmospheres of WDs heavier than He settle into the star's core and out of sight (Paquette *et al.* 1986; Koester 2009). Despite their strong gravitational fields, up to 50% of WDs are 'polluted' and exhibit heavy elements in their atmospheres (Zuckerman *et al.* 2003, 2010; Koester *et al.* 2014). This pollution is exogenous to the star; these heavy elements originate from vaporized rocks, torn apart by the gravitational field of the WD (Debes & Sigurdsson 2002; Jura 2003). Some polluted WDs exhibit infrared excesses, evidencing surrounding debris disks (e.g. Jura *et al.* 2007). By observing polluted white dwarf stars, and measuring the abundances of the heavy elements in their atmospheres, one can constrain the cosmochemistry of the rocky bodies that previously orbited them.

Thus far, parent bodies accreting onto polluted white dwarf stars resemble rocky bodies in the Solar System (e.g. Jura & Young 2014). The material accreting onto the white dwarf stars in this work are chondritic or basaltic; these bodies are devoid of metal cores. Doyle *et al.* (2019) showed that oxygen fugacity (f_{O_2}), a specific geochemical parameter that characterizes the overall oxidation state of a rocky body, can be constrained using WD elemental abundance data.

2. Oxygen Fugacity

The partial pressure of O_2, oxygen fugacity, is a thermodynamic quantity that characterizes the level of oxidation in a system. Geochemists use f_{O_2} as a measure of the partial pressure of O_2 in non-ideal thermodynamic systems. Oxygen fugacity can be used to characterize the oxidation state of a system even where O_2 gas may never have existed, as long as a reaction between the constituents of the system and O_2 can be written and quantified thermodynamically. The equilibrium reaction between pure metallic iron (Fe) and pure wüstite (FeO) defines the reference f_{O_2} commonly used in cosmochemical studies. This so-called Iron-Wüstite (IW) reference reaction is

$$\mathrm{Fe} + \frac{1}{2}\mathrm{O_2} \rightleftharpoons \mathrm{FeO}. \tag{2.1}$$

A simple thermodynamic expression relates the oxygen fugacity for any geochemical system containing oxidized and reduced iron to that of the IW reference:

$$0 = \Delta\hat{G}^0 + RT\ln\left(x_{\mathrm{FeO}}\right) - RT\ln\left(x_{\mathrm{Fe}}\right) - \frac{1}{2}RT\ln\left(f_{O_2}\right) \tag{2.2}$$

where $\Delta\hat{G}^0$ is the Gibb's free energy for the reaction in Equation 2.1 at standard state (1 bar, 298 K, pure Fe and FeO), x_{FeO} and x_{Fe} are mole fractions of FeO and Fe in their respective impure mineral phases, T is the temperature of interest and R is the ideal gas constant. Relating the oxygen fugacity of a rock to a reference reaction like Equation 2.1 results in cancellation of specific thermodynamic constants so that one can refer just to the difference in oxygen fugacity relative to the IW reference, ΔIW. In this case all that is required is the mole fraction of FeO in the rock ($x_{\mathrm{FeO}}^{\mathrm{rock}}$) and the mole fraction of Fe in the metal ($x_{\mathrm{Fe}}^{\mathrm{metal}}$), yielding:

$$\Delta\mathrm{IW} \equiv \log\left(f_{O_2}\right)_{\mathrm{rock}} - \log\left(f_{O_2}\right)_{\mathrm{IW}} = 2\log\left(\frac{x_{\mathrm{FeO}}^{\mathrm{rock}}}{x_{\mathrm{Fe}}^{\mathrm{metal}}}\right) \tag{2.3}$$

Following Doyle *et al.*, minerals comprising rocks in our solar system can be thought of as being composed of the oxides SiO_2, MgO, FeO, CaO and Al_2O_3. The relative proportions of these metal oxides reflect the compositions of the major minerals. The ability to measure all major rock-forming elements in a polluted WD affords the opportunity to use the abundances of FeO as a measures of the oxidation states of exoplanetary materials. Polluted WDs with quantifiable concentrations of O, Fe, Mg, Si, Ca and Al can be used to calculate oxygen fugacity, assuming there was at least a small amount of metal present during the formation of the body. By first assigning oxygen to Mg, Si, Al, and Ca to form these oxides, and then assigning the remaining oxygen to Fe, to form FeO, one can calculate the relative amount of oxidized Fe, as FeO, and assign any remaining Fe to metal, representing iron core material, if present. Measurement uncertainties for the polluted WDs are propagated using a Monte Carlo approach with an $n = 1$ bootstrap to obtain a frequency distribution of oxygen fugacities for each WD.

This use of FeO to estimate oxygen fugacities can be illustrated using Solar System bodies with known oxygen fugacities as examples. For instance, Mercury and Earth have very different intrinsic oxygen fugacities, as recorded at the time of core-mantle equilibration (e.g. Zolotov *et al.* 2013); Frost & McCammon (2008). This difference between Mercury and Earth is manifested by how much iron each body contains in its rock. Earth has about 8 wt. % FeO in its mantle rocks and those rocks produced by partial melting of the mantle (e.g., crustal basalt), whereas Mercury is estimated to have about 1 wt. % FeO in its rocks (Rubie *et al.* 2004; Nittler *et al.* 2018). This difference corresponds to a difference in intrinsic oxygen fugacity of 5 orders of magnitude, with Mercury having f_{O_2} values comparable to that of a solar gas, and Earth having an f_{O_2} 10^5 times greater than solar and closer to the IW reference defined by the reaction in Equation (2.1).

3. Oxygen Fugacity of Rocks Accreted by White Dwarfs

The method used to obtain ΔIW values for the rocky material accreted to WDs is detailed more thoroughly in Doyle *et al.* (2019). The method was validated using Solar

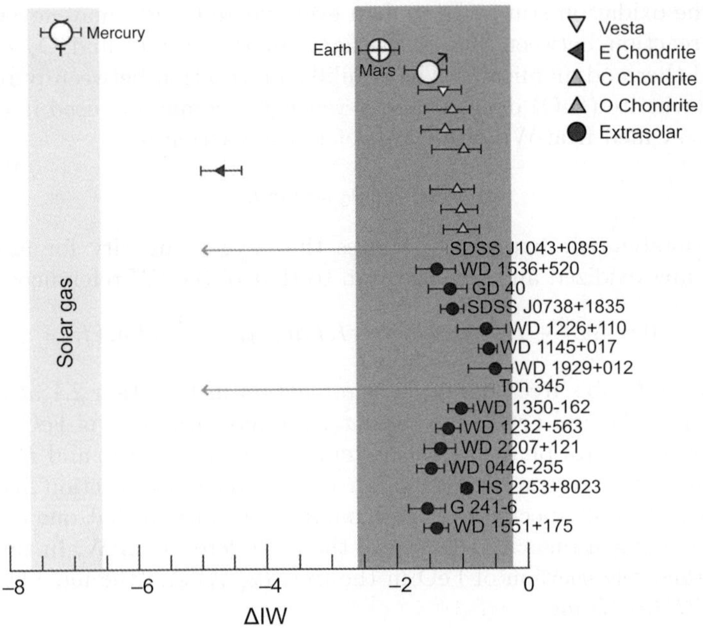

Figure 1. Oxygen fugacity values for rocky bodies accreting onto white dwarf stars. Recovered values for Solar System bodies are shown for comparison. WD abundance data from top to bottom: Melis & Dufour 2017; Farihi *et al.* 2016; Jura *et al.* 2012; Dufour *et al.* 2012; Gansicke *et al.* 2012; Xu *et al.* 2016; Gansicke *et al.* 2012; Swan *et al.* 2019; Xu *et al.* 2019; Wilson *et al.* 2015; Xu *et al.* 2019; Swan *et al.* 2019; Klein *et al.* 2011; Jura *et al.* 2012; Xu *et al.* 2019.

System rocks; the compositions of various Solar System bodies were converted into hypothetical polluted white dwarfs, as if the rocks from these bodies had accreted onto a WD, with typical WD measurement uncertainties. Doyle *et al.* recovered the known intrinsic oxygen fugacities for Mercury, Earth, Mars, Vesta and a range of chondritic bodies, including enstatite and CI chondrites. Figure 1 shows that the Solar System bodies span a range in ΔIW of $\sim$6 dex, which agrees with previous studies showing that Mercury and enstatite meteorites have oxygen fugacity values orders of magnitude lower than those for Earth, Mars, and other chondrite group meteorites (e.g. Cartier & Wood 2019). For the 15 WDs shown in Figure 1, the calculated ΔIW values for the accreted rocks are all similar to those of the oxidized bodies in the Solar System. When metallic iron is abundant, only an upper limit can be constrained for ΔIW, as shown for SDSS J1043+0855 and Ton 345, because the errors in estimating FeO concentrations become unwieldy.

4. Implications

Oxidation. The process that oxidized most rock-forming materials in the Solar System, relative to a hydrogen-rich gas of solar composition, also operated in these other planetary systems. One explanation is that this oxidation occurred early and involved the amount of water present in the disk at the time and place of rock formation (Monteux *et al.* 2018). Another explanation could be aqueous alteration of asteroidal parent bodies, occurring after rock formation but prior to assembly of planets. If the latter case is the correct explanation, then serpentinization, the process of converting anhydrous minerals into hydrous minerals and in so doing releasing H_2 gas, may be the primary oxidation process associated with terrestrial planet formation.

Habitability. Oxygen fugacity is a parameter that is as important as pressure and temperature in determining the dominant mineral assemblages present in planetary interiors.

Along with bulk composition, mineral structure is an important factor in determining how heat is transported through a body, how much water exists in a planet's mantle, and the presence or absence of a magnetic field (Pearson *et al.* 2014; Frost *et al.* 2001; Frost & McCammon 2008; Elkins-Tanton & Seager 2008; Wood *et al.* 2006; Buffett 2000). Additionally, oxygen fugacity and mineral assemblage will play a part in determining what type of volcanism a planet will have, which has implications for the existence of plate tectonics and the composition of the planet's initial atmosphere (e.g. Kolzenburg *et al.* 2018; Schaefer & Fegley 2017). Indeed, many geophysical aspects that are thought to be important for habitability are at least in part determined by oxygen fugacity.

References

Buffett, B. 2000, *Science*, 288, 2007

Cartier, C. & Wood, B. J. 2019, *Elements*, 15, 39

Debes, J. H. & Sigurdsson, S. 2002, *ApJ*, 572, 556

Doyle, A. E., Young, E. D., Klein, B., Zuckerman, B., & Schlichting, H. E. 2019, *Science*, 366, 356

Dufour, P., Kilic, M., Fontaine, G., Bergeron, P., Melis, C., & Bochanski, J. 2012, *ApJ*, 749, 15

Elkins-Tanton, L. T. & Seager, S. 2008, *ApJ*, 688, 628

Farihi, J., Koester, D., Zuckerman, B., Vican, L., Gansicke, B. T., Smith, N., Walth, G., & Breedt, E. 2016, *MNRAS*, 463, 3186

Frost, D. J., Langenhorst, F., & van Aken, P. A. 2001, *Physics and Chemistry of Minerals*, 28, 455

Frost, D. J. & McCammon, C. A. 2008, *Annual Review of Earth and Planetary Sciences*, 36, 389

Gansicke, B. T., Koester, D., Farihi, J., Girven, J., Parsons, S. G., & Breedt, E. 2012, *MNRAS*, 424, 333

Grossman, L., Fedkin, A. V., & Simon, S. B. 2012, *Meteoritics & Planetary Science*, 47, 2160

Jura, M. 2003, *ApJ*, 584, L91

Jura, M., Farihi, J., & Zuckerman, B. 2007, *ApJ*, 663, 1285

Jura, M., Xu, S., Klein, B., Koester, D., & Zuckerman, B. 2012, *ApJ*, 750

Jura, M. & Young, E. D. 2014, *Annual Review of Earth and Planetary Sciences*, Vol 42, 42, 45

Klein, B., Jura, M., Koester, D., & Zuckerman, B. 2011, *ApJ*, 741

Koester, D. 2009, *A&A*, 498, 517

Koester, D., Gansicke, B. T., & Farihi, J. 2014, *A&A*, 566

Kolzenburg, S., Di Genova, D., Giordano, D., Hess, K. U., & Dingwell, D. B. 2018, *Earth and Planetary Science Letters*, 487, 21

Krot, A., Fegley, B., & Lodders, K. 2000, Meteoritical and Astrophysical Constraints on the Oxidation State of the Solar Nebula, ed. V. Mannings, A. P. Boss, & S. S. Russell (University of Arizona Press), 1019–1054

Melis, C. & Dufour, P. 2017, *ApJ*, 834, 1

Monteux, J., Golabek, G. J., Rubie, D. C., Tobie, G., & Young, E. D. 2018, *Space Science Reviews*, 214

Nittler, L. R., Chabot, N. L., Grove, T. L., & Peplowski, P. N. 2018. The chemical composition of Mercury. In S. C. Solomon, L. R. Nittler, B. J. Anderson (Eds.) *Mercury: The View after MESSENGER* (pp. 30–51). Cambridge University Press

Paquette, C., Pelletier, C., Fontaine, G., & Michaud, G. 1986, *ApJS*, 61, 197

Pearson, D. G., Brenker, F. E., Nestola, F., McNeill, J., Nasdala, L., Hutchison, M. T., Matveev, S., Mather, K., Silversmit, G., Schmitz, S., & *et al.* 2014, *Nature*, 507, 221

Rubie, D. C., Gessmann, C. K., & Frost, D. J. 2004, *Nature*, 429, 58

Schaefer, L. & Fegley, B. 2017, *ApJ*, 843, 120

Swan, A., Farihi, J., Koester, D., Hollands, M., Parsons, S., Cauley, P. W., Redfield, S., & Gaensicke, B. T. 2019, *MNRAS*

Wilson, D. J., Gaensicke, B. T., Koester, D., Toloza, O., Pala, A. F., Breedt, E., & Parsons, S. G. 2015, *MNRAS*, 451, 3237

Wood, B. J., Walter, M. J., & Wade, J. 2006, *Nature*, 441, 825

Xu, S., Dufour, P., Klein, B., Melis, C., Monson, N. N., Zuckerman, B., Young, E. D., & Jura, M. A. 2019, *ApJ*, 158

Xu, S., Jura, M., Dufour, P., & Zuckerman, B. 2016, *ApJ*, 816

Zolotov, M. Y., Sprague, A. L., Hauck II, S. A., Nittler, L. R., Solomon, S. C., & Weider, S. Z. 2013, *Journal of Geophysical Research: Planets*, 118, 138

Zuckerman, B., Koester, D., Reid, I. N., & Hunsch, M. 2003, *ApJ*, 596, 477

Zuckerman, B., Melis, C., Klein, B., Koester, D., & Jura, M. 2010, *ApJ*, 722, 725

White Dwarfs as Probes of Fundamental Physics: Tracers of Planetary, Stellar and Galactic Evolution
Proceedings IAU Symposium No. 357, 2019
M. A. Barstow, S. J. Kleinman, J. L. Provencal & L. Ferrario, eds.
doi:10.1017/S1743921320000770

Near-infrared observations of dusty white dwarfs

Laura K. Rogers[1], Siyi Xu[2], Amy Bonsor[1], Simon Hodgkin[1], Kate Y. L. Su[3], Ted von Hippel[4] and Michael Jura†

[1]Institute of Astronomy, University of Cambridge, Madingley Rd, Cambridge, CB3 0HA, UK
email: laura.rogers@ast.cam.ac.uk

[2]Gemini Observatory, 670 N. A'ohoku Place, Hilo, HI 96720, USA

[3]Steward Observatory, University of Arizona, 933 N Cherry Avenue, Tucson, AZ 85721, USA

[4]Department of Physical Sciences, Embry-Riddle Aeronautical University, Daytona Beach, FL 32114, USA

Abstract. Planetary material in the atmospheres of white dwarfs is thought to be scattered inwards from outer planetary systems. Dusty emission in the infrared traces the accretion. As the scattering of many small asteroids is a stochastic process, variability in the infrared emission is predicted. We report a 3 year near-infrared (J, H and K) monitoring campaign of 34 dusty, polluted white dwarfs which aims to search for dust emission variability. We find all white dwarfs have consistent near-infrared fluxes, implying the excess emission is stable. This suggests tidal disruption events which lead to large variabilities are rare and quick (< 1 year) and become stable within a few years. For WD 0408–041, the system that shows both increases and decreases in dust emission over 11 years, our K band data suggest a potential colour change associated with the dust emission that needs further confirmation.

Keywords. white dwarfs, planetary systems, minor planets, asteroids

1. Introduction

Planetary material is found to survive to the white dwarf phase. Observations suggest that between 25% and 50% of white dwarfs have atmospheric pollution of elements heavier than helium (Zuckerman *et al.* 2003, 2010; Koester *et al.* 2014). The rapid gravitational settling times in comparison to the white dwarfs' cooling age implies ongoing accretion from a circumstellar reservoir.

The driving mechanism that results in atmospheric pollution is debated. The favoured theory cites outer planets which perturb asteroids towards the Roche radius of the white dwarf, where they become tidally disrupted, sublimate, and subsequently accrete on to the atmosphere of the white dwarf (Debes & Sigurdsson 2002; Jura *et al.* 2003; Farihi *et al.* 2010; Veras *et al.* 2014). This theory is supported by the discovery of disintegrating planetesimals transiting the polluted and dusty white dwarf WD 1145+017, and the white dwarf ZTF J013906.17+524536.89 (Vanderburg *et al.* 2015; Vanderbosch *et al.* 2019).

G29-38 was the first polluted white dwarf to be discovered to also display an infrared excess (Zuckerman & Becklin 1987). Follow-up observations confirmed its dusty nature (Graham *et al.* 1990). To date, around 40 polluted white dwarfs have a *Spitzer* confirmed infrared excess; this is calculated as between 1.5% and 4% of white dwarfs (e.g. Becklin*et al.* 2005; Kilic *et al.* 2006; Jura *et al.* 2007; Rebassa-Mansergas *et al.* 2019;

† Deceased

Wilson *et al.* 2019). Eight of the white dwarfs with dust also show evidence of circumstellar gas in emission near the same orbital distance as the dust (e.g. Gänsicke *et al.* 2006).

The scattering of planetary bodies that lead to the atmospheric pollution is expected to be a stochastic process (Wyatt *et al.* 2014), with variability predicted on human timescales. Variability in the dusty emission is seen for a few white dwarfs. The white dwarfs SDSS J095904.69–020047.6 (hereafter WD J0959–0200) and SDSS J122859.92+104033.0 (WD 1226+110) demonstrated large drops in dust flux (Xu & Jura 2014; Xu *et al.* 2018), whilst WD 0408–041 (GD56) among others show both increases and decreases in the flux (Farihi *et al.* 2018; Swan *et al.* 2019). The first white dwarf discovered with dust variability was WD J0959–0200. The 3–5 μm flux dropped by 35% between observations separated by a year, and the near-infrared K band flux dropped by 18.5% between observations separated by 8 years (Xu & Jura 2014). The large K band excess for this white dwarf means the dust is hot and close in. The mid-infrared flux drop occurred within a year, implying the variability event was large and became stable quickly.

2. Observations

Motivated by the large infrared variability in WD J0959–0200, we present a study to search for near-infrared variability in 34 dusty white dwarfs. We aim to characterise how the sample varies in the near-infrared in order to better understand how dusty material arrives in the atmospheres of the white dwarfs. We use the Wide Field Camera (WFCAM) on the UK Infrared Telescope (UKIRT) on Mauna Kea, Hawaii. We obtain J, H and K band photometry of our white dwarfs up to 6 times over the 3 year monitoring campaign. We probe short timescales (minutes) between individual frames taken on an observation date, and long timescales (years) between observation dates.

All data obtained with WFCAM were pipeline-processed by the Cambridge Astronomical Survey Unit using standard infrared photometry data reduction steps (CASU, Irwin *et al.* 2004; Dye *et al.* 2006). Further processing using the LIGHTCURVES software† was executed to photometrically calibrate the light curves (Irwin *et al.* 2007).

At each observation date, a number of observations were taken depending on the magnitude of the white dwarf. In order to robustly study the statistics of the photometry, the same analysis method was applied to all dusty white dwarf observations regardless of the magnitude and the number of measurements in the J, H and K bands. Gaussians were fitted to the distribution of photometrically corrected magnitudes from the LIGHTCURVES software. A Markov Chain Monte Carlo approach was implemented to model the magnitude distribution with a Gaussian, from this the mean and standard deviation were inferred. Please see Rogers *et al.* (submitted) for further details.

To ensure that any variability detected was real, background stellar objects were analysed using the same approach for comparison. The wide field of view of WFCAM ensures many background stars are present in the data. The results for the background objects give us a base to determine the observation sensitivity to detecting variability.

3. Results

3.1. *K band variability*

Of the near-infrared bands, the K band is the most sensitive to the dust, therefore, this is where we probe variability. Fig. 1 plots the median value of the magnitude and standard deviation from the Gaussian fit for the white dwarfs and background stars in the K band. All white dwarfs have standard deviations which are consistent with the background stellar objects within $2\,\sigma$. We conclude that from our survey all our objects

† https://github.com/mdwarfgeek/lightcurves

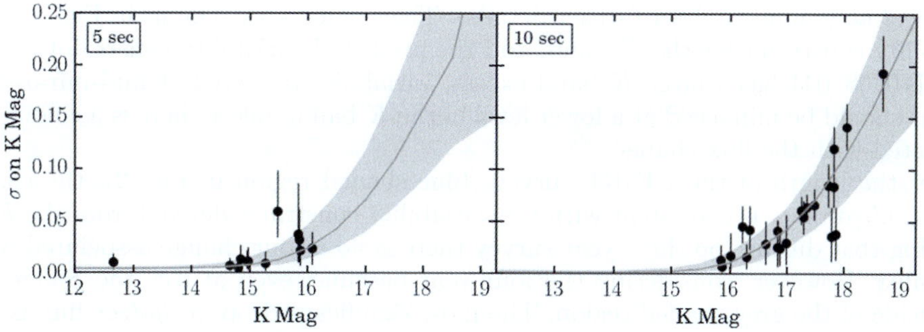

Figure 1. The median K magnitude against the median standard deviation on this magnitude for white dwarfs and background stars in the UKIRT survey. Due to differing observing strategies, the white dwarfs observed with 5 and 10 second frames were analysed separately. The background stars are given by the blue median line, with the 16th and 84th percentile confidence levels for objects in bins of width 0.5 mag. This represents the sensitivity of the survey as a function of magnitude. The white dwarfs are plotted in black with the errors as the 16th and 84th percentiles. This plot is used to distinguish white dwarfs which are variable from those that follow the background distribution.

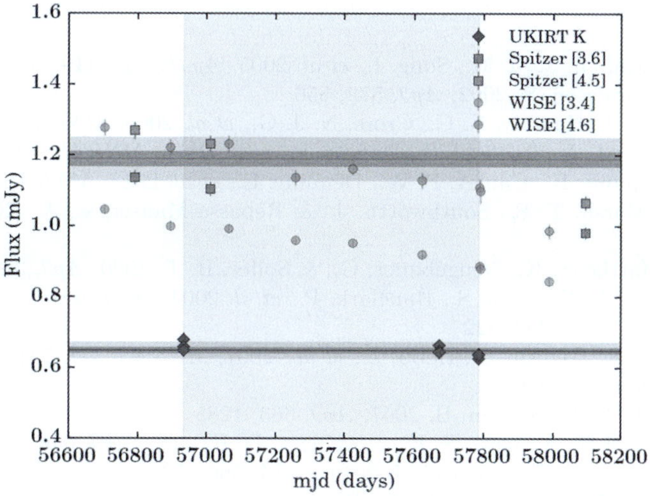

Figure 2. WD 0408–041 light curve of the UKIRT K band data and the *Spitzer*/*WISE* data where Farihi *et al.* (2018) observed a drop in the mid-infrared flux. The grey regions represent the 1 and $3\,\sigma$ median variability constraint for the K band, scaled up to *Spitzer* wavelengths (assuming a constant colour temperature). The blue region shows the period of time that we have UKIRT observations. Figure adapted from Farihi *et al.* (2018).

have stable K band contributions to the dust. This may imply that the tidal disruption events which lead to large variabilities are rare and occur quickly ($<$ 1 year), then once they occur they become stable within a few years.

3.2. *WD 0408–041*

WD 0408–041 has previously shown dust variation at 3–5 μm demonstrating brightening and dimming with peak-to-peak changes of 20% over 11.2 yrs (Farihi *et al.* 2018); a four year data set is shown in Fig. 2. We use the UKIRT K band standard deviation to predict the expected variability at 4.5 μm by assuming the dust emission has a constant colour temperature (i.e., the change is due to the amount of dust, not the location, see

Rogers *et al.* (submitted) for further details). The shaded grey regions in Fig. 2 show the variability constraint for the K band, and the predicted variability constraint at 4.5 μm. As WD 0408–041 has a large K band excess, variability observed at mid-infrared wavelengths should be mimicked at a lower level in the K band, unless there is a colour change associated with the flux change.

Over the length of the UKIRT survey (blue shaded region in Fig. 2), the change in *Spitzer* 4.5 μm flux is consistent with the variability constraint derived from the K band, implying that during the three year survey there is no colour change associated with the variability. However, considering the four year baseline, even at $3\,\sigma$, the *Spitzer* fluxes lie outside of the grey shaded region. The most significant drop in *Spitzer* flux is around 58100 days. We do not have simultaneous K band observations, with the closest K band measurement being over a year previous to this. There is previous evidence that large scale changes can occur on time-scales of less than a year (Xu & Jura 2014), so changes can be large and occur quickly. We conclude that a further K band measurement is required to understand whether the K band flux also dropped around 58100 days, or if there is a colour change associated with this variability. If the latter is true, it may explain why the K band appears to be stable for all objects in the UKIRT monitoring campaign.

References

Becklin, E. E., Farihi, J., Jura, M., Song, I., *et al.* 2005, *ApJL*, 632, 119

Debes, J. H. & Sigurdsson, S. 2002, *ApJ*, 572, 556

Dye, S., Warren, S. J., Hambly, N. C., Cross, N. J. G., *et al.* 2006, *MNRAS*, 372, 1227–1252

Farihi, J., Barstow, M. A., Redfield, S., Dufour, P., *et al.* 2010, *MNRAS*, 404, 2123–2135

Farihi, J., van Lieshout, R., Cauley, P. W., Dennihy, E., *et al.* 2018, *MNRAS*, 481, 2601–2611

Gänsicke, B. T., Marsh, T. R., Southworth, J., & Rebassa-Mansergas, A. 2006, *Science*, 314, 1908–1910

Graham, J. R., Matthews, K., Neugebauer, G., & Soifer, B. T. 1990, *ApJ*, 357, 216–223

Irwin, M. J., Lewis, J., Hodgkin, S., Bunclark, P., *et al.* 2004, *International Society for Optics and Photonics*, 5493, 411–422

Irwin, J., Hodgkin, S., Aigrain, S., Hebb, L., *et al.* 2007, *MNRAS*, 377, 741–758

Jura, M. 2003, *ApJL*, 584, 91

Jura, M., Farihi, J., & Zuckerman, B. 2007, *ApJ*, 663, 1285

Kilic, M., von Hippel, T., Leggett, S. K., & Winget, D. E. 2006, *ApJ*, 646, 474

Koester, D., Gänsicke, B. T., & Farihi, J. 2014, *A&A*, 566, A34

Rebassa-Mansergas, A., Solano, E., Xu, S., Rodrigo, C., *et al.* 2019, *MNRAS*, 489, 3990–4000

Swan, A., Farihi, J., & Wilson, T. G. 2019, *MNRAS: Letters*, 484, 109–113

Vanderburg, A., Johnson, J. A., Rappaport, S., Bieryla, A., *et al.* 2015, *Nature*, 526, 546

Vanderbosch, Z., Hermes, J. J., Dennihy, E., Dunlap, B. H., *et al.* 2019, arXiv:1908.09839

Veras, D., Leinhardt, Z. M., Bonsor, A., & Gänsicke, B. T. 2014, *MNRAS*, 445, 2244–2255

Wilson, T. G., Farihi, J., Gänsicke, B. T., & Swan, A. 2019, *MNRAS*, 487, 133–146

Wyatt, M. C., Farihi, J., Pringle, J. E., & Bonsor, A. 2014, *MNRAS*, 439, 3371–3391

Xu, S. & Jura, M. 2014, *ApJL*, 792, 39

Xu, S., Su, K. Y., Rogers, L. K., Bonsor, A., *et al.* 2018, *ApJ*, 866, 108

Zuckerman, B. & Becklin, E. E. 1987, *Nature*, 330, 138

Zuckerman, B., Koester, D., Reid, I. N., & Hünsch, M. 2003, *ApJ*, 596, 477

Zuckerman, B., Melis, C., Klein, B., Koester, D., *et al.* 2010, *ApJ*, 722, 725

White Dwarfs as Probes of Fundamental Physics: Tracers of Planetary, Stellar and Galactic Evolution
Proceedings IAU Symposium No. 357, 2019
M. A. Barstow, S. J. Kleinman, J. L. Provencal & L. Ferrario, eds.
doi:10.1017/S1743921320000204

The search for planet and planetesimal transits of white dwarfs with the Zwicky Transient Facility

Keaton J. Bell

NSF Astronomy and Astrophysics Postdoctoral Fellow and DiRAC Institute Fellow,
Department of Astronomy, University of Washington, Seattle, WA 98195, USA
email: keatonb@uw.edu

Abstract. Planetary materials orbiting white dwarf stars reveal the ultimate fate of the planets of the Solar System and all known transiting exoplanets. Observed metal pollution and infrared excesses from debris disks support that planetary systems or their remnants are common around white dwarf stars; however, these planets are difficult to detect since a very high orbital inclination angle is required for a small white dwarf to be transited, and these transits have very short (minute) durations. The low odds of catching individual transits could be overcome by a sufficiently wide and fast photometric survey. I demonstrate that, by obtaining over 100 million images of white dwarf stars with 30-second exposures in its first three years, the Zwicky Transient Facility (ZTF) is likely to record the first exoplanetary transits of white dwarfs, as well as new systems of transiting, disintegrating planetesimals. In these proceedings, I describe my project strategy to discover these systems using the ZTF data.

Keywords. planetary systems, white dwarfs, surveys

1. Introduction

White dwarf stars (WDs) represent the final evolutionary stage of $>97\%$ of stars (e.g., Williams *et al.* 2009), including the host stars of every known transiting exoplanet. The present population of planets orbiting WDs reveals the fates of these planets that missions like *Kepler* have shown to be ubiquitous around main sequence stars (e.g., Hsu *et al.* 2019). By virtue of its high volumetric survey speed (Bellm 2016), the Zwicky Transient Facility (ZTF) holds the promise of detecting the first members of this important exoplanet population, the details of which will inform our theories of planetary migration and survival in post-main-sequence circumstellar environments.

Despite the fact that no intact planets orbiting WDs have been previously identified, we have compelling observational evidence that such planets are common. The high gravities of WDs cause metals to diffuse quickly below their photospheres (Koester & Wilken 2006), yet 27–50% of WDs are metal polluted (Koester *et al.* 2014), implying that accretion of metals is ongoing. Chemical abundance analyses support that these metals come from rocky planetary debris (e.g., Gänsicke *et al.* 2012). Furthermore, debris disks are detected via infrared excesses from $\approx 2\%$ of these systems (Rebassa-Mansergas *et al.* 2019). Recently, disintegrating planetesimals have even been discovered to orbit a couple of specific metal-polluted WDs (Vanderburg *et al.* 2015; Manser *et al.* 2019). Some of my own high speed photometry from McDonald Observatory of the transiting debris system WD 1145+017 is displayed in Figure 1.

Detecting planets around WDs is observationally very challenging since transits of these roughly Earth-sized stars have short ($\sim$ minute) durations and require nearly

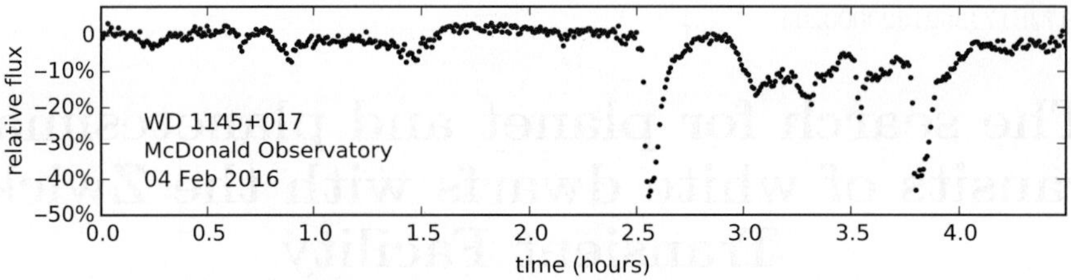

Figure 1. WD 1145+017 light curve from McDonald Observatory covering one orbital period.

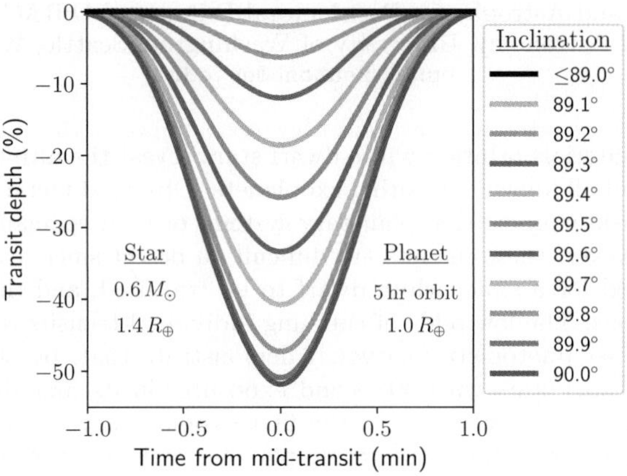

Figure 2. Model light curves of a 1 $R_\oplus$ planet transiting a WD at different inclination angles, simulated with the BATMAN package (Kreidberg 2015) accounting for 30-second ZTF exposures.

edge-on alignments. Still, transits of WDs can be very deep (even occulting), producing unmistakable signals in well-timed images. Figure 2 explores the transit profile of an Earth-sized planet in a 5-hour orbit around a typical 0.6 $M_\odot$ WD at very high inclination angles.

Previous searches for such dips in samples of ∼1000 WDs have yielded no transit detections (from, e.g., WASP, Faedi *et al.* 2011; Pan-STARRS, Fulton *et al.* 2014; *K2*, van Sluijs & Van Eylen 2018; DECam, Dame *et al.* 2019; GALEX, Rowan *et al.* 2019). Discovering the first planets to transit WDs will require orders of magnitude larger sample sizes. Cortés & Kipping (2019) and Lund *et al.* (2018) have simulated the expected yield for WD transits after 10 years of Large Synoptic Survey Telescope (LSST) operations (ca. 2033), anticipating ∼1000 detections. Because LSST is a deep survey, opportunities to follow-up candidate WD transits with large telescopes will be extremely limited. The magnitude range of ZTF, on the other hand, is readily accessible to more modest follow-up facilities, and I demonstrate in these proceedings that the current potential for discovering WD transits with the modern ZTF survey has been largely overlooked.

2. Detectability of Planet Transits with ZTF

Thanks to *Gaia* astrometry, we now know the identities (Gentile Fusillo *et al.* 2019) of most of the ≈220,000 WDs in the ZTF survey footprint (declination > -31 deg) brighter than the median 5σ survey depth of $g = 20.8$ mag, each of which will receive a median ≈ 870 observations over three years (Bellm 2014). I use the BATMAN package (Kreidberg 2015) to synthesize transit light curves for planets that uniformly and isotropically populate the period range of the continuously habitable zones (4.5 hr $< P_{\rm orb} <$ 32 hr; Agol 2011) around typical 0.6 $M_\odot$ WDs (e.g., Tremblay *et al.* 2013). A transit detection is

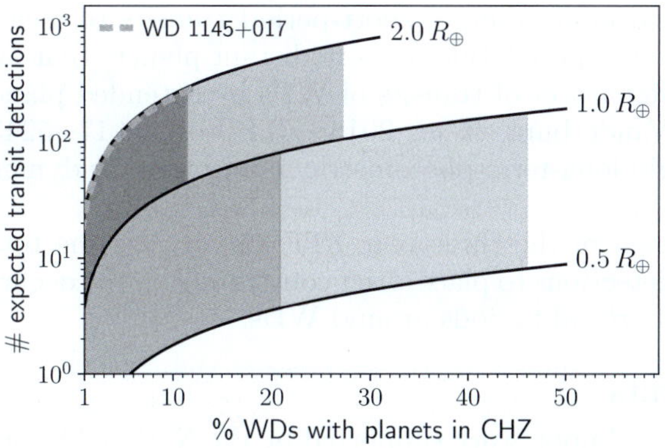

Figure 3. Simulated number of transit detections expected from the three-year ZTF survey for different percentages of WDs hosting 0.5, 1, and 2 $R_\oplus$ planets in the continuously habitable zone. Regions are shaded below the 95% (light) and 68% (dark) confidence upper limits on planet occurrence from van Sluijs & Van Eylen (2018). The dashed line shows the expected detections of disintegrating planetesimals transiting WDs based on the first such system, WD 1145+017.

considered significant when the measured flux decrease exceeds a 5σ threshold, where σ is the photometric error achieved through the magnitude range observed by ZTF (Bellm *et al.* 2019).

Figure 3 displays the expected WD planet yield from ZTF. The solid lines show the number of 5σ transit detections for a given fraction of WDs that host planets of radii 0.5, 1, and 2 $R_\oplus$ in their continuously habitable zones. The shaded regions are consistent with recent constraints based on ~1000 non-detections from *K2* (van Sluijs & Van Eylen 2018). If as few as 1% of WDs host 1–2 $R_\oplus$ planets in their continuously habitable zones, ZTF will detect $\gtrsim 10$ exoplanet transits during its 3-year survey.

ZTF is also expected to discover many new examples of disintegrating planetesimal systems transiting WDs like WD 1145+017, which was found in *K2* data (Vanderburg *et al.* 2015). Using the ground-based light curve of this system obtained with 30-second exposures from the McDonald Observatory 2.1-meter Otto Struve Telescope (Figure 1) as a template, I include estimates of the number of similar systems that ZTF will detect with the dashed line in Figure 3. In fact, the second such transiting planetesimal system has already been discovered from the first public release of ZTF data (ZTF J013906.17+524536.89; Vanderbosch *et al.* 2019).

3. Project Strategy

This work is the topic of the project "Discovering the First Exoplanets Around White Dwarf Stars with the Zwicky Transient Facility" that was granted a three-year NSF Astronomy and Astrophysics Postdoctoral Fellowship award #1903828 (PI: Bell). The project is designed with the following strategy:

(*a*) Use the Astronomy eXtensions for Spark (AXS; Zečević *et al.* 2019) tools to efficiently identify candidate transits of likely WDs in the large ZTF data set (including both public and proprietary partnership data).

(*b*) After validating the data quality of ZTF measurements that are consistent with significant brightness decreases, obtain follow-up high-speed photometry from Apache Point Observatory (APO), confirming the transits and resolving their shapes and periods.

(*c*) Obtain spectra of the WD hosts at APO to characterize stellar atmospheric parameters ($T_{\rm eff}$, $\log g$) and to rule out binarity for targets that exhibit light curves consistent with binary eclipses.

(*d*) Begin long-term monitoring of short-period transits to measure potential timing variations that probe the population of more distant planets that are unlikely to transit.

(*e*) Also identify signatures of transits of WDs by extended planetesimal debris (such as WD 1145+017, Vanderburg *et al.* 2015; ZTF J013906.17+524536.89, Vanderbosch *et al.* 2019), and begin long-term photometric monitoring for dynamical variations (e.g., Gänsicke *et al.* 2016).

(*f*) At the conclusion of the three-year ZTF survey, use the total number of transit detections and non-detections to place strict constraints on the occurrence rates of planets of different sizes and orbital periods around WDs.

Acknowledgements

This material is based upon work supported by the National Science Foundation under Award No. 1903828.

References

Agol, E. 2011, *ApJL*, 731, L31
Bellm, E. 2014, *The Third Hot-wiring the Transient Universe Workshop*, 27
Bellm, E. C. 2016, *PASP*, 128, 084501
Bellm, E. C., Kulkarni, S. R., Graham, M. J., *et al.* 2019, *PASP*, 131, 018002
Cortés, J. & Kipping, D. 2019, *MNRAS*, 488, 1695
Dame, K., Belardi, C., Kilic, M., *et al.* 2019, *MNRAS*, 490, 1066
Faedi, F., West, R. G., Burleigh, M. R., Goad, M. R., & Hebb, L. 2011, *MNRAS*, 410, 899
Fulton, B. J., Tonry, J. L., Flewelling, H., *et al.* 2014, *ApJ*, 796, 114
Gänsicke, B. T., Aungwerojwit, A., Marsh, T. R., *et al.* 2016, *ApJL*, 818, L7
Gänsicke, B. T., Koester, D., Farihi, J., *et al.* 2012, *MNRAS*, 424, 333
Gentile Fusillo, N. P., Tremblay, P.-E., Gänsicke, B. T., *et al.* 2019, *MNRAS*, 482, 4570
Hsu, D. C., Ford, E. B., Ragozzine, D., *et al.* 2019, *AJ*, 158, 109
Koester, D., Gänsicke, B. T., & Farihi, J. 2014, *A&A*, 566, A34
Koester, D. & Wilken, D. 2006, *A&A*, 453, 1051
Kreidberg, L. 2015, *PASP*, 127, 1161
Lund, M. B., Pepper, J. A., Shporer, A., *et al.* 2018, arXiv e-prints, arXiv:1809.10900
Manser, C. J., Gänsicke, B. T., Eggl, S., *et al.* 2019, *Science*, 364, 66
Rebassa-Mansergas, A., Solano, E., Xu, S., *et al.* 2019, *MNRAS*, 489, 3990
Rowan, D. M., Tucker, M. A., Shappee, B. J., *et al.* 2019, *MNRAS*, 486, 4574
Tremblay, P.-E., Ludwig, H.-G., Steffen, M., & Freytag, B. 2013, *A&A*, 559, A104
Vanderbosch, Z., Hermes, J. J., Dennihy, E., *et al.* 2019, arXiv e-prints, arXiv:1908.09839
Vanderburg, A., Johnson, J. A., Rappaport, S., *et al.* 2015, *Nature*, 526, 546
van Sluijs, L. & Van Eylen, V. 2018, *MNRAS*, 474, 4603
Williams, K. A., Bolte, M., & Koester, D. 2009, *ApJ*, 693, 355
Zečević, P., Slater, C. T., Jurić, M., *et al.* 2019, *AJ*, 158, 37

White Dwarfs as Probes of Fundamental Physics: Tracers of Planetary, Stellar and Galactic Evolution
Proceedings IAU Symposium No. 357, 2019
M. A. Barstow, S. J. Kleinman, J. L. Provencal & L. Ferrario, eds.
doi:10.1017/S174392132000023X

Searching for low-mass companions around white dwarfs and subdwarfs from Kepler field

Jurek Krzesinski[1] and Adam Blokesz[2]

[1]Astronomical Observatory of the Jagiellonian University, ul. Orla 171, 30-244 Krakow, Poland
email: jk@oa.uj.edu.pl

[2]Pedagogical University of Cracow, ul. Podchorazych 2, 30-084 Cracow, Poland
email: abl@astro.as.up.krakow.pl

Abstract. Knowing the late stages of the stellar evolution is crucial for understanding the fate of planets around subdwarfs and white dwarfs. Simulations by Staff *et al.* (2016) show that exoplanets engulfed in the extending stellar envelope will quickly spiral down onto the parent star. Therefore, we do not expect to find planets on close by orbits to the subdwarfs (Blokesz *et al.* 2019) or white dwarfs. However, the recent observation of planetary debris around WD 1145+017 white dwarf suggests, there might exists planets farther away from these stars. Using binarograms, O-C diagrams and Fourier transform for the Kepler space telescope data, we investigate a problem of missing planets around white dwarfs in binary systems, single white dwarfs and subdwarfs type B. The last ones, being the only stars which (due to the lack of hydrogen) go directly to the white dwarf cooling track after their red giant phase.

Keywords. (stars:) planetary systems, (stars:) subdwarfs, (stars:) white dwarfs, (stars:) binaries: general

1. Introduction

Most of the stars have their exoplanetary systems around them. During the evolution, the host stars go through the red giant (RG), horizontal branch and asymptotic giant (AGB) phases on the HR diagram before they become white dwarfs. However, even though white dwarfs (WD) and subdwarfs are considered to be the best objects for exoplanet searches, no exoplanets have been found around any of them. There were reports on a discovery of exoplanets around the horizontal and extreme horizontal branch stars in the past ten years, but all of them awaited the counterpart papers denying the existence of exoplanets.

The lack of exoplanets around WDs suggests that planets are lost after their hosts stars leave the main sequence. The hydrodynamic simulations by Staff *et al.* (2016) show, that for 10 M_J planet it takes less than 3 years to spiral down onto 3.5 $M_\odot$ host star, if the planet gets engulfed by the stellar envelope during the RG, and less than 100 years during the AGB phases. Therefore we can expect, there are no nearby survivors to the host stars. Any planet which had survived the RG/AGB phase, should be further away than the RG/AGB radius of the host star.

Recent observation of the disintegrating planetary debris around the WD 1145+017 (Vanderburg & Rappaport 2018) could be the indirect indication for the planetary presence around white dwarfs. If present, the exoplanets further away from a white dwarf could perturb asteroid orbits, so the rocky material falls onto the WD.

Table 1. The list of analyzed objects

KIC ID#	Type	Mass in $M_\odot$	ref.	remarks
KIC 8626021	DBV	0.55	1	single
KIC 9472174	sdBV+dM	0.48 + 0.12	2	eclipsing binary, P_{orb} = 0.12576528 d
KIC 7975824	sdB+WD	0.47 + 0.59	3	eclipsing binary, P_{orb} = 0.40375026 d

Notes: [1](Giammichele *et al.* 2018), [2](Zola & Baran 2013), [3](Bloemen *et al.* 2011)

2. Search and analysis techniques

To find out how far away from the WD (or subdwarf) the exoplanets might have survived, we began our search for planets around extreme horizontal branch stars, WDs, as well as binary stars (the common envelope phase would also affect the exoplanetary systems evolution). One of the most important stars for our investigations are B-type subdwarfs (sdB). These stars lost most of the hydrogen from their envelope before the helium flash, therefore after the RG phase they will go directly to the WD cooling track omitting the entire AGB phase. As such, sdBs allow to track the fate of the exoplanetary systems (if present) right after the host star RG phase.

Our analysis techniques were based on the O-C diagrams for eclipcing binary systems, Fourier Transform (FT) of the light curves and binarograms. The "binarogram" term was introduced by Ballona (2014). In short, this method is designed to detect the time-delay effect in the most efficient way using pulsation frequencies as a clock. The method was successfully tested on delta Scuti stars (about ~1.5 as massive as the Sun). The lowest mass limit of an object, which could be detected with this method, was estimated to be ~2 M_J. Because sdB and WD masses are smaller than the delta Scuti stars, we might be able to detect lower-mass objects.

For our analysis we used Kepler space telescope data (the primary mission) of several sdBs, WDs and eclipsing binaries. This report outlines the results obtained for three objects listed in Table 1.

3. Light curves analysis

KIC 8626021: a pulsating helium atmosphere WD (DBV) showing a rich pulsating frequency spectrum in its light curve FT. The short cadence (SC) Kepler date was analyzed using the binarogram method. Due to the variability in the pulsation mode amplitudes and frequencies of the star, the resulting binarograms were contaminated by a number of artifact orbital frequencies. Therefore to rule out the artifacts, binarograms were calculated for different sets of frequencies (taken from Giammichele *et al.* 2018) as well as two halfs of the data. The real orbital signal should be independent on the data and frequency sets used for calculations and it should have the same frequency and amplitude with the accuracy corresponding to the binarogram noise. The noise and the 4 σ detection thresholds in binarograms were calculated similar way as the noise in FT.

As a result, all signals visible within 0.001–3.0 c/d orbital frequency range of the KIC 8626021 binarogram were ruled out and classified as artifacts.

KIC 9472174: since the star is an eclipsing binary (composed of the red dwarf and sdBV stars, see Table 1), we used the eclipses to calculate times of minima and the O-C diagram. The star was analyzed this way before by Baran *et al.* (2015), where mid-eclipse times were calculated using Kwee-van Woerden method (KW) (Kwee & van Woerden 1956), after prewhitening pulsation modes of the sdBV from the SC Kepler data. From their O-C diagram shape (Fig. 1), authors inferred the presence of the ~2 M_J mass object on the 0.9 AU orbit around the system. The orbital period of the putative exoplanet was estimated to be around 412 days (0.0024 c/d).

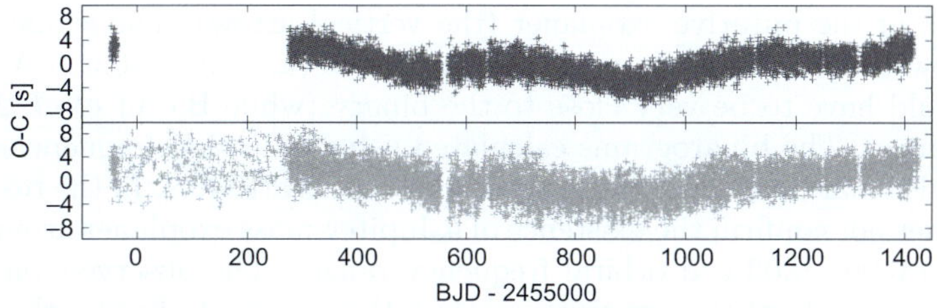

Figure 1. KIC 9472174 O-C diagrams: Baran *et al.* (2015) (top) and our (bottom). Sparse points (near 0–200 BJD) in the figure correspond to the LC data.

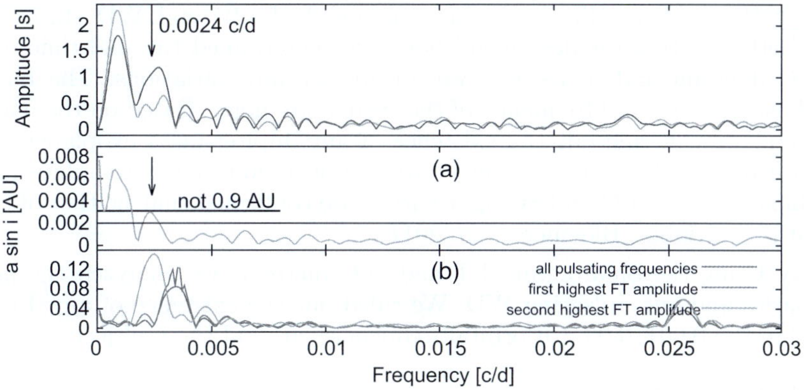

Figure 2. KIC 9472174. Top - FT of the O-C diagram shown in Fig. 1 (violet - Baran *et al.* 2015; yellow - our O-C diagram). Bottom - binarograms calculated using a) - the orbital frequency of the KIC 9472174 binary and b) - pulsation frequencies of the sdBV component. The vertical arrows show the orbital frequency of the putative exoplanet.

Since the SC data cover the light curve minima with 11-13 points only, we decided to use a light curve template and the least square method, to fit this template to the KIC 9472174 data and calculate minima times this way. The template of the binary light curve was prepared by phasing and averaging the whole light curve. In addition to SC data, we also used the long cadence (LC) light curve of the binary to calculate mid-eclipse times. In this case however, the template was composed of the multiple of 10 phased and averaged LC data (to compensate for the poor data point coverage of the light curve). The resulting O-C diagram is presented at the bottom of Fig. 1, where the sparse pints in the figure correspond to the LC data. We did not prewhitened pulsation frequencies from the data, therefore we think, we observe larger scatter in our O-C diagram (comparing to the O-C scatter in the upper panel of Fig. 1).

Thanks to the length of the Kepler data, the number of minima points in the O-C diagram is large enough to calculate the Fourier transform (FT) out of the O-C curve. This allows to determine the amplitude and frequency of the O-C variations from the FT amplitude spectra, instead of guessing the best model which would fit these variations. A top panel of Fig. 2 shows the FT of both Baran *et al.* (2015) (violet) and our (yellow) O-C diagrams. It appears, the signal amplitude near the orbital frequency of the putative exoplanet (marked with the vertical arrows) is ~twice lower in the FT of our O-C than Baran *et al.* (2015) one.

The bottom panels of Fig. 2 present binarograms calculated using a) - the orbital frequency of the KIC 9472174 binary, b) - frequencies of the pulsating modes of the sdBV component. Neither of the two binarograms show signals near the orbital frequency (nor the amplitude) of the putative exoplanet (the vertical arrow). The signal visible in the panel a) (below the arrow) has a very low amplitude (a×sin(i) ∼0.003 AU), meaning - the planet would have to be very close to the binary (while Baran *et al.* 2015 gives the ∼0.9 AU distance). The binarograms calculated using pulsation frequencies in the panel b) show several signals, but all of them were ruled out as artifacts. In the result we neither found any planet nor confirm the existence of a Jupiter-mass exoplanet around this binary system (within 0.001–0.03 c/d orbital frequency range). The observed variations in the O-C diagram, needs further investigations but they are not due to the exoplanetary presence. At the time of writing, we can only speculate, the variations might be due to spots on the dM component or amplitude modulations of the pulsation modes of the sdBV star.

KIC 7975824: another eclipsing binary composed of sdB and WD stars. Using the template method fitting (as described above), we determined the mid-time of eclipses. In the O-C diagram and in its FT, we do not see any variations. The binarograms calculated using the orbital frequency of the system are noisy. Based on them and on the O-C diagram, we can rule out the existence of any Jupiter-mass exoplanet within the 0.001–0.03 c/d orbital frequency range. After subtraction of the template light curve from the binary system LC and SC light curves, we confirmed that neither of the binary component is a variable (Bloemen *et al.* 2011).

Summary Using the binarogram, FT and O-C methods we analyzed the light curve of two binaries and one pulsating WD. We ruled out the existence of any Jupiter-mass planet within the 0.001–0.03 c/d orbital frequencies around these objects.

Acknowledgements

This work was supported by the National Science Centre, Poland. The project registration number is 2017/25/B/ST9/00879.

References

Baran, A.-S., Zola, S., Blokesz, A., Ostensen, R.-H., & Silvotti, R. 2015, *A&A*, 577, A146
Ballona, L. 2014, *MNRAS*, 443, 1946
Blokesz, A., Krzesinski, J., & Kedziora-Chudczer L. 2019, *A&A*, 627, A86
Bloemen, S., Marsh, T.-R, Ostensen, R.-H., *et al.* 2011, *MNRAS*, 410, 1787
Giammichele, N., Charpinet, S., Fontaine, G., *et al.* 2018, *Nature*, 554, 73
Kwee, K. K. & van Woerden, H. 1956, *Bull. Astron. Inst. Neth.*, 9, 252
Staff, J., De Marco, O., Wood, P., Galaviz, P., & Passy, J.-C. 2016, *MNRAS*, 458,832
Zola, S. & Baran, A. 2013, *Central Eur. Astrophys. Bull.*, 37, 227
Vanderburg, A. & Rappaport, S.-A. 2018, *Handbook of Exoplanets (New York: Springer International Publishing AG)*

White Dwarfs as Probes of Fundamental Physics: Tracers of Planetary, Stellar and Galactic Evolution
Proceedings IAU Symposium No. 357, 2019
M. A. Barstow, S. J. Kleinman, J. L. Provencal & L. Ferrario, eds.
doi:10.1017/S1743921320000447

Variation of fundamental constants and white dwarfs

Susana J. Landau[1,2]

[1]Departamento de Física, Facultad de Ciencias Exactas y Naturales, Universidad de Buenos Aires and IFIBA, Ciudad Universitaria - Pab. I, Buenos Aires 1428, Argentina
email: `slandau@df.uba.ar`

[2]CONICET, Godoy Cruz 2290, 1425 Ciudad Autónoma de Buenos Aires, Argentina

Abstract. Theories that attempt to unify the four fundamental interactions and alternative theories of gravity predict time and/or spatial variation of the fundamental constants of nature. Different versions of these theories predict different behaviours for these variations. As a consequence, experimental and observational bounds are an important tool to check the validity of such proposals. In this paper, we review constraints on the possible variation of the fundamental constants from astronomical observations and geophysical experiments designed to test the constancy of the fundamental constants of nature over different timescales. We also focus on the limits that can be obtained from white dwarfs, which can constrain the variation of the constants with the gravitational potential.

Keywords. quasars: absorption lines, cosmology: theory, cosmic microwave background, supernovae: general , stars: white dwarfs

1. Introduction

Our present knowledge of the properties of nature is based on two fundamental theories: General Relativity (GR) which describes gravitational interaction and the Standard Model of Elementary Particles which depicts the electromagnetic, strong and weak interactions. In this way, all physical phenomena at low energies can be described by the solutions of the equations of both theories. However, there is something missing in this picture, namely, there are 20 parameters in these equations that are not given by the theories and must be determined by experiments. These parameters are the ones that we call the fundamental constants and examples of them are the masses of elementary particles like quarks and leptons; the gauge coupling constants of the electromagnetic, weak and strong interactions; the gravitational constant G_N, the velocity of light c and the Higgs vacuum expectation value $< v >$. In turn, the principle of equivalence on which GR is based, requires the invariance of these quantities against changes in position, time and reference system. The gauge coupling constants of $U(1)$, $SU(2)$ and $SU(3)$ are related to the fine structure constant α, the QCD energy scale Λ_{QCD} and the Fermi coupling constant G_F through the following equations:

$$\alpha^{-1}(E) = \frac{5}{2}\alpha_1^{-1}(E) + \alpha_2^{-1}(E) \tag{1.1}$$

$$\Lambda_{QCD} = E\exp\left[-\frac{2\pi}{7}\alpha_3^{-1}(E)\right] \tag{1.2}$$

$$G_F = \frac{\pi\ \alpha_2(M_W)}{\sqrt{2}M_W^2} \tag{1.3}$$

where M_W refers to the mass of the W boson. Furthermore, gyromagnetic factors of atoms g_i are also considered fundamental constants and for the purpose of this paper we will consider the proton's mass m_p as a fundamental constant.

Since the Large Number hypothesis was formulated by Dirac in 1937, the variation of the fundamental constants has been the subject of numerous research papers. Dirac (1937) observed a remarkable coincidence: the dimensionless ratio between the gravitational and the electromagnetic force between a proton and an electron is of order 10^{-39} while the time it takes for the light to pass through a hydrogen atom is just 3 orders of magnitude smaller. This remarkable coincidence led him to formulate the Large Number Hypothesis according to which one or more constants are simple functions of the age of the universe. Furthermore, Dirac proposed that the variation of G_N is of the form: $G_N \sim t^{-1}$. Simultaneously, Milne (1937) also considered a possible variation in the G_N but with a different dependence with time : $G_N \sim t$. Following Dirac's proposal, Teller (1948) suggested that the fine structure constant is also a function of cosmological time and in particular he proposed the following dependence: $\alpha \sim (\log t)^{-1}$. Later, Pochoda & Schwarzschild (1964), showed that Dirac's proposal cannot explain the Sun's observed age and luminosity. In an attempt to rescue Dirac's idea, Gamow (1967) proposed that while G_N remains constant, α varies as a linear function of time. However, Dyson (1967) ruled out Gamow's proposal using the abundance of long lived β decayers in meteorites. At the same time, Bahcall & Schmidt (1967) reached the same conclusion using quasar absorption system spectra. The interest in time and spatial variation of fundamental constants got renewed when the attempt to unify the four interactions of nature resulted in the development of multidimensional theories such as Kaluza-Klein theories (Kaluza 1921; Klein 1926; Marciano 1984), string theories (Maeda 1988; Barr & Mohapatra 1988; Damour & Polyakov 1994; Damour *et al.* 2002) and related brane theories (Youm 2001; Palma *et al.* 2003). In Kaluza-Klein theories, the variation of the fundamental constants is related to the cosmological evolution of the radius of the extra dimensions, while in the case of superstrings and branes, the variation of the vacuum expectation value of a non-massive scalar field (for example the dilaton in string theories) is responsible for such variation. On the other hand, scalar-tensor theories of gravity are natural frameworks to study the variation of G_N. In these theories, a scalar field couples to the Ricci scalar and its dynamics provides the physical mechanism for the variation in G_N (Jordan 1959; Brans & Dicke 1961). More recently, Moffat (2006) proposed an alternative theory of gravity, which attempts to explain observational data from a large number of astrophysical scenarios without the need to include dark matter. In the latter, G_N is also allowed to vary with space and/or time. Following a different path of research, Bekenstein (1982) proposed a theoretical framework to study the fine structure constant variability based on general assumptions: covariance, gauge invariance, causality and time-reversal invariance of electromagnetism, as well as the idea that the Planck-Wheeler length $\left(10^{-33}\text{cm}\right)$ is the shortest scale allowable in any theory. Similar phenomenological frameworks based on first principles were also proposed by other authors (Barrow *et al.* 2002; Olive & Pospelov 2002; Chamoun *et al.* 2001; Barrow & Magueijo 2005). In these phenomenological frameworks, the physical mechanism responsible for such variation is a scalar field which is added to the theory.

The experimental research can be grouped into astronomical and local methods. The latter ones include geophysical methods such as the natural nuclear reactor that operated about 1.8 10^9 years ago in Oklo, Gabon, and laboratory measurements such as detailed comparisons of several atomic clock frequencies with different atomic number. The astronomical methods are based mainly in the analysis of spectra from high-redshift quasar absorption systems. Moreover, data from type Ia supernovae also provide limits on the possible variation in α. Besides, primoridal nucleosyntesis and the Cosmic Microwave

Backdground (CMB) provide constraints on the the variation of the fundamental constants in the early universe. Moreover, white dwarf spectra and the mass-radius relation in white dwarfs also provide stringent constraints on the variation in α and the proton to electron mass $\mu = \frac{m_p}{m_e}$. It was mentioned above that in most of the theories that predict variation of the fundamental constants, the physical mechanism for such variation is the dynamics of a scalar field. Near massive objects, like a white dwarf, the effect of the scalar field can change. Therefore, constraints on the fundamental constants from white dwarfs, can also be regarded as limits on those variations in terms of a variation in the gravitational potential. In such a way, the constraints on $\frac{\Delta\alpha}{\alpha}$ can be translated in terms of a dependence on a dimensionless gravitational potential:

$$\frac{\Delta\alpha}{\alpha} \simeq k_\alpha^{(1)} \Delta\Phi + k_\alpha^{(2)} (\Delta\Phi)^2 \tag{1.4}$$

Similar, constraints on the possible variation of the proton to electron mass μ can be expressed as follows:

$$\frac{\Delta\mu}{\mu} \simeq k_\mu^{(1)} \Delta\Phi + k_\mu^{(2)} (\Delta\Phi)^2 \tag{1.5}$$

In Equations 1.4 and 1.5 Φ refers to the gravitational potential and the coefficients $k_\alpha^{(i)}$, $k_\mu^{(i)}$ can be determined by experimental or observational data. In atomic clock experiments, the difference in the gravitational potential is of order $\Delta\Phi \sim 10^{-10}$, while the difference between the gravitational potential in a white dwarf and the respective one on Earth is $\Delta\Phi \sim 10^{-4} - 10^{-5}$. This is the main reason to affirm that white dwarf spectra are the ideal probe for a relationship between the fundamental constants like α and μ and strong gravitational fields. In this article, we will describe bounds on the fundamental constants from geophysical experiments and astronomical observations with a focus on those that can be obtained from white dwarfs. In section 2 we describe the bounds from geophysical data like the Oklo nuclear reactor and atomic clocks. We also discuss the bounds on the relation between the fundamental constants and the gravitational potential that can be obtained with this method. In section 3 we review the bounds from quasar absorption systems and discuss the status of the claimed variation in α from the Many Multiplet Method. Besides, bounds from primordial nucleosynthesis and the Cosmic Microwave Background are discussed in Section 4. Furthermore, bounds from type Ia supernovae data are described in Section 5. A detailed discussion on the bounds that can be obtained from white dwarf data on the possible variation of α, μ and G_N is described in Section 6. We also provide a description on the bounds on G_N that can be obtained from the Lunar Laser Ranging experiment (see Section 7) and helioseismology (see Section 8). Finally, in Section 9 we discuss our conclusions.

2. Bounds from geophysical data

2.1. *The Oklo Phenomenon*

One of the most stringent limits on time variation of the fine structure constant α follows from an analysis of isotope ratios in the natural uranium fission reactor that operated 1.8×10^9 years ago at the present day site of the Oklo mine in Gabon, Africa. From an analysis of nuclear and geochemical data, the operating conditions of the reactor could be reconstructed and the thermal neutron capture cross sections of several nuclear species measured. In particular, a shift in the lowest lying resonance energy level in $^{149}\mathrm{Sm} : \Delta = E_r^{149(\mathrm{Oklo})} - E_r^{149(\mathrm{now})}$ can be derived from a shift in the neutron capture cross section of the same nucleus (Damour & Dyson 1996). The shift in Δ can be translated into a bound on a possible difference between between the values of α and G_F during the Oklo phenomenon and their value now:

$$\Delta = 10^6 \mathrm{eV}\frac{\Delta\alpha}{\alpha} + 5.6\mathrm{eV}\frac{\Delta G_F}{G_F} \tag{2.1}$$

Several authors have analyzed the Oklo data to put constraints on the possible variation of the fundamental constants (Petrov *et al.* 2006; Davis & Hamdan 2015). The most stringent limit was established by Davis & Hamdan (2015):

$$\frac{\Delta\alpha}{\alpha} < 1.1 \times 10^{-8} \tag{2.2}$$

2.2. *Atomic Clocks*

Atomic clocks provide the most stringent bounds on the variation of fundamental constants. The method consists of comparing atomic clock frequencies of different transitions, which in turn have different dependence with the fundamental constants. For example, the dependence of a hyperfine transition frequency with the fundamental constants can be expressed as follows:

$$\nu_{\mathrm{Hyp}} \sim \alpha^2 g_i \frac{m_e}{m_p} R_\infty c F_{REL}(\alpha Z) \tag{2.3}$$

where g_i is gyromagnetic factor, R_∞ is Rydberg's constant, m_e and m_p are the electron's and proton's mass respectively and F_{REL} is the relativistic contribution to the energy. In such way, clocks based on hyperfine transitions in alkali atoms with different atomic number Z can be used to set bounds on $\alpha^k \frac{\mu_{A_1}}{\mu_{A_2}}$ where k depends on the frequencies measured and μ_{A_i} refers to the nuclear magnetic moment of each atom (Prestage *et al.* 1995; Marion *et al.* 2003; Guéna *et al.* 2012).

On the other hand, an optical transition frequency has the following dependence on α:

$$\nu_{opt} \sim R_\infty B F_i(\alpha) \tag{2.4}$$

where B is a numerical constant assumed not to vary in time and $F_i(\alpha)$ is a dimensionless function of α that takes into account level shifts due to relativistic effects. Thus, comparing an optical transition frequency with a hyperfine transition frequency can be used to set bound on $\alpha^k \frac{m_e}{m_p}\frac{\mu_A}{\mu_B}$ (Le Targat *et al.* 2013; Huntemann *et al.* 2014; Falke *et al.* 2014). Furthermore, the most stringent bound on α variation using this method was obtained by Rosenband *et al.* (2008) and it follows from a comparison between two optical transition frequencies. On the other hand, the energy difference between two adjacent rotational levels in a diatomic molecule is proportional to $\frac{1}{Mr^2}$, r being the bond length and M the reduced mass. Morevover, the vibrational transition of the same molecule has, in first approximation, a $\sqrt{M}$ dependence. In this way, it follows that the vibro-rotational molecular transitions are proportional to $\frac{1}{\sqrt{M}}$. Therefore, comparing vibro-rotational transitions with an hyperfine transition gives information about the variaton of $g_{\mathrm{hyp}}\sqrt{\frac{m_e}{m_p}}\alpha^k$. Shelkovnikov *et al.* (2008) used this latter method to compare the vibro-rotational transition of the SF_6 molecule with the ^{133}Cs hyperfine transition. Their results can also be used to constrain $k_\mu^{(1)}$ from Eq. 1.5. In this experiment, the difference in the gravitational potential in the Earth due to the Sun between aphelion and perihelion was estimated to be $\Delta\Phi = 10^{-10}$, and therefore the bound on $\frac{\Delta\mu}{\mu}$ yieds:

$$k_\mu^{(1)} < 4 \times 10^{-4} \tag{2.5}$$

On the other hand, comparison of radio-frequency transitions between nearly degenerate, opposite parity excited states in atoms allows us to establish bounds on the possible variation in α due to the large relativistic corrections of opposite sign for the opposite-parity levels. This method was used by Leefer *et al.* (2013) to establish a bound in α from the comparison of frequency transitions in two isotopes of atomic dysprosium (Dy). This bounds can also be used to set constraints on $k_{\alpha}^{(1)}$ in Equation 1.4. In fact, since the estimated difference in the gravitational potential between aphelion and perihelion for this experiment is $\Delta\Phi = 3.3 \times 10^{-10}$, the bound on $\frac{\Delta\alpha}{\alpha}$ yieds:

$$k_{\alpha}^{(1)} = (-5.5 \pm 5.2) \times 10^{-7} \tag{2.6}$$

Finally, combining all available laboratory data Godun *et al.* (2014) obtain:

$$\frac{\dot{\alpha}}{\alpha} = (0.7 \pm 2.1) \times 10^{-17}\text{yr}^{-1} \tag{2.7}$$

$$\frac{\dot{\mu}}{\mu} = (0.2 \pm 1.1) \times 10^{-16}\text{yr}^{-1} \tag{2.8}$$

3. Bounds from Quasar Absorption Systems

Quasar absorption systems present ideal laboratories in which to test for possible variations of the fundamental constants. The relationship between the wavelength observed in quasar spectra (λ_{obs}) and the ones measured in the laboratory (λ_{lab}) can be expressed:

$$\lambda_{\text{obs}} = \lambda_{\text{lab}}\,(1 + z)\,. \tag{3.1}$$

Several methods have been developed to test the possible variation of the fundamental constants, among them, it is important to mention the following: i) Alkali Doublet Method, ii) Many Multiplet Method, iii) Comparison of molecular spectra with laboratory spectra, iv) Comparison of radio spectra with optical spectra, v) Comparison of radio spectra with molecular spectra and vi) Conjugate Lines Method. Some of them such as the Akali Doublet Method rely on the comparison between the observed wavelengths in the quasar and the one measured in the laboratory. Others, are based on the comparison of absorption redshifts due to different transitions that happen in the same absorption cloud. We will not discuss all methods in this article. We will focus on i) the Many Multiplet Method which is so far the method that allows us to place the more stringent constraints on $\frac{\Delta\alpha}{\alpha}$from quasar spectra and ii) the comparison between molecular and optical spectra which provides stringent constraints on the variation of μ.

The Many Multiplet Method was proposed by Webb *et al.* (1999) and relies on the comparison of transitions with different atomic masses in the same absorption cloud. In short, the Many Multiplet Method consist of adjusting the Voigt profiles of several absorption lines, including besides the usual fit parameters: column density, Doppler width, redshift, a possible variation of the fine structure constant. In this way, the method allows us to gain an order of magnitude in sensibility with respect to previously reported data. Using this method and data provided by the Keck telescope, Webb *et al.* (1999) claimed a detection of a variation in α. However, an independent analysis performed with observations made with UVES at the Very Large Telescope (VLT) provided null results (Srianand *et al.* 2004). Contrary to previous results, a third analysis, this time with VLT/UVES archival data also indicated a variation in α, but now with α increasing with redshift (Webb *et al.* 2011; King *et al.* 2012). These results, led the authors to suggest a spatial dipole-type variation in α. The weighted mean of the 293 measurements obtained with data from both Keck and VLT telescopes by the group of Webb et al at $0.3 < z < 3.1$ is:

$$\left(\frac{\Delta\alpha}{\alpha}\right) = -2.16 \pm 0.86 \times 10^{-6}. \tag{3.2}$$

On the other hand, more recently, a reanalysis of systematic errors with new techniques showed that there is no compelling evidence for any variation in α from quasar data

(Whitmore & Murphy 2015). However, it should be noted that in all those mentioned analyses the data acquisition procedures were far from ideal, in particular, regarding the key issue of wavelength calibration. Trying to confirm the above mentioned results was the main motivation for the ESO UVES Large Program, for which key improvements in the data acquisition procedures were implemented (Bonifacio *et al.* 2014). Furtheremore, other dedicated measurements of the variation in α with the quasar method were performed recently, by several authors (Agafonova *et al.* 2011; Bainbridge & Webb 2017; Evans *et al.* 2014; Kotuš *et al.* 2017; Molaro *et al.* 2013; Songaila & Cowie 2014). In particular Murphy *et al.* (2016) developed a method to test a variation in α which is not influenced by long-range distortions. The weighted mean of all recent dedicated measurements which span the redshift range $1 < z < 2.4$ is (Martins 2017):

$$\left(\frac{\Delta\alpha}{\alpha}\right) = -0.64 \pm 0.65 \times 10^{-6}, \tag{3.3}$$

showing no variation in α. Nevertheless, Murphy *et al.* (2016) reached the conclusion that their quasar sample is too small to rule out the dipole model. On the other hand, it should be noted that all above mentioned measurements were performed with spectrographs such as UVES, HARPS or Keck-HIRES, which are far from optimal for this type of measurement. Therefore, more precise measurements using the new generation of high-resolution spectrograph, like ESPRESSO for the VLT and E-ELT-HIRES for the E-ELT, are expected to improve significantly the precision of the data.

Varshalovich & Levshakov (1993) developed a method for constraining the variation in $\mu = \frac{m_p}{m_e}$ which is based on the fact that wavelengths of electron-vibro-rotational lines depend on the reduced mass of the molecules, with different dependence for different transitions. In such way, it is possible to distinguish the cosmological redshift of a line from the shift caused by a variation in μ. The wavelength of a molecular line at redshift $z_{\rm abs}$ can be expressed as:

$$\lambda_i = \lambda_i^{\rm lab}\,(1 + z_{\rm abs})\left(1 + K_i \frac{\Delta\mu}{\mu}\right) \tag{3.4}$$

where $\lambda_i^{\rm lab}$ is the wavelength measured in the laboratory and K_i is a coefficient for the molecular band. Several authors have used this method to place stringent constraints on the variation in μ at $0.6 < z < 4.2$ (Kanekar 2011; Bagdonaite *et al.* 2013; van Weerdenburg *et al.* 2011; Rahmani *et al.* 2013; Daprà *et al.* 2015; Albornoz Vásquez *et al.* 2014; Bagdonaite *et al.* 2015). Measurements at low redshift were obtained with radio/mm observations and its weighed mean yields $\frac{\Delta\mu}{\mu} = (-0.24 \pm 0.09) \times 10^{-6}$ (Martins 2017) while the high redshift sample proceeds from UV/optical observations with weighed mean $\frac{\Delta\mu}{\mu} = (2.9 \pm 1.9) \times 10^{-6}$ (In table 2 we show the weighed mean of the complete data set). There is weak evidence for a variation in μ from both samples, even though it should be stressed that the sign of the variation is different at high and low redshift. Besides, Bagdonaite *et al.* (2013b) obtain the most stringent limit from observations of metanol transitions at $z = 0.89 : \frac{\Delta\mu}{\mu} = (-1 \pm 1.8) \times 10^{-7}$.

4. Bounds from the early universe

4.1. *Cosmic Microwave Background*

The Cosmic Microwave Background (CMB) radiation is one of the best tools to study the early universe because it provides information about the physical conditions in the Universe just before decoupling of matter and radiation. Possible variation in α and the electron mass m_e affect the Thompson scattering cross section $\sigma_T = \frac{8\pi}{3}\frac{\hbar^2 c^2}{m_e^2}\alpha^2$ and the ionization fraction. The main effect in the ionization fraction can be expressed as follows:

$$x_e \simeq \left(\frac{m_e}{k_B T}\right)^{\frac{3}{2}} \exp - \left(\frac{B_1}{k_B T}\right) \tag{4.1}$$

where $B_1 = \frac{\alpha^2 m_e c^2}{2}$ is the Hydrogen binding energy. The effect of a possible variation in α and/or m_e on the CMB doppler peaks is a shift in the position of the peaks and modification in the height of the peaks. From the comparison of the theoretical prediction with observational data obtained by the Planck Collaboration *et al.* (2015), bounds on the possible variation in α and/or m_e can be obtained. Assuming indepedent variations in α and m_e, Hart & Chluba (2018) obtain:

$$\left(\frac{\Delta\alpha}{\alpha}\right) = (0.7 \pm 2.5) \times 10^{-3} \quad \text{and} \quad \left(\frac{\Delta m_e}{m_{e0}}\right) = (3.9 \pm 7.4) \times 10^{-3} \tag{4.2}$$

where the last bound was obtained considering also Baryon Acoustic Oscilation (BAO) data in the statistical analysis. On the other, hand if joint variations of α and m_e are assumed and BAO data are considered in the analysis, the limits obtained are:

$$\left(\frac{\Delta\alpha}{\alpha}\right) = (1.1 \pm 2.6) \times 10^{-3} \quad \text{and} \quad \left(\frac{\Delta m_e}{m_{e0}}\right) = (5.6 \pm 8) \times 10^{-3} \tag{4.3}$$

4.2. *Big Bang Nucleosynthesis (BBN)*

The study of the light nuclei production during the first three minutes of the universe is another important tool to study the early universe. In short, the abundances of light nuclei depend on the following physical parameters: freeze-out time of the weak interactions, neutron-proton mass diference, binding energies of the light nuclei and cross sections of the reactions producing D, ^{4}He, ^{7}Li, ^{6}Li, Tritium, ^{3}He, and ^{7}Be, which, in turn are functions of the fundamental constants α, the Higgs vacuum expectation value $< v >$ and G_N. It should be stressed that unlike other methods described in this paper, it is necessary to asume a theoretical model for the strong interaction within this analysis. Mosquera & Civitarese (2017) considered the Argonne potential for the nucleon-nucleon interaction. From the comparison with the observed abundances of D, ^{4}He and ^{7}Li they obtained:

$$\frac{\Delta\alpha}{\alpha} = -0.022 \pm 0.006 \quad \text{and} \quad \frac{\Delta < v >}{< v >} = 0.042 \pm 0.01 \tag{4.4}$$

whereas if the Bonn potential is assumed, the analysis yielded:

$$\frac{\Delta\alpha}{\alpha} = -0.022^{+0.003}_{+0.004} \quad \text{and} \quad \frac{\Delta < v >}{< v >} = 0.036 \pm 0.007 \tag{4.5}$$

On the other hand, the value of G_N determines the expansion rate of the universe and thus, the relevant time scales for the weak and nuclear reactions. As a consequence, assuming a variation in G_N at the time of the BBN, translates in turn into a variation of the light element abundances with respect to the ones predicted by the standard cosmological model. Copi *et al.* (2004) and Bambi *et al.* (2005) compared the theoretical predictions for the light nuclei abundances with the present abundances of D, ^{4}He and ^{7}Li and established the following bound:

$$\frac{\dot{G}}{G} = (0.05 \pm 0.35) \times 10^{-12} \text{yr}^{-1} \tag{4.6}$$

5. Bounds from supernovae type Ia

Type Ia supernovae (SNe Ia) are among the most energetic and interesting phenomena in our universe. Furthermore, the spectra and light curves of normal SNe Ia are very homogeneous in such a way that they are considered one of the best standard candles known today. All this makes them suitable astronomical objects to test the possible variation of fundamental constants. The homogeneity of the light curve is basically caused by the homogeneity of the nickel mass produced during the supernova outburst. This effect is primary determined by the value of the Chandrasekhar mass which in turn depends on the value of the fundamental constants G_N, α or the velocity of light c as follows:

$$M_{\rm ch} \sim \left(\frac{\hbar c}{G_N}\right)^{3/2} \sim \left(\frac{e^2}{\alpha^2 G_N}\right)^{3/2} \tag{5.1}$$

In addition, the possible variation of the fundamental constants with cosmological time would also change the cosmic evolution and therefore affect the luminosity distance relation. Gaztañaga *et al.* (2002) performed an analysis allowing only for the variation in G_N and showed that the latter is typically several times smaller than the change produced by the corresponding variation of the Chandrasekhar mass. Furthermore, these authors used data from the Supernova Cosmology Project to put the following bound:

$$\frac{\dot{G}_N}{G_N} < 10^{-11}{\rm yr}^{-1} \tag{5.2}$$

Garcia-Berro *et al.* (2006) considered that the variation in G_N follows the predictions of scalar-tensor theories and reached similar conclusions. On the other hand, Kraiselburd *et al.* (2015) analised the dependence of type Ia supernovae explosions on α, including both the change in the Chandrasekhar mass and the the dependence of the mean opacity of the expanding supernovae photosphere. Furthermore, motivated by the results of the quasar data, they considered a spatial dipole-type variation for α. From the comparison of supernovae data from the Union 2.1 compilation, they obtained:

$$\frac{\Delta\alpha}{\alpha} = (2.1 \pm 0.8) \times 10^{-2} \tag{5.3}$$

In a posterior work, Negrelli *et al.* (2018) considered the variation in c as the source of the variation in the Chandrasekhar mass. In addition, they also included the change of the energy release during the explosion in their analysis. Comparing the theoretical predictions for a spatial dipole-type variation in c, with the Union 2.1 and JLA supernovae type Ia data, they established:

$$\frac{\Delta c}{c} = (-3.7 \pm 0.8) \times 10^{-2} \tag{5.4}$$

6. Bounds from White Dwarfs

6.1. *Bounds on α and μ*

Hot white dwarf stars are ideal laboratories to probe for a relationship between the fundamental constants and strong gravitational fields. Hot white dwarfs with masses comparable to the sun and radii comparable to Earth generate strong gravitational fields and are typically bright with numerous absorption lines. The relationship between the laboratory wavelengths ($\lambda_{\rm lab}$) and those observed near a white dwarf ($\lambda_{\rm WD}$) is:

$$\frac{\Delta\lambda}{\lambda} = \frac{\lambda_{\rm WD} - \lambda_{\rm lab}}{\lambda_{\rm lab}} = z_{\rm WD} - Q_\alpha \frac{\Delta\alpha}{\alpha}(1 + z_{\rm WD}), \tag{6.1}$$

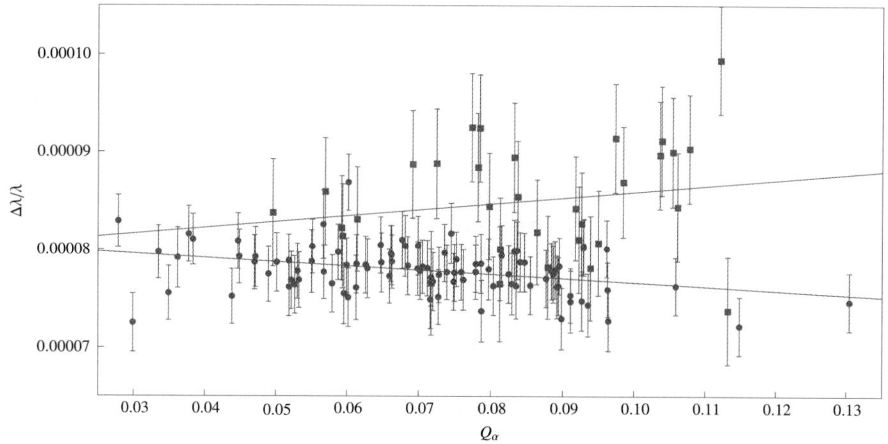

Figure 1. $\frac{\Delta\lambda}{\lambda}$ vs. Q_α for transitions in FeV (blue circles) and NiV (red squares). The slope of the lines give $\frac{\Delta\alpha}{\alpha} = (4.2 \pm 1.6) \times 10^{-5}$ for FeV and $\frac{\Delta\alpha}{\alpha} = (-6.1 \pm 5.8) \times 10^{-5}$ for NiV, respectively. Reprinted with permission from Physical Review Letters 111, 010801 (2013). Copyright(2013) by the American Physical Society.

where $Q_\alpha = 2q/\omega_0$ is the relative sensitivity of the transition frequency to a variation in α. Berengut *et al.* (2013) observed the wavelength shift in 96 quadruply ionized iron and 32 quadruply ionized nickel absorption features from the white dwarf star G191-B2B and derived separate limits for each metal. In all measurements line calibration was the expected culprit for the observed signal. Both results (see Figure 1) are inconsistent with each other at the 1.6 sigma level, the likely reason being related to uncertainties in laboratory wavelength measurements and systematics in the wavelength calibration. More recently, Bainbridge *et al.* (2017) presented preliminary results of a similar analysis incorporating improvements such as; i) the use of new laboratory wavelengths; ii) measurements of a sample of objects rather than a single object ; and iii) the use of robust techniques from quasar absorption systems (the Many Multiplet method) in the data analysis method. Therefore, future measurements with improved data analysis methods could provide more stringent bounds.

On the other hand, a possibe variation in μ would be manifested as shifts in the observed wavelengths of molecular hydrogen (H_2) when compared to laboratory wavelengths as follows

$$\frac{\lambda_i^{WD}}{\lambda_i^{\text{lab}}} = (1 + z_{WD})(1 + \frac{\Delta\mu}{\mu} K_i) \tag{6.2}$$

where λ_i^{WD} represents the H_2 transition wavelength observed in white dwarf spectra, λ_i^{lab} is a corresponding wavelength measured in the laboratory and K_i are the sensitivity coefficients. Bagdonaite *et al.* (2014) applied this method to the spectrum of the white dwarf star G29–38 with a potential of $2 \times 10^4 \Phi_{\text{earth}}$ to obtain:

$$\Delta\mu/\mu = (-5.8 \pm 3.8_{\text{stat}} \pm 0.3_{\text{sys}}) \times 10^{-5} \tag{6.3}$$

while for the white dwarf GD133 with a potential of $10^4 \Phi_{\text{earth}}$ the resulting bound was:

$$\Delta\mu/\mu = (-2.7 \pm 4.7_{\text{stat}} \pm 0.2_{\text{sys}}) \times 10^{-5} \tag{6.4}$$

We have discussed in Section 1 the theoretical motivation for considering varying fundamental constants and the inclusion of scalar fields as the physical mechanism for that variation in most theoretical models. Therefore, constraints on $\Delta\mu/\mu$ can be interpreted in terms of a dependence on a dimensionless gravitational potential. Applying Eq. 1.5 to the above bounds yields the following limit:

$$k_\mu^{(2)} < 1 \times 10^3 \tag{6.5}$$

which is several orders of magnitude more stringent than the ones obtained from Earth-based experiments. And we remember the reader that the constraint on $k_\mu^{(1)}$ is obtained from atomic clock experiments (see Section 2.2).

On the other hand, Magano *et al.* (2017) obtained bounds on the possible variation of the fundamental constants from the mass-radius relation in white dwarfs. Unlike the analyses performed for other bounds described in this paper†, it is necessary to assume a theoretical model for the relation between the variation of the electron and nucleon masses and α. For this, the authors assume the expressions given by the generic class of unification models proposed by Coc *et al.* (2007):

$$\frac{\Delta m_e}{m_e} = \frac{1}{2}(1+S)\frac{\Delta\alpha}{\alpha} \tag{6.6}$$

$$\frac{\Delta m_N}{m_N} = [0.8R + 0.2(1+S)]\frac{\Delta\alpha}{\alpha} \tag{6.7}$$

where m_N refers to the nucleon masses, and R and S depend of the specific unification model considered. Using these relations, the mass continuity and hidrostatic equations can be expressed for the case of varying constants as follows:

$$\frac{dm'}{dr} = m_0 r^2 {x_F}^3 \tag{6.8}$$

$$\frac{dx_F}{dr} = -K_1(1+\beta)\ \frac{m'}{r^2}\frac{\sqrt{1+x_F^2}}{x_F} \tag{6.9}$$

$$m = K_2(1+\gamma)\ m' \tag{6.10}$$

where $K_1 = \frac{16}{3\pi(2q)^2}\frac{R_\odot}{m_0}\frac{c^3}{G\hbar^3}\alpha_e\alpha_N$, $K_2 = \frac{8}{3\pi(2q)8}\frac{{R_\odot}^3}{M_\odot m_0}\frac{c^5}{G^2\hbar}{\alpha_e}^{\frac{3}{2}}{\alpha_N}^{\frac{1}{2}}$, $\alpha_i = \frac{Gm_i^2}{\hbar c}$, $x_F = \frac{p_F}{m_e c}$, p_F is the Fermi momentum, m_0 is a dimensionless constant and q is the number of electrons per nucleon. In the above equations, β and γ enclose the dependence of the above equations with the fundamental constants and are given by the following expressions:

$$\beta = \left[\frac{9}{5}R + \frac{8}{5}(1+S)\right]\frac{\Delta\alpha}{\alpha}, \gamma = \left[\frac{4}{5}R + \frac{23}{10}(1+S)\right]\frac{\Delta\alpha}{\alpha}.$$

Results of the numerical integration of the above equations from the original work are shown in Figure 2, for the cases: i)$\beta = 0$ and $\gamma = 0$ (standard model), ii) $\beta = \pm 0.01$ and iii)a more simple polytropic model; toghether with present data for the mass-radius relation. Next, a statistical analysis was performed to estimate bounds on α from observational data. For each star i in the catalog, Magano *et al.* (2017) chose a value of $M_\star$ to minimize the following quantity:

$$ {\chi_i}^2(M_\star) = \frac{(M_\star - M_i)^2}{\sigma_{M,i}^2} + \frac{(R_{th}(M_\star) - R_i)^2}{\sigma_{R,i}^2}, \tag{6.11}$$

where M_i, $\sigma_{M,i}$, R_i, and $\sigma_{R,i}$ are the mass and radius of the ith star and their respective uncertainties, and $R_{th}(M)$ is the theoretical prediction. Thus, the total value of χ^2 to

† with the exception of primordial nucleosynthesis (see Section 4.2 where a model for the strong interactions has to be assumed)

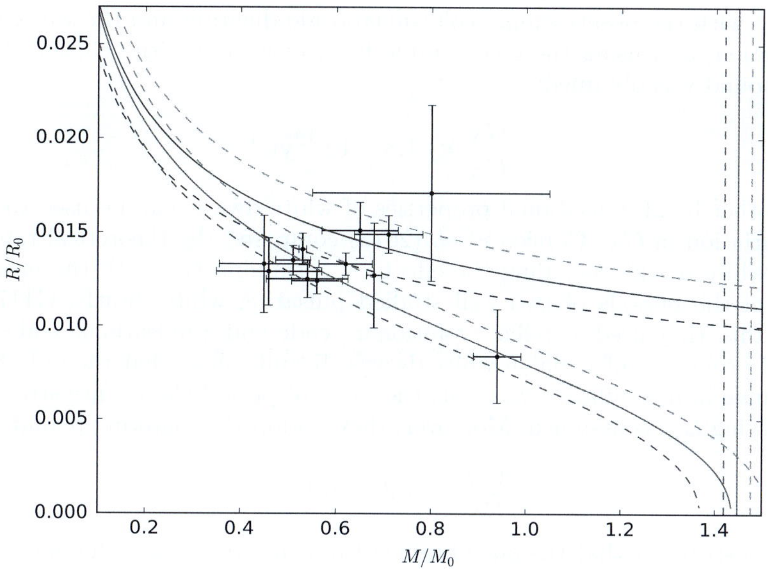

Figure 2. Solid red line for the standard model, with the nearby darker and lighter dashed lines corresponding to $\beta \pm = 0.01$. Solid blue line shows the nonrelativistic limit of the polytropic model, and darker and lighter dashed lines for $x = \pm 0.1$. $x = \left[\frac{4}{3}R + \frac{5}{6}(1+S)\right]\frac{\Delta\alpha}{\alpha}$. The black points with error bars correspond to the data. R_0 and M_0 refer to the Sun's radius and mass respectively. Reprinted with permission from Physical Review D96, 083012 (2017). Copyright(2017) by the American Physical Society.

be calculated was $\chi^2 = \sum_{i=1}^{N} \chi_i{}^2(\hat{M}_i)$ with $\hat{M}_i$ as the value of M that minimizes the corresponding χ_i^2. From the statistical analysis, the following constraints were obtained:

$$\beta = 0.012 \pm 0.032\,, \tag{6.12}$$

$$\gamma = 0.006 \pm 0.060 \tag{6.13}$$

If the typical values suggested in Coc *et al.* 2007 were assumed: $R \sim 30$ and $S \sim 160$, allowing a 10% uncertainty in each of them, the following bound was established:

$$\frac{\Delta\alpha}{\alpha} = (2.7 \pm 9.1) \times 10^{-5} \tag{6.14}$$

Finally, it should be noted, that improvements in mass and radius measurements, such as those expected from the Gaia satellite, will allow us to obtain more stringent constraints.

6.2. *Bounds on G_N*

White dwarfs can also be used to put independent limits on a possible variation in G_N. The most important reason for this is that white dwarfs have very long evolutionary timescales. A possible variation of G_N over cosmological ages changes the temperature in the central regions of white dwarf progenitors. As a consequence, the thermonuclear rates, luminosities and main sequences lifetimes of these stars are also modified. Furthermore, white dwarf cooling times are also affected by a change in G_N. García-Berro *et al.* (2011) calculated the effects of a varying G_N on the main sequence ages using a stellar evolutionary code. Furthermore, employing modified white dwarf cooling ages in order to account for a varying G_N, they built the white dwarf luminosity function for the old, metal-rich Galactic open cluster NGC 6791. Comparing the theoretical predictions for a varying G_N with the observational color-magnitude diagram and the white dwarf luminosity function, and using the distance modulus measured with eclipsing binaries, the following bound was obtained:

$$\frac{\dot{G}_N}{G_N} < -1.8 \times 10^{-12} \mathrm{yr}^{-1} \tag{6.15}$$

On the other hand, pulsational properties of white dwarfs can be used to constrain a possible variation in G_N. Córsico *et al.* (2013) compared the theoretical rates of period change of white dwarfs including the effects of a running G_N with the measured rates of change of the periods of two well studied pulsating white dwarfs, G117-B15A and R548. For this, they used a stellar evolutionary code and a pulsational code to compute pulsational properties of variable white dwarfs. Results show that the pulsation periods do not change in a significant way, but the rates of period change are strongly affected when a varying G_N is assumed. Moreover, they obtain the following bound

$$\frac{\dot{G}_N}{G_N} < -1.3 \times 10^{-10} \mathrm{yr}^{-1} \tag{6.16}$$

which is less stringent that the one obtained from the white dwarf luminosity function.

7. Bounds from the Lunar Laser Ranging experiment

The Lunar Laser Ranging (LLR) experiment, has provided high-precision values of the Earth-Moon distance, through measurements of round trip travel times of laser pulses between stations on the Earth and retroreflectors on the Moon. Results of this experiment have been used to test General Relativity, the Strong Equivalence Principle, metric parameters, preferred frame effects and the temporal variation of the gravitational constant G_N. Hofmann *et al.* (2010) updated the Institut fur Erdmessung (IfE) LLR model taking the effect of a fluid lunar core into consideration. In this way, they reduced the uncertainty in the current variation in G_N by a factor of 2, obtaining:

$$\frac{\dot{G}_N}{G_N} = (0.7 \pm 3.8) \times 10^{-13} \mathrm{yr}^{-1} \tag{7.1}$$

8. Bounds from Helioseismology

Guenther *et al.* (1998) considered solar models including the possible variation of the gravitational constant G_N during the solar lifetime. Furthermore, they compared the p-mode oscilation spectra of those models with observations from the Global Oscillation Network Group (GONG) instrument and Birmingham Solar Oscillation Network (BiSON). As a result of their research, they obtain the following bound:

$$\frac{\dot{G}_N}{G_N} < 1.6 \times 10^{-12} \mathrm{yr}^{-1} \tag{8.1}$$

9. Summary and Conclusions

In this paper we have described the most important present bounds on the possible variaton of the fundamental constants α, μ and G_N. We also have mentioned the dependence of some observables on other fundamental constants such as the velocity of light c, the Fermi constant G_F and the Higgs vaccum expectation value $< v >$. In order to have a comprehensive picture, Tables 1, 2 and 3 show a summary of the bounds that can be obtained with each of the methods described in this review, together with the time interval for which the change in the fundamental constants was measured. It is important to stress, that bounds on the variation in α and μ from white dwarf spectra, are

Table 1. Bounds on $\frac{\Delta\alpha}{\alpha}$. Columns: (1) Method considered (2) mean value and the corresponding 1–σ error, (3) the time interval for which the variation was measured, 4) Reference. The bound obtained from white dwarf data constraints the possible variation in α with the gravitational potential.

Source	$\frac{\Delta\alpha}{\alpha} \pm \sigma$	Δt **(yr)**	**Reference**
Atomic Clocks	$(2.1 \pm 6.3) \times 10^{-17}$	13.8×10^9	Godun *et al.* (2014)
Oklo	$< 1.1 \times 10^{-8}$	1.8×10^9	Davis & Hamdan (2015)
Quasars	$(-0.64 \pm 0.65) \times 10^{-6}$	$(8–11) \times 10^9$	Martins (2017)
CMB	$(0.7 \pm 2.5) \times 10^{-3}$	13.8×10^9	Hart & Chluba (2018)
BBN	$(-2.2 \pm 0.6) \times 10^{-2}$	13.8×10^9	Mosquera & Civitarese (2017)
Supernovae type Ia	$(2.1 \pm 0.8) \times 10^{-2}$	$(0.1–9) \times 10^{10}$	Kraiselburd *et al.* (2015)
White Dwarfs	$(2.7 \pm 9.1) \times 10^{-5}$		Magano *et al.* (2017)

Table 2. Bounds on $\frac{\Delta\mu}{\mu}$. Columns: (1) Method considered (2) mean value and the corresponding 1–σ error, (3) the time interval for which the variation was measured, 4) Reference. The bound obtained from white dwarf data constraints the possible variation in μ with the gravitational potential.

Source	$\frac{\Delta\mu}{\mu} \pm \sigma$	Δt **(yr)**	**Reference**
Atomic Clocks	$(0.2 \pm 2.1) \times 10^{-16}$	2–4	Godun *et al.* (2014)
Quasars	$(-0.23 \pm 0.1) \times 10^{-6}$	$(6.4–12.4) \times 10^9$	Martins (2017)
CMB	$(3.9 \pm 7.4) \times 10^{-3}$	13.8×10^9	Hart & Chluba (2018)
White Dwarfs	$(-5.8 \pm 4.1) \times 10^{-5}$		Bagdonaite *et al.* (2014)

Table 3. Bounds on $\frac{\dot{G}_N}{G_N}$. Columns: (1) Method considered (2) mean value and the corresponding 1–σ error, (3) the time interval for which the variation was measured, 4) Reference.

Source	$\frac{\dot{G}_N}{G_N} \pm \sigma$ **in units of** 10^{-12}	Δt **(yr)**	**Reference**
BBN	(0.05 ± 0.35)	13.8×10^9	Copi *et al.* (2004); Bambi *et al.* (2005)
LLR	(0.07 ± 0.38)	40	Hofmann *et al.* (2010)
Supernovae	< 10	5×10^9	Gaztañaga *et al.* (2002)
Helioseismology	< -1.6	4.6×10^9	Guenther *et al.* (1998)
White Dwarfs	< -1.8	7.7×10^9	García-Berro *et al.* (2011)

constraints of changes with the gravitational potential (which could also be regarded as a spatial variation), but they do not refer under any point of view to a time variation of fundamental constants. Finally, we would like to emphasize, the most important aspects that were discussed in this review:

(*a*) There is no evidence for a time or space variation of the fundamental constants.

(*b*) The most stringent limits on variation of the fundamental constants are provided by atomic clocks experiments:

(*c*) White dwarf stars are ideal laboratories to test for a relationship between the fine-structure constant or the proton-to-electron-mass and strong gravitational fields. Current constraints are derived from molecular hydrogen measurements and the mass-radius relationship while future measurements of Fe V and Ni V spectra in white dwarfs could provide more stringent bounds.

Acknowledgments

S.L. is grateful to IAU, IUPAP and the University of Leicester for financial support to attend the IAU 357 Symposium. S.L. is also supported by the National Agency for the Promotion of Science and Technology (ANPCYT) of Argentina grant PICT-2016-0081; and by grant UBACYT 20020170100129BA (2018) from Buenos Aires University.

References

Agafonova, I. I., Molaro, P., Levshakov, S. A., *et al.* 2011, *A&A*, 529, A28
Albornoz Vásquez, D., Rahmani, H., Noterdaeme, P., *et al.* 2014, *A&A*, 562, A88
Bagdonaite, J., Daprà, M., Jansen, P., *et al.* 2013, *Physical Review Letters*, 111, 231101
Bagdonaite, J., Jansen, P., Henkel, C., *et al.* 2013, *Science*, 339, 46
Bagdonaite, J., Salumbides, E. J., Preval, S. P., *et al.* 2014, *Physical Review Letters*, 113, 123002
Bagdonaite, J., Ubachs, W., Murphy, M. T., *et al.* 2015, *Physical Review Letters*, 114, 071301
Bahcall, J. N. & Schmidt, M. 1967, *Physical Review Letters*, 19, 1294
Bainbridge, M. B. & Webb, J. K. 2017, *MNRAS*, 468, 1639
Bainbridge, M., Barstow, M., Reindl, N., *et al.* 2017, *Universe*, 3, 32
Bambi, C., Giannotti, M., & Villante, F. L. 2005,*Physical Review D*, 71, 123524
Barr, S. M. & Mohapatra, P. K. 1988, *Physical Review D*, 38, 3011
Barrow, J. D. & Magueijo, J. 2005, *Physical Review D*, 72, 043521
Barrow, J. D.,Sandvik, H. B. & Magueijo, J. 2002, *Physical Review D*, 65, 063504
Bonifacio, P., Rahmani, H., Whitmore, J. B., *et al.* 2014, *Astronomische Nachrichten*, 335, 83
Brans, C. & Dicke, R. H. 1961, *Physical Review*, 124, 925
Bekenstein, J. D. 1982, *Physical Review D*, 25, 1527
Berengut, J. C., Flambaum, V. V., Ong, A., *et al.* 2013, *Physical Review Letters*, 111, 010801
Chamoun, N., Landau, S. J. & Vucetich, H. 2001, *Physics Letters B*, 504, 1
Coc, A., Nunes, N. J., Olive, K. A., *et al.* 2007, *Physical Review D*, 76, 023511
Copi, C. J., Davis, A. N., & Krauss, L. M. 2004, *Physical Review Letters*, 92, 171301
Córsico, A. H., Althaus, L. G., García-Berro, E., *et al.* 2013, *JCAP*, 2013, 032
Damour, T. & Dyson, F. 1996, *Nuclear Physics B*, 480, 37
Damour, T., Piazza, F., & Veneziano, G. 2002, *Physical Review Letters*, 89, 081601
Damour, T. & Polyakov, A. M 1994, *Nuclear Physics B*, 95, 10347
Daprà, M., Bagdonaite, J., Murphy, M. T., *et al.* 2015, *MNRAS*, 454, 489
Davis, E. D. & Hamdan, L. 2015, *Physical Review C*, 92, 014319
Dirac, P. A. M. 1937, *Nature*, 139, 323
Dyson, F. 1967, *Physical Review Letters*, 19, 1291
Evans, T. M., Murphy, M. T., Whitmore, J. B., *et al.* 2014, *MNRAS*, 445, 128
Falke, S., Lemke, N., Grebing, C., *et al.* 2014, *New Journal of Physics*, 16, 073023
Gamow, G. 1967, *Physical Review Letters*, 19, 759
Garcia-Berro, E., Kubyshin, Y., Loren-Aguilar, P., *et al.* 2006, *International Journal of Modern Physics D*, 15, 1163
García-Berro, E., Lorén-Aguilar, P., Torres, S., *et al.* 2011, *JCAP*, 2011, 021
Gaztañaga, E., García-Berro, E., Isern, J., *et al.* 2002, *Physical Review D*, 65, 023506
Godun, R. M., Nisbet-Jones, P. B. R., Jones, J. M., *et al.* 2014, *Physical Review Letters*, 113, 210801
Guéna, J., Abgrall, M., Rovera, D., *et al.* 2012, *Physical Review Letters*, 109, 080801
Guenther, D. B., Krauss, L. M., & Demarque, P. 1998, *ApJ*, 498, 871
Hofmann, F., Müller, J., & Biskupek, L. 2010, *A&A*, 522, L5
Huntemann, N., Lipphardt, B., Tamm, C., *et al.* 2014, *Physical Review Letters*, 113, 210802
Hart, L. & Chluba, J. 2018, *MNRAS*, 474, 1850
Jordan, P. 1959, *Zeitschrift fur Physik*, 157, 112
Kaluza, Th. 1921, *Sitzungber. Preuss. Akad. Wiss.K*, 1, 966
Kanekar, N. 2011, *ApJ Letters*, 728, L12
King, J. A., Webb, J. K., Murphy, M. T., *et al.* 2012, *MNRAS*, 422, 3370
Klein, O. 1926, *Z. Phys.*, 37, 895
Kotuš, S. M., Murphy, M. T., & Carswell, R. F. 2017, *MNRAS*, 464, 3679
Kraiselburd, L., Landau, S. J., Negrelli, C., *et al.* 2015, *Astrophysics Space and Science*, 357, 4
Leefer, N., Weber, C. T. M., Cingöz, A., *et al.* 2013, *Physical Review Letters*, 111, 060801
Le Targat, R., Lorini, L., Le Coq, Y., *et al.* 2013, *Nature Communications*, 4, 2109
Maeda, K. 1988, *Modern Physics. Letters A*, 3, 243
Marciano, W. J. 1984, *Physical Review Letters*, 52, 489

Magano, D. M. N., Vilas Boas, J. M. A., & Martins, C. J. A. P. 2017, *Physical Review D*, 96, 083012

Marion, H., Pereira Dos Santos, F., Abgrall, M., *et al.* 2003, *Physical Review Letters*, 90, 150801

Martins, C. J. A. P. 2017, *Reports on Progress in Physics*, 80, 126902

Milne, E. 1937, *Proceeding Royal Society A*, 158, 324

Moffat, J. W. 2006, *JCAP*, 2006, 004

Molaro, P., Centurión, M., Whitmore, J. B., *et al.* 2013, *A&A*, 555, A68

Mosquera, M. E. & Civitarese, O. 2017, *Physical Review C*, 96, 045802

Murphy, M. T., Malec, A. L., & Prochaska, J. X. 2016, *MNRAS*, 461, 2461

Negrelli, C., Kraiselburd, L., Landau, S., *et al.* 2018, *International Journal of Modern Physics D*, 27, 1850099

Olive, K. A. & Pospelov, M. 2002, *Physical Review D*, 65, 085044

Palma, G. A.,Brax, P., Davis, A. C. & van de Bruck, C. 2003, *Physical Review D*, 68, 123519

Petrov, Y. V., Nazarov, A. I., Onegin, M. S., *et al.* 2006, *Physical Review C*, 74, 064610

Planck Collaboration, Ade, P. A. R., Aghanim, N., *et al.* 2015, *A&A*, 580, A22

Prestage, J. D., Tjoelker, R. L., & Maleki, L. 1995, *Physical Review Letters*, 74, 3511

Pochoda, P. & Schwarzschild, M. 1964, *ApJ*, 139, 587

Rahmani, H., Wendt, M., Srianand, R., *et al.* 2013, *MNRAS*, 435, 861

Rosenband, T., Hume, D. B., Schmidt, P. O., *et al.* 2008, *Science*, 319, 1808

Songaila, A., & Cowie, L. L. 2014, *ApJ*, 793, 103

Shelkovnikov, A., Butcher, R. J., Chardonnet, C., *et al.* 2008, *Physical Review Letters*, 100, 150801

Srianand, R., Chand, H., Petitjean, P., *et al.* 2004, *Physical Review Letters*, 92, 121302

Teller, E. 1948, *Physical Review D*, 73, 801

van Weerdenburg, F., Murphy, M. T., Malec, A. L., *et al.* 2011, *Physical Review Letters* , 106, 180802

Varshalovich, D. A. & Levshakov, S. A. 1993, *Soviet Journal of Experimental and Theoretical Physics Letters*, 58, 237

Webb, J. K., Flambaum, V. V., Churchill, C. W., *et al.* 1999, *Physical Review Letters*, 82, 884

Webb, J. K., King, J. A., Murphy, M. T., *et al.* 2011, *Physical Review Letters*, 107, 191101

Whitmore, J. B. & Murphy, M. T. 2015, *MNRAS*, 447, 446

Youm, D. 2001, *Physical Review D*, 64, 085011

White Dwarfs as Probes of Fundamental Physics: Tracers of Planetary, Stellar and Galactic Evolution
Proceedings IAU Symposium No. 357, 2019
M. A. Barstow, S. J. Kleinman, J. L. Provencal & L. Ferrario, eds.
doi:10.1017/S1743921320000745

Clues to the origin and properties of magnetic white dwarfs

Adela Kawka

International Centre for Radio Astronomy Research - Curtin University
GPO Box U1987, Perth, WA 6845, Australia
email: adela.kawka@curtin.edu.au

Abstract. A significant fraction of white dwarfs possess a magnetic field with strengths ranging from a few kG up to about 1000 MG. However, the incidence of magnetism varies when the white dwarf population is broken down into different spectral types providing clues on the formation of magnetic fields in white dwarfs. Several scenarios for the origin of magnetic fields have been proposed from a fossil field origin to dynamo generation at various stages of evolution. Offset dipoles are often assumed sufficient to model the field structure, however time-resolved spectropolarimetric observations have revealed more complex structures such as magnetic spots or multipoles. Surface mapping of these field structures combined with measured rotation rates help distinguish scenarios involving single star evolution from other scenarios involving binary interactions. I describe key observational properties of magnetic white dwarfs such as age, mass, and field strength, and confront proposed formation scenarios with these properties.

Keywords. white dwarfs, stars: magnetic fields, stars: atmospheres

1. Introduction

Following the discovery of a magnetic field in 78 Virginis (Babcock 1947), Blackett (1947) contemplated the presence of stronger magnetic fields in white dwarfs assuming magnetic flux conservation throughout evolution. The strong circular polarization spectrum of the first known magnetic white dwarf, Grw+70° 8247 (Kemp *et al.* 1970), implied a longitudinal field measurement of $\approx 10^7$ G, which was nearly a thousand times the field strength observed in Ap stars at the time. Following the original discovery, many more magnetic white dwarfs were found with strong magnetic fields ($B \gtrsim 10^7$ G) including G195-19 as the prototype for the cool, continuum-like, helium-rich magnetic white dwarfs (DCP, Angel & Landstreet 1971) and G99-37 as the prototype for carbon-enriched magnetic white dwarfs (DQP but originally classified as DGp, Greenstein 1969, Landstreet & Angel 1971). Early observations already demonstrated that magnetic fields are present across all spectral types. Although a search for lower strength magnetic fields ($\lesssim 5$ MG) carried out by Angel *et al.* (1981) proved negative, the number of white dwarfs with magnetic fields greater than 1 MG steadily grew by 1995 to 40 objects (see Schmidt & Smith 1995). Spectropolarimetric measurements proved capable of reaching below 1 MG field strength with the discovery of two low-field white dwarfs (0.1 MG) in a large survey of 170 objects (Schmidt & Smith 1994, 1995). With even larger surveys, such as the Sloan Digital Sky Survey (SDSS) and with larger telescopes and more sensitive instruments, the number of magnetic white dwarfs has grown to over 600 (see, e.g., Ferrario *et al.* 2015) but still includes only 18 objects with fields below 1 MG (Kawka & Vennes 2012, Landstreet *et al.* 2017).

This paper aims to summarise our current understanding of the magnetic white dwarf population and the methods used to detect and determine magnetic fields in white dwarfs

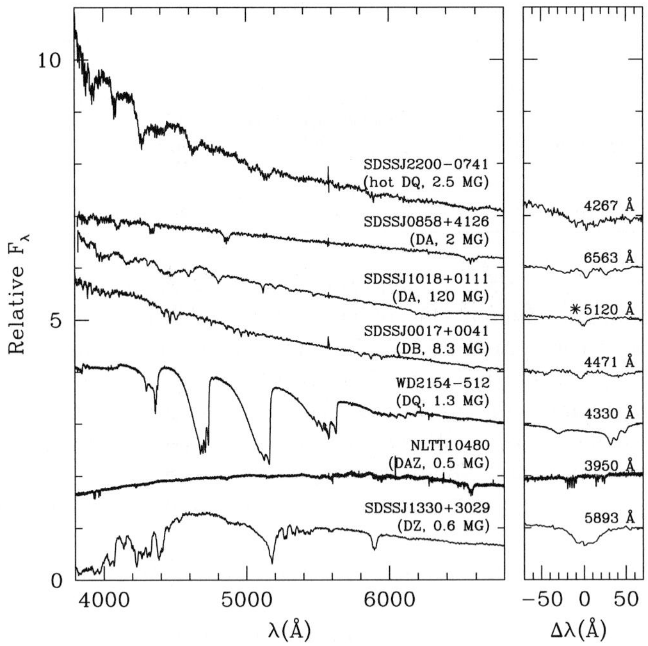

Figure 1. Example spectra showing the different spectral classes for which magnetism has been detected. The left panel shows the optical spectra. The left panel highlights a particular feature that is representative of the spectral type which is centred on the given wavelength. The absorption line at 5120 Å, marked with a star, in SDSSJ1018+0111 is an example of a stationary feature. In this case it is a Zeeman component of Hβ.

(Section 2). The properties of the magnetic white dwarf population are presented in Section 3 and the various models for the origin of magnetic fields in white dwarfs are presented in Section 4. Finally, Section 5 exposes current challenges in modelling magnetic atmospheres in particular the effect of magnetic fields on convective motion and the atmospheric structure.

2. Measuring magnetic field strengths

There are two complementary ways to detect and measure magnetic fields. The most straight forward follows the detection of Zeeman splitted lines. Depending on the atomic structure and magnetic field strength, the Zeeman effect can be calculated using distinct methods. Fig. 1 shows examples of the different classes of magnetic white dwarfs. The magnetic DC white dwarfs, which have featureless spectra, are not included in the plot. The magnetic DQ shown is WD 2154-512 which exhibits CH molecular bands in its spectrum (shown in the left panel of Fig. 1 and centred at 4330 Å) and which were used in measuring its magnetic field strength from spectropolarimetry (Berdyugina *et al.* 2007). The figure also shows the different regimes of the Zeeman effect on various elements.

The normal Zeeman effect occurs when a spectral line of an atom with zero spin ($S=0$) splits into three components when immersed in a magnetic field. In this case, we can calculate the magnetic field strength using:

$$\Delta\lambda = \frac{eB\lambda^2}{4\pi m_e c^2} \approx 4.67\times 10^{-7}\lambda^2 B \tag{2.1}$$

where λ is the wavelength in Å, B is the magnetic field in MG, e is the electron charge (e.s.u.), m_e is the electron rest mass, and c is the speed of light.

Table 1. Calculations of Zeeman splitted lines.

Element	B Range (G)	Spectral range	Reference
Hydrogen	$0-5\times10^{12}$	Lyman, Balmer, Paschen, Brackett	Schimeczek & Wunner (2014)
Helium	$0-2.35\times10^{11}$	HeI	Becken & Schmelcher (2001)
Helium	$2.35\times10^{11}-2.35\times10^{13}$	HeI	Al-Hujaj & Schmelcher (2003)
Sodium	$0-4.7\times10^{7}$	$n=3-7$	González-Férez & Schmelcher (2003)
Calcium	$10^{6}-10^{8}$	CaII H&K	Kemic (1975)

The anomalous Zeeman effect depends on the electron spin and occurs in atoms with non-zero spin. The splitting in this regime can be complex and is observed at lower field strength where the spin and orbital angular momenta remain coupled. In general, the energy level of an electron can be described by three quantum numbers, J which is the total angular momentum, L which is the orbital angular momentum, and S which is the spin orbital momentum. When an atom is placed in a magnetic field its levels split into $2J+1$ components. These are defined by the magnetic quantum number $m = -J, -J+1, ..., J-1, J$ and the Zeeman shifted lines can be calculated using:

$$\Delta\lambda = \frac{eB\lambda^2}{4\pi m_e c^2}(g_l m_l - g_u m_u) \approx 4.67\times10^{-7}\lambda^2 B(g_l m_l - g_u m_u) \tag{2.2}$$

where m_u and m_l are the magnetic quantum numbers of the upper and lower levels, respectively. The permitted transitions are defined by $\Delta m = 0, \pm1$. Here, $\Delta m = 0$ defines the π components and $\Delta m = \pm1$ the σ components. The Landé factors for the upper and lower levels are g_u and g_l, respectively and they can be calculated assuming LS coupling:

$$g = 1 + \frac{J(J+1) - L(L-1) + S(S+1)}{2J(J+1)}, \tag{2.3}$$

The LS coupling is generally valid for lighter atoms such as sodium, magnesium, aluminium and calcium. For heavier atoms like iron LS coupling is no longer valid and other coupling needs to be applied. The Vienna Atomic Line Database† provides atomic data including Landé factors for light and heavy elements.

The Paschen-Back effect occurs when the m_l degeneracy is removed. In this regime the spin and orbital angular momenta decouple and the anomalous Zeeman effect reverts to the normal or linear Zeeman effect and Eqn. 2.1 can be used. The linear Zeeman splitting of hydrogen atoms falls into this regime.

As the magnetic field strength increases, the quadratic effect begins to take over. In this regime, the l degeneracy is also removed and shifts in the π components and asymmetry in the line profiles are observed. Increasing the field strength will cause even more complex Zeeman patterns and for more detail on these higher field regimes see Ferrario *et al.* (2020). Landi Degl'Innocenti & Landolfi (2004) present conveniently written Zeeman splitting formulae for the low field regime and quadratic effect. For further reading consult Herzberg (1945). Table 1 lists references to available calculations of line shifts and transition probabilities for hydrogen and helium at stronger magnetic fields, as well as some heavier elements such as sodium, calcium and carbon. For helium, the calculations of Becken & Schmelcher (2001) and Al-Hujaj & Schmelcher (2003) supersede the calculations of Kemic (1974) and allowed the measurement of magnetic field strengths of strongly magnetic DB white dwarfs (Jordan *et al.* 1998; Wickramasinghe *et al.* 2002). Measuring magnetic field strengths at high fields ($\gtrsim 100$ MG) can be difficult due to the complexity of the Zeeman pattern. Also, the magnetic field on a white dwarf surface is not uniform and can vary significantly, up to a factor of two even if a simple dipole is

† http://vald.astro.uu.se/ (Ryabchikova *et al.* 2015).

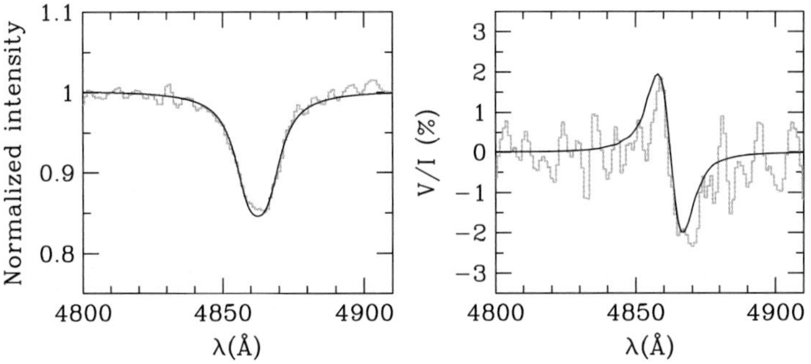

Figure 2. Flux (left) and circular polarization (right) spectra of NLTT 2219. The flux spectrum (grey) is compared to the best-fitting Hβ line profile (black). The circular polarization spectrum is compared to a model polarization spectrum at $B_l = -97$ kG (Kawka & Vennes 2012).

assumed. This can smear out most of the absorption features, however some line components become stationary where the wavelength goes through a minimum or maximum as a function of the field strength. These stationary components can be observed in spectra even when the field varies considerably (see Fig 1).

Spectropolarimetric measurements add additional geometric information following a magnetic field detection. To date, most spectropolarimetric observations have measured the circular polarization, the Stokes parameter V. The longitudinal field strength (B_l in G) can be determined using:

$$v = \frac{V}{I} = \frac{dI}{d\lambda} \frac{4.67 \times 10^{-13} \lambda^2 B_l}{I} \tag{2.4}$$

This equation provides for a determination of the longitudinal field strength from the measured polarization V and the intensity spectrum I for stars where the Zeeman splitting is smaller than the width of the line profile. Several low field DA white dwarfs have been recently detected using this technique (e.g., Kawka & Vennes 2012; Landstreet & Bagnulo 2019). Fig. 2 shows the Hβ line profile of the weakly magnetic NLTT 2219 as well as the circular polarization spectrum showing the polarized σ components.

Spectropolarimetry has also been used to detect and measure magnetic fields in the carbon polluted DQ white dwarfs. The first DQ white dwarf that showed circular polarization was G99-37 (Angel & Landstreet 1974) followed by LP 790-29 (Liebert *et al.* 1978). Estimating a magnetic field strength is still difficult for DQ white dwarfs because of their relatively broad molecular carbon bands. However, Angel & Landstreet (1974) were able to obtain an estimate of the magnetic field of G99-37 using CH molecular bands, a rare feature among DQ white dwarfs. Berdyugina *et al.* (2007) and Vornanen *et al.* (2010) further developed the method involving CH molecular bands to measure magnetic fields and determined a longitudinal field strength of 7.3 ± 0.3 MG for G99-37. The same method enabled them to identify the second magnetic DQ with CH bands (GJ 841B) with a field strength $B_l = 1.3 \pm 0.5$ MG.

Currently, there are several instruments capable of spectropolarimetry. These include the FOcal Reducer and low dispersion Spectrograph (FORS2) at the European Southern Observatory (Appenzeller *et al.* 1998) located at the Paranal Observatory in Chile, the Echelle SpectroPolarimetric Device for the Observation of Stars (ESPaDOnS: Donati *et al.* 2006) on the Canada-France-Hawaii Telescope (CFHT) located on Mauna Kea in Hawaii, the Intermediate-dispersion Spectrograph and Imaging System (ISIS) on the William Herschel Telescope (WHT) located on La Palma in the Canary islands, Spain and the Robert Stobie Prime Focus Imaging Spectrograph (RSS) on the South African Large Telescope near Sutherland in South Africa.

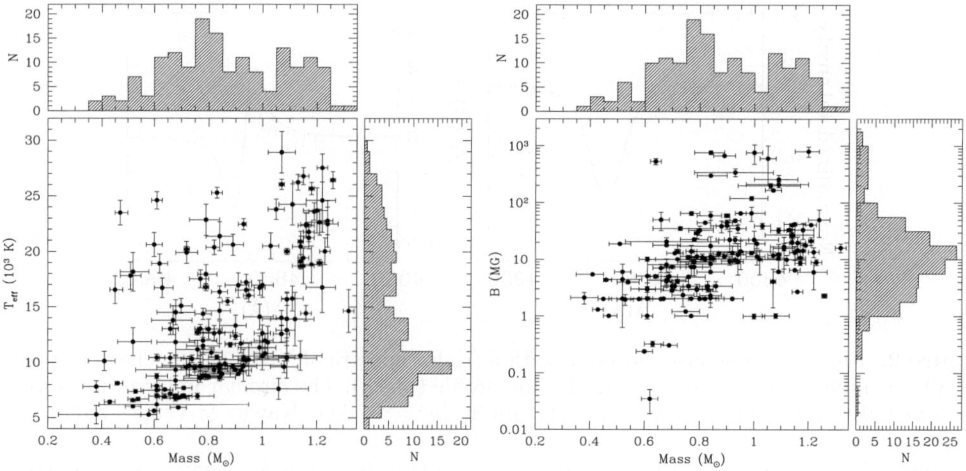

Figure 3. Mass versus temperature distribution (left) and mass versus magnetic field strength distribution (right) of magnetic DA white dwarfs from SDSS DR12.

3. Population properties

To investigate the properties of magnetic white dwarfs as a population, the sample of known magnetic DA white dwarfs was cross-correlated with SDSS DR12 photometric measurements (Alam *et al.* 2015) and Gaia DR2 parallax measurements (Lindegren *et al.* 2018). The effective temperature and mass were determined by fitting the SDSS photometric measurements with synthetic magnitudes that were set to the stellar distance using Gaia parallaxes and the stellar radius R by applying mass-radius relations from Benvenuto & Althaus (1999). Fig. 3 shows the distribution of temperature and magnetic field strength versus mass for a sample of known magnetic white dwarfs with available SDSS photometric measurements and Gaia parallaxes. The mass distribution shows that there are two peaks with the main one at $\sim 0.8\,M_\odot$ and the second broader one at $\sim 1.1\,M_\odot$. The average mass of these magnetic white dwarfs is $0.87\,M_\odot$, which is higher than the average mass of $0.6\,M_\odot$ for non-magnetic white dwarfs (Tremblay *et al.* 2019). A higher incidence of magnetism in hot massive white dwarfs discovered by the Extreme Ultraviolet Explorer (EUVE) had already been reported by Vennes (1999). There appears to be a lack of cool magnetic white dwarfs with a high mass, which is most likely due to their intrinsic faintness and a shortage of cool white dwarfs in past SDSS colour selections.

A relationship appears to exist between the mass and magnetic field strength, with more massive white dwarfs having stronger magnetic fields, however the correlation coefficient is only 0.43 and therefore should be viewed with caution. Using population synthesis calculations under the merger scenario for the generation of fields (see Section 4), Briggs *et al.* (2018) predicted that more massive magnetic white dwarfs should have weaker fields than less massive magnetic white dwarfs, which contradicts the weak correlation observed in Fig. 3. In their simulations they show that when an asymptotic giant branch (AGB) star merges with a late stage star, it produces stronger fields than when an AGB star merges with an earlier main-sequence star. However, they also note that double degenerate mergers would not follow this trend and instead would result in more massive white dwarfs with stronger field strengths and faster rotation than magnetic white dwarfs that merged in a common envelope phase.

Using the SDSS DR2/Gaia sample of magnetic white dwarfs, we can investigate potential correlations between the cooling age of the white dwarf and the magnetic field

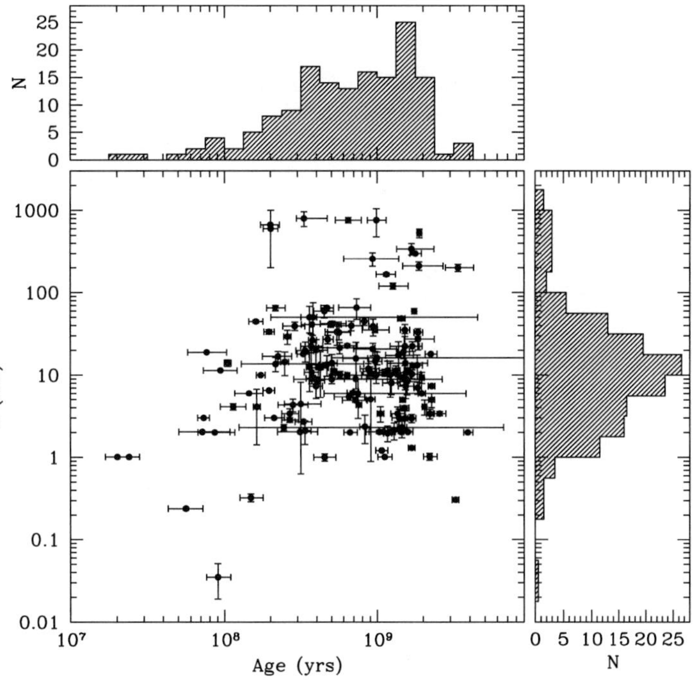

Figure 4. Cooling age versus magnetic field strength distribution of the magnetic DA white dwarfs from SDSS DR12.

strength and find out if there is any evidence for magnetic field decay or evolution. Fig. 4 shows no obvious correlation between the cooling age versus the magnetic field strength. The cross-correlation coefficient is only 0.14 indicating that the two properties are not correlated. Ferrario *et al.* (2015) reached the same conclusions when they compared the magnetic field strength with the effective temperature.

Many magnetic white dwarfs are faster rotators than their non-magnetic counterparts. Most rotational periods in magnetic white dwarfs were determined from photometric time series (e.g., Brinkworth *et al.* 2013), although some were discovered using polarimetry or spectropolarimetry. Photometric and spectropolarimetric measurements revealed that the massive, high field white dwarf RE J0317-853 (EUVE J0317-85.5) to be a very fast rotator with a period of 725 seconds (Barstow *et al.* 1995; Ferrario *et al.* 1997; Vennes *et al.* 2003). It remained the fastest rotating white dwarf until photometric timeseries of hot DQs revealed them to have rotational periods as short as 5 minutes (e.g., Dunlap *et al.* 2010; Dufour *et al.* 2011). Fig. 5 shows that most of the magnetic white dwarfs have rotation periods shorter than 10 hours with a distribution peaking at 2–3 hours. This can be compared to the rotational period distribution for non-magnetic white dwarfs determined from pulsation studies showing a mean rotation period of 35 hours (Hermes *et al.* 2017).

Availability of photometric and astrometric data from large scale surveys has finally allowed us to better understand the properties of the magnetic white dwarf population. Detailed, case by case, spectropolarimetric investigations remain essential to understand the diverse properties of individual magnetic white dwarfs.

4. Origin of magnetic fields

The fossil field origin for the existence of magnetic white dwarfs remains a viable model because the magnetic flux of Ap/Bp stars is similar to that of magnetic white

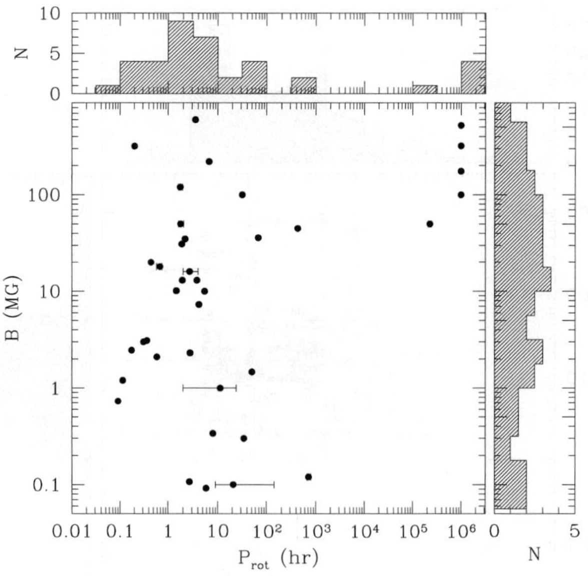

Figure 5. Rotational velocity versus magnetic field strength distribution for known magnetic white dwarfs for which a rotational period was measured.

dwarfs. Assuming flux conservation ($B_{\rm WD}/B_{\rm MS} = (R_{\rm MS}/R_{\rm WD})^2$) the magnetic fields of Ap/Bp stars can evolve into the strongest fields observed in white dwarfs (Tout *et al.* 2004). Angel *et al.* (1981) compared the space density of magnetic white dwarfs to the space density of magnetic Ap/Bp stars and proposed that there is a sufficient number of Ap/Bp stars to generate all known magnetic white dwarfs. Kawka & Vennes (2004) revisited this comparison and showed that if Ap/Bp stars evolve to high field magnetic white dwarfs ($B \gtrsim 10$ MG), additional progenitors are required for white dwarfs with lower field strengths. Wickramasinghe & Ferrario (2005) proposed that white dwarfs with $B \lesssim$ a few ten MG could form from main-sequence stars with $M \gtrsim 4.5\,M_\odot$ and low fields of $\approx 10-100$ G. However, Aurière *et al.* (2007) showed that a magnetic desert exists below 300 G for A and B type stars, which eliminates that possibility.

One of the main difficulties with the fossil field origin is that there are no known close, detached magnetic white dwarfs paired with a main-sequence star (Liebert *et al.* 2015), although thousands of close, detached but non-magnetic white dwarf plus M dwarf spectroscopic binaries have been identified in SDSS (Rebassa-Mansergas *et al.* 2016).

To address this conundrum, Tout *et al.* (2008) and Wickramasinghe *et al.* (2014) proposed that magnetic fields in isolated white dwarfs are created during a common envelope phase experienced by the progenitor close binary system. There are likely variations on the magnetic field generation model proposed by Tout *et al.* (2008). Potter & Tout (2010) and Wickramasinghe *et al.* (2014) showed that a dynamo mechanism within the common envelope can generate magnetic fields with the strongest fields arising in mergers that are differentially rotating near break-up.

Nordhaus *et al.* (2011) also proposed a variation on the model, where the low mass star is tidally disrupted by the pre-white dwarf during the common envelope phase, resulting in the formation of an accretion disc. A dynamo is then generated in the disc creating the magnetic field which is then transferred to the degenerate core via accretion.

García-Berro *et al.* (2012) showed that magnetic fields can also be created during the merger of two white dwarfs. When the white dwarfs merge a hot, convective and differentially rotating corona is created which produces a dynamo and, thus, the magnetic field. However, using a population synthesis of binary stars, Briggs *et al.* (2015) showed that the majority of magnetic fields in white dwarfs would have formed within a common envelope with less than 1% being formed during the merger of two white dwarfs.

Isern *et al.* (2017) proposed a model where low magnetic fields ($B \lesssim 0.1$ MG) are generated by phase separation when the white dwarf begins to crystallize. They showed that when white dwarfs reach sufficiently low temperatures that the core begins crystallizing, phase separation of the main elements occurs. This leads to an unstable, convective liquid mantle on top of a solid core producing a dynamo which creates a magnetic field.

It is possible that the range of observed magnetic fields stem from several distinct processes. By studying the various classes of magnetic white dwarfs, we may be able to place constraints on the specific processes that could lead to a particular class of magnetic white dwarfs.

4.1. *Double degenerate mergers*

Evidence of magnetic fields being formed in double degenerate mergers has been found in common proper motion binaries. EUVE J0317-85.5 is a hot ($T_{\rm eff} = 33\,000$ K), strongly magnetic ($B = 450$ MG) and massive white dwarf ($M = 1.3\,M_\odot$) in common proper motion with the non-magnetic white dwarf LB 9802 (Barstow *et al.* 1995; Ferrario *et al.* 1997; Vennes *et al.* 2003). EUVE J0317-85.5 being significantly more massive than LB 9802 would have had a more massive main-sequence progenitor and therefore it should have a much longer cooling age compared to LB 9802. However, EUVE J0317-85.5 is much hotter than LB 9802, and therefore its cooling age is also shorter. This strongly suggests that EUVE J0317-85.5 is the result of a merger (Ferrario *et al.* 1997). Another possible example is the common proper motion binary SDSS J150746.49+521002.1 plus SDSS J150746.80+520958.0. SDSS J1507+5209 is magnetic and more massive than SDSS J1507+5210, however their effective temperatures are similar (Dobbie *et al.* 2012). Since they are in a common proper motion binary, their total ages should be the same, however the total age of the magnetic white dwarf is much shorter than the total age of the non-magnetic companion. Therefore SDSS J1507+5209 is most likely the product of a merger and thus the original system was a triple system.

A small number of common proper motion binaries provide evidence to support that some magnetic fields in white dwarfs are formed through mergers. The key property is the discrepancy in the total ages of the magnetic white dwarf and non-magnetic white dwarf if single star evolution is assumed for both stars.

Spectroscopic follow-up of newly identified common proper motion white dwarf pairs may add additional cases and shed more light on the origin of magnetic fields.

4.2. *Incidence of magnetism*

The incidence of magnetism varies between different spectral classes, but it is also affected by the population sampling criteria. Taking the population of magnetic white dwarfs as a whole, then in magnitude limited surveys such as those of Palomar-Green or SDSS (Schmidt & Smith 1995; Kepler *et al.* 2013) the incidence is as low as $\sim 5\%$, while in volume limited surveys, the incidence is much higher at $13-20\%$ (Kawka *et al.* 2007). Table 2 lists the many classes of magnetic white dwarfs including the individual prototypes and the incidence within specific spectral classes.

4.2.1. Common proper motion binaries

It was already mentioned that there are no close, detached magnetic white dwarfs paired with a main sequence star, however there are a few magnetic white dwarfs in

Table 2. Incidence of magnetism among different classes of white dwarfs.

Spectral Type	Prototype	Reference	Fraction (%)	Reference
DAH	Grw+70°8247	Kemp *et al.* (1970)	4 ± 1.4	Schmidt & Smith (1995)
DAZH (cool)	G77–50	Farihi *et al.* (2011)	~ 50	Kawka *et al.* (2019)
DBH	GD229	Swedlund *et al.* (1974)	~ 1.5	Kawka (2018)
DCP	G195–19	Angel & Landstreet (1971)	~ 5	Putney (1997)
DZH (cool)	LHS2534	Reid *et al.* (2001)	21.6 ± 3.3	Hollands (2017)
DQH (hot)	SDSS J1337+0026	Dufour *et al.* (2007)	~ 70	Dufour *et al.* (2013)
DQH (cool)	G99–37	Landstreet & Angel (1971)	~ 4	Vornanen *et al.* (2013)

common proper motion with a main sequence star. In these systems the stars are assumed to have been born at the same time and remain distant enough from each other to avoid interaction and thereby follow single star evolution. An examination of the incidence of magnetic white dwarfs in such systems within the Solar neighbourhood ($d < 20$ pc) reveals that magnetic white dwarfs in wide binaries are not that rare. Holberg *et al.* (2013) identified 11 so-called Sirius-like systems within 20 pc of the Sun. These are systems where the white dwarf is paired with a main sequence star with a spectral type of K or earlier. Of these 11 binary systems, one has a magnetic white dwarf: WD 1009–184 is a cool, polluted helium rich white dwarf with a weak surface averaged field of ≈ 300 kG (Bagnulo & Landstreet 2019). Therefore, this sets an incidence of $\sim 9\%$, comparable to the incidence of single magnetic white dwarfs. More recently, Hollands *et al.* (2018) revisited the 20 pc sample with Gaia and identified a total of 30 white dwarfs that are in a wide binary system with a main sequence companion (including M dwarfs) and 3 of these are magnetic. These include WD 1009–184, along with the cool DA WD2150+591 (Landstreet & Bagnulo 2019) and the cool DQ WD2153–512 (Vornanen *et al.* 2010). This corresponds to an incidence of $\sim 10\%$. If we exclude systems that are likely to interact or have already interacted, then this incidence would be higher. The formation of magnetic fields in white dwarfs that are in wide binaries can be assumed to follow the same paths as those followed by isolated magnetic white dwarfs.

4.2.2. Polluted white dwarfs

Magnetic fields have been detected in all spectral classes of white dwarfs. Most of the known magnetic white dwarfs have hydrogen-rich atmospheres, similar to non-magnetic white dwarfs. Some spectral classes of white dwarfs display significantly higher incidence of magnetism than the general white dwarf population. Cool, polluted white dwarfs show an enhanced incidence of magnetism. Kawka & Vennes (2014) and Kawka *et al.* (2019) showed that $\approx 50\%$ of cool ($T_{\rm eff} < 6000$ K), polluted hydrogen-rich DAZ white dwarfs are magnetic with field strengths below 1 MG. Similarly, Hollands *et al.* (2015) and Hollands (2017) showed that $21.6 \pm 3.3\%$ of cool ($T_{\rm eff} < 8000$ K) polluted helium-rich DZ white dwarfs are magnetic with field strengths ranging from ≈ 0.3 MG up to ≈ 30 MG. Only two DAZ white dwarfs with an effective temperature above 6000 K have magnetic fields, NLTT 53908 which is only slightly warmer at 6250 K (Kawka & Vennes 2014) and WD 2105-820 which is much hotter at $T_{\rm eff} = 10800$ K (Landstreet *et al.* 2012). In the case of DZ white dwarfs, Hollands (2017) found that there are no known magnetic DZ white dwarfs warmer than 8000 K. The lack of warmer counterparts to the magnetic, polluted white dwarfs provides constraints on the origin of the magnetic field in these stars. Some of these possibilities are discussed below.

One possible scenario which may explain the presence of magnetic fields in this particular class involves a gaseous planet colliding with an older white dwarf creating differential rotation which would generate a relatively weak magnetic field (Farihi *et al.* 2011; Kawka *et al.* 2019). The time-scale for a post main sequence planetary system to dynamically stabilize is relatively short at about 100 Myr (Debes & Sigurdsson 2002). Therefore, at the old ages encountered in cool DAZs and DZs (>1 Gyr) a destabilizing agent is required to create a planetary or asteroidal debris disc causing atmospheric pollution and contributing to the creation of a magnetic field. Farihi *et al.* (2011) proposed that planets and asteroids could be destabilized by a close encounter with another star. These cool white dwarfs have been orbiting the Galaxy for billions of years and Farihi *et al.* (2011) estimated that a stellar encounter has a 50% probability of occurring every 0.5 Gyr. Therefore, the older the white dwarf, the more likely it will encounter a star that is close enough to perturb the orbit of the outer planets. If one of these planets is a gaseous type then this would create the magnetic field while the perturbed rocky planets and asteroids would pollute the white dwarf. The pairing of these events would explain the dual nature of polluted, magnetic white dwarfs.

Hollands (2017) suggested that cool, polluted and magnetic white dwarfs could have been generated in the cores of giant stars following Kissin & Thompson (2015). In this scenario, a dynamo is generated within the giant core during the shell burning phases. They showed that angular momentum pumped inward in combination with the absorption of a planet or a binary companion could generate a dynamo at the boundary between the radiative and convective envelopes. This could create magnetic fields of up to about 10^7 G. However this magnetic field could remain buried below a non-magnetized envelope for up to about 1–2 Gyr. Since the creation of the magnetic field is aided by the engulfment of a planet, it is likely that other planets or asteroids could survive and with time migrate closer toward the white dwarf and be accreted, polluting the white dwarf atmosphere.

4.2.3. Hot DQ white dwarfs

The class of hot DQs shows the highest incidence of magnetism among the various classes of white dwarfs. Approximately 70% of hot DQs are magnetic (Dufour *et al.* 2013). Hot DQs have temperatures ranging from 18 000 K up to 24 000 K and have an atmosphere dominated by carbon (Dufour *et al.* 2008). Coutu *et al.* (2019) showed that they are all massive ($0.8-1.2$ $M_\odot$) and about half of these stars have been found to be photometrically variable with periods ranging from about 5 minutes up to 2.1 days. These variations have been attributed to the rotation of the white dwarfs (Dunlap & Clemens 2015; Williams *et al.* 2016). These properties point toward hot DQ white dwarfs being the outcome of mergers of white dwarfs. Coutu *et al.* (2019) showed that the hot DQ sequence extends toward lower temperatures. These cooler objects are also massive and they can be distinguished from normal mass (~ 0.6 $M_\odot$) DQ white dwarfs by having a higher carbon abundance. Coutu *et al.* (2019) also showed that the distribution of transverse velocities of these objects is larger than for other classes of objects with similar temperatures, suggesting that they must come from an older population. If these massive DQs formed in a double degenerate merger, this would provide the necessary delay to make these stars appear younger than they really are. One star where we clearly see a discrepancy between the total age assuming single star evolution and the age based on kinematics is LP 93-21, which is a massive DQ with halo like orbit (Kawka *et al.* 2020) and is thus highly unlikely to be the unburnt remnant of a subluminous Type Ia supernova explosion (e.g., Vennes *et al.* 2017) as proposed by Ruffini & Casey (2019). It is likely that the majority of these cooler massive DQs will also harbour a magnetic field and be fast rotators. The number of these massive DQs is small compared to other classes of white dwarfs, however it is in agreement with the conclusion of Briggs *et al.* (2015) who predicted that less than about one in four hundred mergers involve double degenerate mergers.

The different properties of the various classes of magnetic white dwarfs, along with their respective incidences of magnetism allows links to be established with potential formation scenarios.

5. Structure of magnetic atmospheres

Magnetic fields have the potential to affect the transport of energy within a stellar atmosphere, and therefore the atmospheric structure as well. Field variations across the stellar surface add more complexities to this problem.

5.1. *Convective energy transport*

As white dwarfs cool, they develop convection zones with a depth increasing with age (Fontaine *et al.* 2013). White dwarfs with hydrogen-rich atmospheres begin to develop a convection zone at about 14 000 K, whereas white dwarfs with helium-rich atmospheres develop convective zones much earlier at about 30 000 K. As white dwarfs cool, these convection zones become the main energy transport mechanism, however it is possible that if a magnetic field is present, this transport of energy can be affected. Suppression of convection by magnetic fields was first investigated by D'Antona & Mazzitelli (1975). Furthermore, Wickramasinghe & Martin (1986) assumed that magnetic fields can provide resistance to large scale motions and therefore convective energy transport was suppressed in their atmospheric models.

Recently, Valyavin *et al.* (2014) revisited the idea that magnetic fields can suppress convection to explain the higher incidence of magnetism among cool white dwarfs (Liebert & Sion 1979; Valyavin & Fabrika 1999). They argued that suppression of convection causes a slow down in white dwarf cooling for white dwarfs below 12 000 to 14 000 K. Tremblay *et al.* (2015) conducted radiation magnetohydrodynamic simulations and showed that convection can be inhibited in a white dwarf atmosphere by magnetic fields as low as 50 kG, but that the cooling age is not affected until the convective zone couples with the degenerate core. This occurs at about 6000 K for higher mass white dwarfs ($M \sim 1\,M_\odot$) and at lower temperatures for lower mass white dwarfs.

Empirical support for suppression of convection was provided by Bédard *et al.* (2017) and Gentile Fusillo *et al.* (2018) using the weakly magnetic white dwarf WD 2105–820 (Landstreet *et al.* 2012). The effective temperature of WD 2105–820 is about 10 000 K and, therefore, it is cool enough to develop a shallow convective zone. Bédard *et al.* (2017) showed that using radiative models they were able to match the photometrically and spectroscopically derived effective temperatures, as well as match the spectroscopically determined distance with the measured parallax. However, convective models showed discrepancies between the spectroscopically and photometrically determined temperatures, as well as the distances obtained from the spectroscopic solutions and the measured parallax. Non-magnetic white dwarfs with similar temperatures to WD 2105–820 show good agreement between spectroscopically and photometrically determined temperatures when using convective models. Additional support for the inhibition of convection was presented by Gentile Fusillo *et al.* (2018), who showed that consistent results were obtained from fitting ultraviolet and optical spectra when using purely radiative models, whilst fitting with convective models resulted in inconsistent results. Again, non-magnetic white dwarfs with similar temperatures show consistent results when convective models are used.

Cooler magnetic white dwarfs have deeper and thicker convection zones. Kawka *et al.* (2019) analyzed the cool ($\approx$5200 K), magnetic white dwarf NLTT 7547 using convective and radiative models. Their analysis showed that the convective models provide a better fit to the Balmer lines compared to radiative models. They also showed that by suppressing convection in the model atmospheres, the temperature gradient steepens sharply at larger optical depths and higher pressure. The model atmospheres also show a downturn in the density at higher pressure which would give rise to a Rayleigh-Taylor instability. They concluded that convective models are preferred in cool and magnetic white dwarfs. Bédard *et al.* (2017) also analyzed a sample of cool (6600–8510 K) and magnetic white dwarfs and showed that in these white dwarfs, consistent results between spectroscopic and photometric solutions, as well as distances inferred from spectroscopy and parallax, were achieved with convective models. As a result, they concluded that it may become more difficult to suppress convection in cooler white dwarfs where convective energy transport becomes more important.

A small number of cool, average mass and magnetic DQs are known (e.g., Schmidt *et al.* 1999; Vornanen *et al.* 2010). The carbon in the atmospheres of these DQs is dredged up from the core by the deep convection zone (Pelletier *et al.* 1986). If convection is suppressed by the magnetic field, then we should not find any magnetic DQ white dwarfs since there is no convective motion to bring up the carbon to the surface. There are two possibilities to explain the existence of these cool and magnetic DQs. The first is that the carbon is not dredged up but it has been present in the atmosphere from the outset and as such they could be the descendants of hot DQs. In this case, they should also be massive. The two prototypical magnetic DQs, LHS2229 and LP 790-29, have $T_{\rm eff} < 5000$ K and $M \lesssim 0.7\,M_\odot$ (Blouin *et al.* 2019). They are also not known to vary rapidly. The second explanation is the same as for the cool magnetic DA white dwarfs, that is the convection zone is too deep and therefore the magnetic field cannot suppress it.

5.2. *Magnetic field structure*

The magnetic field structure in a white dwarf can provide clues to how the magnetic field was created. Complex magnetic fields are predicted for white dwarfs that formed in a merger (García-Berro *et al.* 2012). Simpler fields that are approximately dipolar may be an attribute of white dwarfs that are descendants of Ap or Bp stars (Braithwaite & Spruit 2004). When modelling the magnetic field geometry in white dwarfs, a centred or offset dipole is often assumed. However, rotating white dwarfs provide the means to study the field structure and reveal a diversity in the field geometry. Depending on the magnetic field structure the Zeeman splitted lines may vary as a function of the rotational phase and reveal how the magnetic field strength varies on the white dwarf surface. In addition to measuring the strength of the magnetic field, circular and linear polarization can provide the direction of the magnetic field. With a series of circular polarization spectra covering the rotation period of 0.243 days, Landstreet *et al.* (2017) showed that WD 2047+372 can be modelled by a simple dipole with a weak field of 91.8 ± 0.8 kG. However, their analysis of WD 2359–434 revealed a more complex field structure which required a combination of a dipolar field and a non-aligned quadrupolar field.

Detailed modelling of rotating white dwarfs with stronger fields has resulted in even more complex structures. The high field, massive and hot white dwarf EUVE J0317–855 has a field of 185 MG with a magnetic spot of 425 MG (Burleigh *et al.* 1999; Vennes *et al.* 2003). Another white dwarf with a magnetic spot is WD 1953-011 (Maxted *et al.* 2000; Valyavin *et al.* 2008) which was revealed over its rotational period of 1.4418 days (Brinkworth *et al.* 2005). Euchner *et al.* (2006) used Zeeman tomography to show that the magnetic field of PG 1015+014 is a combination of three individually offset non-aligned dipoles.

Our study of the structure of magnetic white dwarf atmospheres offers computational challenges. Complete suppression of the convection zone appears unlikely at lower temperatures. Complex surface field distributions are observed in several white dwarfs but such observations are challenging and impractical in most cases.

6. Summary

With the growth of large scale photometric, spectroscopic and astrometric surveys we have been able to study the magnetic white dwarf population as a whole in greater detail. The growing sample of white dwarfs has revealed that the incidence of magnetism varies across the spectral classes, with the class of hot DQs showing the highest incidence at about 70%. There is evidence that these objects are a result of double degenerate mergers, which is one of the proposed scenarios for the creation of magnetic fields.

There are several proposed scenarios for the creation of magnetic fields in white dwarfs, with the merging scenario being favoured, although many challenges remain. It is also likely that magnetic fields in white dwarfs may be created by several distinct processes.

The magnetic field structure across the surface can be complex. We have some observational evidence that shows that the white dwarf atmospheric structure is affected by the presence of a magnetic field, such as the suppression of convection. This provides computational challenges in modelling white dwarf atmospheres in the presence of complex magnetic fields.

Acknowledgements

AK thanks S. Vennes, L. Ferrario and D.T. Wickramasinghe for stimulating discussions and the Science Organizing Committee for the kind invitation to present this review. The International Centre for Radio Astronomy Research is a joint venture between Curtin University and the University of Western Australia, funded by the state government of Western Australia and the joint venture partners.

References

Alam, S. *et al.* 2015, *ApJS*, 219, 12

Al-Hujaj, O.-A. & Schmelcher, P. 2003, *PhRvA*, 68, 053403

Angel, J. R. P., Borra, E. F., Landstreet, J. D. 1981, *ApJS*, 45, 457

Angel, J. R. P. & Landstreet, J. D. 1971, *ApJ* Letters, 164, L15

Angel, J. R. P. & Landstreet, J. D. 1974, *ApJ*, 191, 457

Appenzeller, I. *et al.* 1998, *Msngr*, 94, 1

Aurière, M. *et al.* 2007, *A&A*, 475, 1053

Babcock, H. W. 1947, *ApJ*, 105 105

Bagnulo, S. & Landstreet, J. D. 2019, *A&A*, 630, A65

Barstow, M. A., Jordan, S., O'Donoghue, D., Burleigh, M. R., Napiwotzki, R., Harrop-Allin, M. K. 1995, *MNRAS*, 277, 971

Becken, W. & Schmelcher, P. 2001, *PhRvA*, 63, 053412

Bédard, A., Bergeron, P., Fontaine, G. 2017, *ApJ*, 848, 11

Benvenuto, O. G. & Althaus, L. G. 1999, *MNRAS*, 303, 30

Berdyugina, S. V., Berdyugin, A. V., Piirola, V. 2007, *PhRvL*, 99, 091101

Blackett, P. M. S. 1947, *Nature*, 159, 658

Blouin, S., Dufour, P., Thibeault, C., Allard, N. F. 2019, *ApJ*, 878, 63

Braithwaite, J. & Spruit, H. C. 2004, *Nature*, 431, 819

Briggs, G. P., Ferrario, L., Tout, C. A., Wickramasinghe, D. T., Hurley, J. R. 2015, *MNRAS*, 447, 1713

Briggs, G. P., Ferrario, L., Tout, C. A., Wickramasinghe, D. T. 2018, *MNRAS*, 478, 899

Brinkworth, C. S., Burleigh, M. R., Lawrie, K., Marsh, T. R., Knigge, C. 2013, *ApJ*, 773, 47

Brinkworth, C. S., Marsh, T. R., Morales-Rueda, L., Maxted, P. F. L., Burleigh, M. R., Good, S. A. 2005, *MNRAS*, 357, 333

Burleigh, M. R., Jordan, S., Schweizer, W. 1999, *ApJ* Letters, 510, L37

Coutu, S., Dufour, P., Bergeron, P., Blouin, S., Loranger, E., Allard, N. F., Dunlap, B. H. 2019, *ApJ*, 885, 74

D'Antona, F. & Mazzitelli, I. 1975, *A&A*, 42, 127

Debes, J. H. & Sigurdsson, S. 2002, *ApJ*, 572, 556

Dobbie, P. D. *et al.* 2012, *MNRAS*, 421, 202

Donati, J.-F., Catala, C., Landstreet, J. D., Petit, P. 2006, *ASP Conf. Ser.* Vol. 358, 362

Dufour, P., Liebert, J., Fontaine, G., Behara, N. 2007, *Nature*, 450, 522

Dufour, P., Béland, S., Fontaine, G., Chayer, P., Bergeron, P. 2011, *ApJ* Letters, 733, L19

Dufour, P., Fontaine, G., Liebert, J., Schmidt, G. D., Behara, N. 2008, *ApJ*, 683, 978

Dufour, P., Vornanen, T., Bergeron, P., Fontaine, G., Berdyugin, A. 2013, *ASP Conf. Ser.* Vol. 469, 167

Dunlap, B. H., Barlow, B. N., Clemens, J. C. 2010, *ApJ* Letters, 720, L159

Dunlap, B. H. & Clemens, J. C. 2015, *ASP Conf. Ser.* Vol. 493, 547

Euchner, F., Jordan, S., Beuermann, K., Reinsch, K., Gänsicke, B. T. 2006, *A&A*, 451, 671

Farihi, J., Dufour, P., Napiwotzki, R., Koester, D. 2011, *MNRAS*, 413, 2559

Ferrario, L., de Martino, D., Gänsicke, B. T. 2015, *SSRv*, 191, 111

Ferrario, L., Vennes, S., Wickramasinghe, D. T., Bailey, J. A., Christian, D. J. 1997, *MNRAS*, 292, 205

Ferrario, L., Wickramasinghe, D. T., Kawka, A. 2020, *AdSpR*, Ed., S. Balman, 66, 1025

Fontaine, G., Brassard, P., Charpinet, S., Randall, S. K., Van Grootel, V. 2013, *ASP Conf. Ser.* Vol. 479, 211

García-Berro, E. *et al.* 2012, *ApJ*, 749, 25

Gentile Fusillo, N. P., Tremblay, P.-E., Jordan, S., Gänsicke, B. T., Kalirai, J. S., Cummings, J. 2018, *MNRAS*, 473, 3693

González-Férez, R. & Schmelcher, P. 2003, *EPJD*, 23, 189

Greenstein, J. L. 1969, *ApJ*, 158, 281

Hermes, J. J. *et al.* 2017, *ApJS*, 232, 23

Herzberg, G. 1945, *Atomic Spectra and Atomic Structure*, New York: Dover

Holberg, J. B., Oswalt, T. D., Sion, E. M., Barstow, M. A., Burleigh, M. R. 2013, *MNRAS*, 435, 2077

Hollands, M. A. 2017, *PhD Thesis*, Univ. Warwick

Hollands, M. A., Gänsicke, B. T., Koester, D. 2015, *MNRAS*, 450, 681

Hollands, M. A., Tremblay, P.-E., Gänsicke, B. T., Gentile-Fusillo, N. P., Toonen, S. 2018, *MNRAS*, 480, 3942

Isern, J., García-Berro, E., Külebi, B., Lorén-Aguilar, P. 2017, *ApJ* Letters, 836, L28

Jordan, S., Schmelcher, P., Becken, W., Schweizer, W. 1998, *A&A*, 336, L33

Kawka, A. 2018, *CoSka*, 48, 228

Kawka, A. & Vennes, S. 2004, *Proc. IAU Symp.* 224, 879

Kawka, A. & Vennes, S. 2012, *MNRAS*, 425, 1394

Kawka, A. & Vennes, S. 2014, *MNRAS*, 439, L90

Kawka, A., Vennes, S., Ferrario, L. 2020, *MNRAS*, 491, L40

Kawka, A., Vennes, S., Ferrario, L., Paunzen, E. 2019, *MNRAS*, 482, 5101

Kawka, A., Vennes, S., Schmidt, G. D., Wickramasinghe, D. T., Koch, R. 2007, *ApJ*, 654, 499

Kemic, S. B. 1974, *JILA Pub.* 1154

Kemic, S. B. 1975, *Ap&SS*, 36, 459

Kemp, J. C., Swedlund, J. B., Landstreet, J. D., Angel, J. R. P. 1970, *ApJ* (Letters), 161, L77

Kepler, S. O. *et al.* 2013, *MNRAS*, 429, 2934

Kissin, Y. & Thompson, C., 2015, *ApJ*, 809, 108

Landi Degl'Innocenti, E. & Landolfi M., 2004, *Polarization in spectral lines*, Vol. 307, Astrophysics and Space Library, Kluwer Academic Publishers, Dordrecht

Landstreet, J. D. & Angel, J. R. P. 1971, *ApJ* Letters, 165, L67

Landstreet, J. D. & Bagnulo, S. 2019, *A&A*, 623, A46

Landstreet, J. D. & Bagnulo, S. 2019, *A&A*, 628, A1

Landstreet, J. D., Bagnulo, S., Valyavin, G. G., Fossati, L., Jordan, S., Monin, D., Wade, G. A. 2012, *A&A*, 545, A30

Landstreet, J. D., Bagnulo, S., Valyavin, G., Valeev, A. F. 2017, *A&A*, 607, A92

Liebert, J., Angel, J. R. P., Stockman, H. S., Beaver, E. A. 1978, *ApJ*, 225, 181

Liebert, J., Ferrario, L., Wickramasinghe, D. T., Smith, P. S. 2015, *ApJ*, 804, 93
Liebert, J. & Sion, E. M. 1979, *ApL*, 20, 53
Lindegren, L., *et al.* 2018, *A&A*, 616, A2
Maxted, P. F. L., Ferrario, L., Marsh, T. R., Wickramasinghe, D. T. 2000, *MNRAS*, 315, L41
Nordhaus, J., Wellons, S., Spiegel, D. S., Metzger B. D., Blackman E. G. 2011, *PNAS*, 108, 3135
Pelletier, C., Fontaine, G., Wesemael, F., Michaud, G., Wegner, G. 1986, *ApJ*, 307, 242
Potter, A. T. & Tout, C. A. 2010, *MNRAS*, 402, 1072
Putney, A. 1997, *ApJS*, 112, 527
Rebassa-Mansergas, A. *et al.* 2016, *MNRAS*, 458, 3808
Reid, I. N., Liebert, J., Schmidt, G. D. 2001, *ApJ* Letters, 550, L61
Ruffini, N. J. & Casey, A. R. 2019, *MNRAS*, 489, 420
Ryabchikova, T., Piskunov, N., Kurucz, R. L., Stempels, H. C., Heiter, U., Pakhomov, Y., Barklem, P. S. 2015, *PhyS*, 90, 054005
Schimeczek, C. & Wunner, G. 2014, *ApJS*, 212, 26
Schmidt, G. D. & Smith, P. S. 1994, *ApJ* (Letters), 423, L63
Schmidt, G. D. & Smith, P. S. 1995, *ApJ*, 448, 305
Schmidt, G. D., Liebert, J., Harris, H. C., Dahn, C. C., Leggett, S. K. 1999, *ApJ*, 512, 916
Swedlund, J. B., Wolstencroft, R. D., Michalsky, J. J., Kemp, J. C. 1974, *ApJ* Letters, 187, L121
Tout, C. A., Wickramasinghe, D. T., Ferrario, L. 2004, *MNRAS*, 355, L13
Tout, C. A., Wickramasinghe, D. T., Liebert, J., Ferrario, L., Pringle, J. E. 2008, *MNRAS*, 387, 897
Tremblay, P.-E., Cukanovaite, E., Gentile Fusillo, N. P., Cunningham, T., Hollands, M. A. 2019, *MNRAS*, 482, 5222
Tremblay, P.-E. *et al.* 2015, *ApJ*, 812, 19
Valyavin, G. & Fabrika, S. 1999, *ASP Conf. Ser.* Vol. 169, 206
Valyavin, G., Wade, G. A., Bagnulo, S., Szeifert, T., Landstreet, J. D., Han, I., Burenkov, A. 2008, *ApJ*, 683, 466
Valyavin, G. *et al.* 2014, *Nature*, 515, 88
Vennes, S. 1999, *ApJ*, 525, 995
Vennes, S., Schmidt, G. D., Ferrario, L., Christian, D. J., Wickramasinghe, D. T., Kawka, A. 2003, *ApJ*, 593, 1040
Vennes, S., Nemeth, P., Kawka, A., Thorstensen, J. R., Khalack, V., Ferrario, L., Alper, E. H. 2017, *Science*, 357, 680
Vornanen, T., Berdyugina, S. V., Berdyugin, A. V., Piirola, V. 2010, *ApJ* Letters, 720, L52
Vornanen, T., Berdyugina, S. V., Berdyugin, A. 2013, *A&A*, 557, A38
Wickramasinghe, D. T. & Ferrario, L. 2005, *MNRAS*, 356, 1576
Wickramasinghe, D. T. & Martin, B. 1986, *MNRAS*, 223, 323
Wickramasinghe, D. T., Schmidt, G., Ferrario, L., & Vennes, S. 2002, *MNRAS*, 332, 29
Wickramasinghe, D. T., Tout, C. A., Ferrario, L. 2014, *MNRAS*, 437, 675
Williams, K. A., Montgomery, M. H., Winget, D. E., Falcon, R. E., Bierwagen, M. 2016, *ApJ*, 817, 27

White Dwarfs as Probes of Fundamental Physics: Tracers of Planetary, Stellar and Galactic Evolution
Proceedings IAU Symposium No. 357, 2019
M. A. Barstow, S. J. Kleinman, J. L. Provencal & L. Ferrario, eds.
doi:10.1017/S1743921320000769

A new look at magnetic white dwarfs

François Hardy[1], Patrick Dufour[1] and Stefan Jordan[2]

[1]Département de Physique, Université de Montréal, C.P. 6128, Succursale Centre-Ville, Montréal, Québec, H3C 3J7, Canada
email: hardy@astro.umontreal.ca

[2]Astronomisches Rechen-Institut am Zentrum für Astronomie, Universität Heidelberg, Germany

Abstract. We present a homogeneous analysis of a large sample of magnetic white dwarf stars (with SDSS, PanSTARRS and Gaia data) using state-of-the-art magnetic atmosphere models and fitting techniques. We discuss the properties of the sample as well as the implication on our understanding of the nature and evolution of such objects.

Keywords. stellar: white dwarf, magnetic field, atmospheric modelling

1. Introduction

It is estimated that about 10 to 20% of white dwarfs are magnetic at some level (Kawka & Vennes 2005; Liebert *et al.* 2003; Kepler *et al.* 2015). Thanks mainly to the *Sloan Digital Sky Survey* (SDSS), their number has increased significantly in the last 15 years or so, from about 70 known objects at the turn of the last millennium to now over 600 (Ferrario *et al.* 2015; Kepler *et al.* 2015). Yet, despite these observational developments, there still remain a lot of open questions concerning the origin of magnetic white dwarfs (MWDs). Possible progenitors include mergers, binary evolution, and fossil field evolution contributing in proportions that still need to be quantified appropriately. While there is some evidence that MWDs are more massive than their non-magnetic counterparts, this is based on a relatively small number of object (and very few with a parallax measurement), and relatively ancient atmosphere models.

In fact, most of the studies of MWDs can be traced back to the work of Wickramasinghe & Martin in the '70s and '80s (hydrogen/helium line splitting with the Zeeman effect and off-centered dipole geometry field), Jordan in the '90s (use of tables calculated by the Tübingen group to model the line splitting of hydrogen and helium under arbitrary magnetic field intensities, and the decomposition of magnetic field geometries into finite elements), with a few additional contributions from Bergeron, Vennes and Kawka (see Ferrario *et al.* 2015 and references therein). The latest analysis of a large sample, that of Külebi *et al.* (2009), is already a decade old, a time when parallax measurements where few and far between and computer resources much more limited than today.

Here we present preliminary work on magnetic white dwarfs based on our newly developed magnetic atmosphere code. This code is currently the only one, to our knowledge, that can treat line splitting for any chemical composition (stars of spectral type DA, DB, DZ, DQ etc.) while solving magnetized radiative transfer with Stokes parameters (ie. computing the emergent polarized flux) for an arbitrary magnetic field geometry (not limited to an offset dipole).

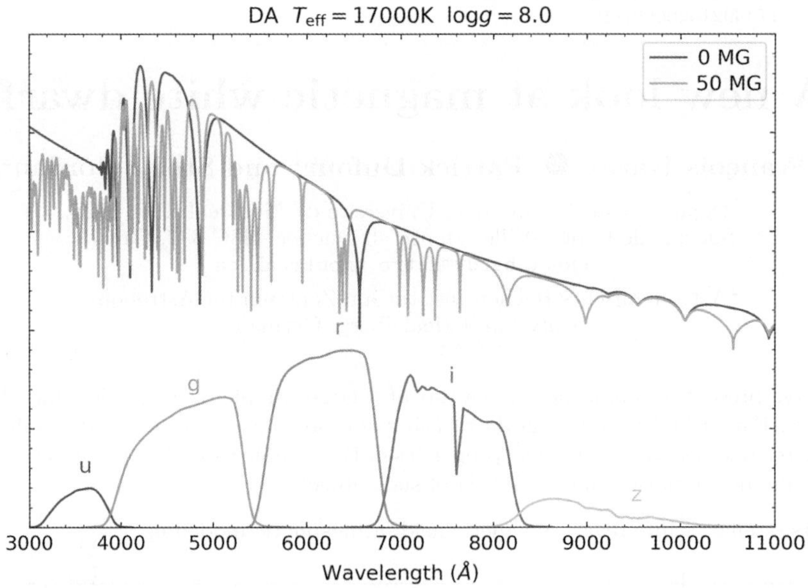

Figure 1. Comparisons of synthetic spectra with 0 and 50 MG magnetic field strengths. Note that the splitting of the lines under a large field provides additional blanketing in adjacent photometric bands (SDSS *ugriz* shown here).

2. Analysis

Atmospheric parameters for magnetic white dwarfs are notoriously difficult to obtain due to the complex line profiles and positions that arise as a result of a possibly variable magnetic strength (and corresponding line splitting) across the surface of the star. Moreover, the reliability of the mass determination from spectral line fitting is somewhat questionable since it is still unclear how to treat Stark and Zeeman broadening simultaneously when the fields are strong. We thus decided to revisit magnetic DA white dwarfs due to the recent release of Gaia data. Indeed, decent estimations can now be obtained from fits to the energy distribution when parallax measurements are available, a situation that was unfortunately possible only for about two dozen objects before the mission. Thanks to the second data release, **289** MWDs now have a parallax measurement, 235 of which are DAs according to the *Montreal White Dwarf Database* (Dufour *et al.* 2017). Of those DAs, we selected a subsample of **127** stars taken from Külebi *et al.* (2009) that had both a Gaia parallax and SDSS photometry and revisit their atmospheric parameters using the photometric method assuming the magnetic field strength and geometry they had determined.

Briefly, the method consists of the simple derivation of the atmospheric parameters of the star by fitting the solid angle ($4\pi\frac{R^2}{D^2}$) and effective temperature from a comparison of the observed flux f_ν to those of our model grid (H_ν, effective temperatures from 5000 to 30,000 K, $\log g$ from 6.5 to 9.5, and a magnetic field intensity, B, form 0 to 500 MG):

$$f_\nu = 4\pi\left(\frac{R^2}{D^2}\right) H_\nu\left(T_{\rm eff}, g, \text{composition}, B\right) \tag{2.1}$$

where D is the distance (calculated from the parallax) and R is the radius of the star. Since D is known, the derived value of R can be converted to mass/surface gravity using the white dwarf mass-radius relationship. Figure 1 shows two synthetic spectra, one at zero-field and another with a (constant) 50 MG magnetic field. This particular example

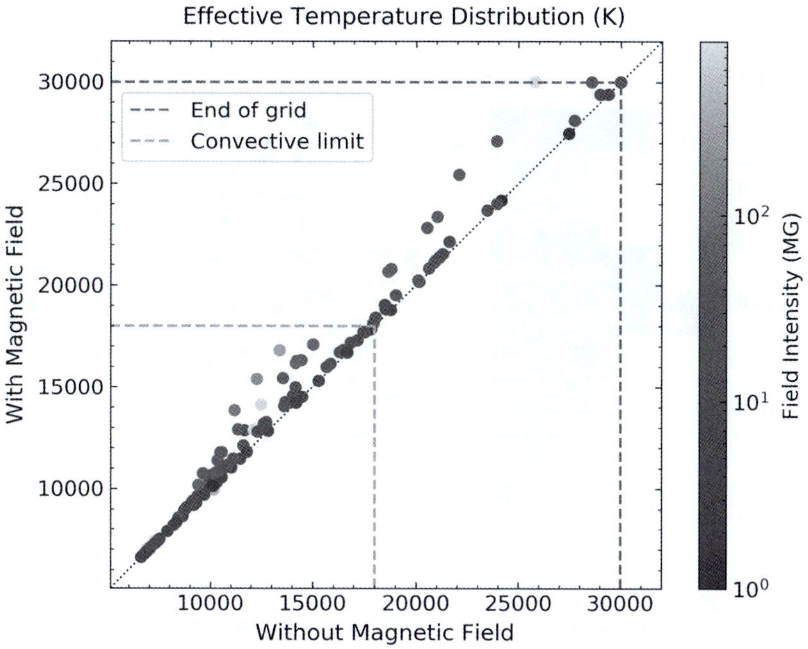

Figure 2. Comparison of effective temperatures derived with and without the inclusion of magnetic field splitting. Fits including the magnetic field yield slightly higher effective temperatures and the strength of this effect is correlated with the intensity of the field.

shows that the u and g bands are heavily affected by the shifted lines at 50 MG. Our approach now explicitly takes into account the impact of the line splitting and blanketing due to the presence of the magnetic field. Taking this into account could thus significantly affect the effective temperature determination for high magnetic field strengths, hence also affecting the radius, and thus mass, determination.

3. Results

Figure 2 compares the effective temperatures obtained assuming no magnetic field to those obtained when assuming the field intensity determined by Külebi *et al.* (2009). The effective temperatures derived when including the magnetic field are generally greater or equal to the non-magnetic ones. The difference in derived effective temperatures is due to the strong magnetic shifting of the lines into different photometric bands (a weak field will only move those spectral lines slightly, not affecting the photometry significantly).

As a consequence of the higher effective temperature, a smaller radius (and thus a larger mass) is needed to fit the solid angle. Figure 3 shows the mass distribution of our sample. Clearly, the shape of the distribution is not centered around $0.6M_{\odot}$, as is usually found in samples of non-magnetic DA white dwarfs (see Genest-Beaulieu & Bergeron 2019 for example). More interestingly, our magnetic DA sample of stars with parallax-derived masses seems to indicate the presence of a correlation between mass and field strength. The average mass of stars with a magnetic field intensity lower than 3 MG is $0.69\,M_{\odot}$ while it is $0.86\,M_{\odot}$ for those with a field higher than 3 MG.

Is this correlation the signature of some different mass loss mechanism depending on the magnetic field strength? Or maybe we are in the presence of two distinct populations (low field/low mass = fossil, high field/high mass = mergers)?

To conclude, we have shown that it is important to take into account the magnetic field splitting for an accurate effective temperature (and thus mass) determination of

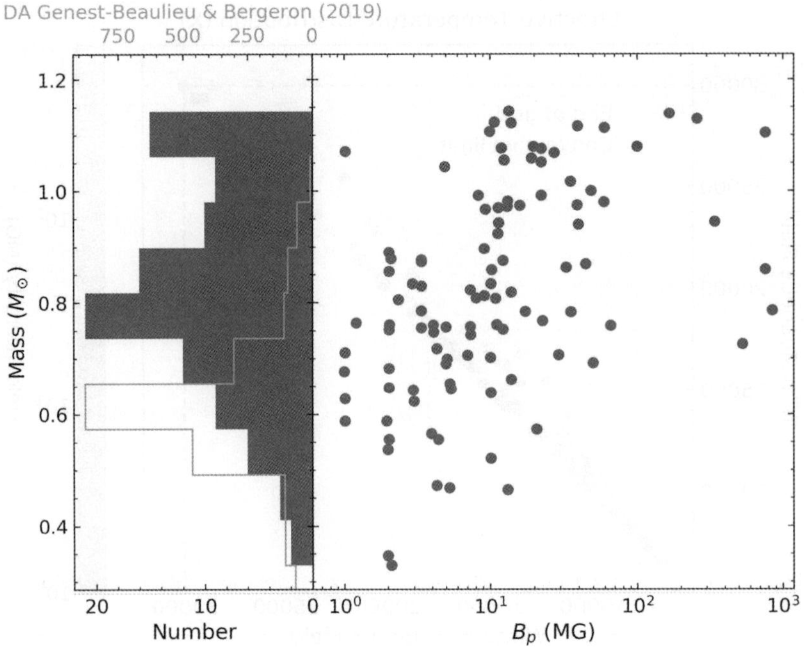

Figure 3. Left: mass distribution of magnetic DAs (blue) compared to non-magnetic DAs (Genest-Beaulieu & Bergeron 2019, orange). Right: Mass vs field intensity for our sample.

magnetic DA white dwarfs. In future work, we plan to extend our study to a larger sample of magnetic DAs as well as to all known magnetic white dwarfs in general, regardless of their spectral type.

References

Bergeron, P., Ruiz, M.-T., & Leggett, S. K. 1992, *ApJ*, 400, 315
Dufour, P., Bergeron, P., Schmidt, G. D., *et al.* 2006, *ApJ*, 651, 1112
Dufour, P., Blouin, S., Coutu, S., *et al.* 2017, 20th European White Dwarf Workshop, 3
Ferrario, L., de Martino, D., & Gänsicke, B. T. 2015, *SSR*, 191, 111
Genest-Beaulieu, C. & Bergeron, P. 2019, *ApJ*, 882, 106
Jordan, S. & Koester, D. 1986, *A&AS*, 65, 367
Jordan, S. 1989, IAU Colloq. 114: White Dwarfs, 333
Jordan, S. 1992, *A&A*, 265, 570
Kawka, A. & Vennes, S. 2005, 14th European Workshop on White Dwarfs, 101
Kawka, A. & Vennes, S. 2014, *MNRAS*, 439, L90
Kepler, S. O., Kleinman, S. J., Pelisoli, I., *et al.* 2010, *American Institute of Physics Conference Series*, 19
Kepler, S. O., Pelisoli, I., Koester, D., *et al.* 2015, *MNRAS*, 446, 4078
Külebi, B., Jordan, S., Euchner, F., *et al.* 2009, *A&A*, 506, 1341
Liebert, J., Harris, H. C., Dahn, C. C., *et al.* 2003, *AJ*, 126, 2521
Martin, B. & Wickramasinghe, D. T. 1981, *MNRAS*, 196, 23
Martin, B. & Wickramasinghe, D. T. 1982, *MNRAS*, 200, 993
Martin, B. & Wickramasinghe, D. T. 1984, *MNRAS*, 206, 407
Wickramasinghe, D. T. 1972, *MNRAS*, 76, 129
Wickramasinghe, D. T. & Martin, B. 1979, *MNRAS*, 188, 165

White Dwarfs as Probes of Fundamental Physics: Tracers of Planetary, Stellar and Galactic Evolution
Proceedings IAU Symposium No. 357, 2019
M. A. Barstow, S. J. Kleinman, J. L. Provencal & L. Ferrario, eds.
doi:10.1017/S1743921320000435

Continuous gravitational waves from magnetized white dwarfs

Surajit Kalita and Banibrata Mukhopadhyay

Department of Physics, Indian Institute of Science, Bangalore 560012, India
emails: surajitk@iisc.ac.in, bm@iisc.ac.in

Abstract. Recent evidence of super-Chandrasekhar white dwarfs (WDs), from the observations of over-luminous type Ia supernovae (SNeIa), has been a great astrophysical discovery. However, no such massive WDs have so far been observed directly as their luminosities are generally quite low. Hence it immediately raises the question of whether there is any possibility of detecting them directly. The search for super-Chandrasekhar WDs is very important as SNeIa are used as standard candles in cosmology. In this article, we show that continuous gravitational waves can allow us to detect such super-Chandrasekhar WDs directly.

Keywords. (stars:) white dwarfs, gravitational waves, stars: magnetic fields, stars: rotation

1. Introduction

Recent detections of gravitational waves (GWs) from LIGO and VIRGO events have opened a new window in astronomy. These GW events were observed when two compact objects in a binary system, such as black holes (BHs) or neutron stars (NSs) or one BH and one NS, merged to produce a single compact object. It is important to note that these events are short lasting. However, there is another type of GW, known as continuous GW, which is continuously emitted with a certain amplitude and frequency. We discuss in this article that isolated rotating white dwarfs (WDs) as well as binary WDs are prominent sources producing continuous GW.

During the past decade, a good number of peculiar type Ia supernovae (SNeIa) have been observed which are over-luminous in nature. It was inferred that these SNeIa must have progenitor WDs with masses $\sim 2.1-2.8 M_\odot$ (Howell *et al.* 2006; Scalzo *et al.* 2010). This immediately indicates the violation of the Chandrasekhar mass-limit (currently accepted value $\sim 1.4 M_\odot$ for non-magnetized static WDs). Many researchers proposed different models to explain these exceedingly massive WDs (e.g. Das & Mukhopadhyay 2013; Kalita & Mukhopadhyay 2018; Ong 2018). In this article, we show that in the presence of magnetic fields and rotation, the mass of WDs can increase significantly and eventually they can efficiently emit continuous gravitational radiation. Once such GW is detected by future space-based detectors, it will provide direct evidence of super-Chandrasekhar WDs.

2. Model of the magnetized rotating white dwarfs

It is well known that GWs have two polarizations, h_+ and $h_\times$. Moreover if an object rotates with an angular frequency Ω, with the rotation axis disaligned with the body's principal axis, it will emit gravitational radiation. For such an object, the polarizations at time t are given by (Zimmermann & Szedenits 1979; Bonazzola & Gourgoulhon 1996)

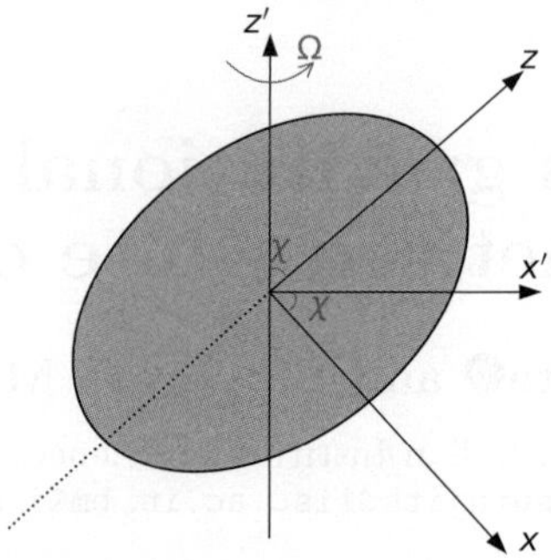

Figure 1. A cartoon diagram of magnetized rotating WD with misalignment between magnetic field and rotation axes.

$$\begin{aligned} h_+ &= h_0 \sin\chi \left[\frac{1}{2} \cos i \sin i \cos\chi \cos\Omega t - \frac{1+\cos^2 i}{2} \sin\chi \cos 2\Omega t \right], \\ h_\times &= h_0 \sin\chi \left[\frac{1}{2} \sin i \cos\chi \sin\Omega t - \cos i \sin\chi \sin 2\Omega t \right], \end{aligned} \tag{2.1}$$

with

$$h_0 = -\frac{6G}{c^4} Q_{z'z'} \frac{\Omega^2}{d}, \tag{2.2}$$

where $Q_{z'z'}$ is the quadrupole moment of the distorted star, χ the angle between the rotation axis z' and the body's third principal axis z, i the angle between the rotation axis of the object and our line of sight, G Newton's gravitational constant and c the speed of light. It has already been shown that a toroidal magnetic field makes a star prolate, whereas a poloidal magnetic field (as well as rotation) deforms it to an oblate shape (Subramanian & Mukhopadhyay 2015). As shown in Figure 1, we assume that the magnetic field is along the $z-$axis, whereas the star is rotating about the $z'-$axis. Using this configuration and from Equation (2.2), the GW amplitude becomes

$$h_0 = \frac{2G}{c^4} \frac{\Omega^2 \epsilon I_{xx}}{d} (2\cos^2\chi - \sin^2\chi), \tag{2.3}$$

where $\epsilon = (I_{zz} - I_{xx})/I_{xx}$ is the ellipticity of the body and I_{xx}, I_{yy}, I_{zz} are the principal moments of inertia of the object about its three principal axes ($x-$, $y-$, $z-$axes respectively). Here we use the *XNS* code (Pili *et al.* 2014) to model WDs. This code solves the axisymmetric equilibrium configuration of stellar structure in general relativity. A detail discussion about all the parameters used to model WDs is given by Kalita & Mukhopadhyay (2019). Moreover, we assume the distance between the WD and the detector to be 100 pc.

3. Structure of rotating magnetized white dwarfs

We show typical cases for rotating WDs with toroidal and poloidal magnetic fields in Figure 2(a) and 2(b) respectively. It is observed that the deformation of the core is more prominent than that of the outer regions. The overall shape of the WDs depends on the combined effects arsing from the rotation and the structure of the magnetic field. It is important to note that the central field could be as large as $\sim 10^{14}$ G even if the surface field is $\sim 10^9$ G (Heyl 2000; Ferrario *et al.* 2015). The stability of these magnetized WDs (B-WDs) is determined constraining the kinetic to gravitational energy ratio as well as the magnetic to gravitational energy ratio (Braithwaite 2009).

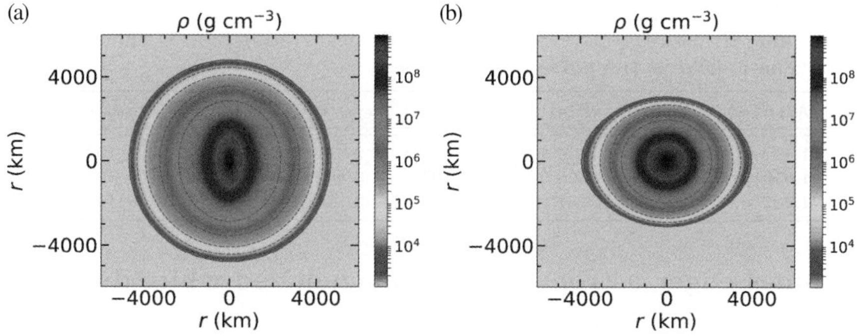

Toroidal magnetic field Poloidal magnetic field

Figure 2. Uniformly rotating B-WDs with (a) $\Omega = 0.02$ rad/s, (b) $\Omega = 1.32$ rad/s. The color code corresponds to density distribution.

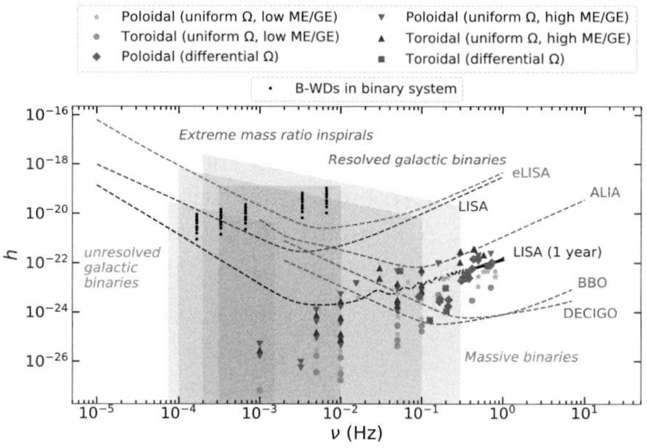

Figure 3. Dimensionless GW amplitude for WDs as a function of frequency with $\chi = 3°$ and $d = 100$ pc.

4. Gravitational radiation from rotating magnetized white dwarfs

The gravitational radiation has been calculated for various stable B-WDs by varying the central density and the magnetic field strength and also considering both uniform and differential rotation. Figure 3 shows the strength of the GW for these B-WDs assuming the misalignment between the rotation and magnetic field axes to be 3°. It is evident from this figure that many of the isolated rotating B-WDs can be detected by future space-based detectors, such as, LISA, DECIGO, BBO, etc. It is also important to note that these B-WDs cannot be detected by LISA directly, while it might be possible to detect some of them with the help of LISA's 1 year integration curve. Moreover, since our galaxy is very rich in binary WDs, they may create some confusion noise (Ruiter *et al.* 2010). However, since the time period of B-WDs is much smaller than the orbital time period of binaries, there is no problem in distinguishing isolated B-WDs from binaries. Some of the massive black hole binaries may create confusion noise too, which can easily be removed with proper source modeling, but this is beyond the scope of this work. A similar exploration has also been carried out in the case of neutron stars (Kalita & Mukhopadhyay 2018).

Table 1. $L_{\rm GW}$ and $L_{\rm EM}$ for WDs considering $\dot{P}=10^{-15}$ Hz s^{-1} and $\chi=3°$. M and R are respectively the mass and mean radius, B_s is the surface magnetic field at the pole of the B-WDs.

$M(M_\odot)$	R(km)	P(s)	B_s(G)	$L_{\rm GW}$(ergs s^{-1})	$L_{\rm EM}$(ergs s^{-1})
1.420	1718.8	1.5	6.12×10^{8}	2.91×10^{35}	5.50×10^{34}
1.640	1120.7	2.0	2.78×10^{9}	3.46×10^{36}	3.03×10^{34}
1.702	1027.2	3.1	4.52×10^{9}	3.17×10^{35}	8.23×10^{33}

5. Multi-messenger astronomy with magnetized white dwarfs

Being quadrupolar in nature, GW is associated with quadrupolar luminosity. Moreover, since the B-WDs have magnetic fields as well as rotation, they behave as rotating dipoles, and they must possess luminosity due to dipole radiation along with gravitational radiation, which means B-WDs have electromagnetic counterparts. The luminosity due to electromagnetic dipole radiation is given by $L_{\rm EM}=4\pi^2_{z'z'}{}^{I}\dot{P}/P^3$, where P is the rotational period of the body which is changing with time as $\dot{P}$. Moreover, the GW luminosity is given by

$$L_{\rm GW}=\frac{G}{c^5}\left\langle \dddot{Q}_{ij}\dddot{Q}_{ij}\right\rangle=\frac{G\Omega^6}{20c^5}(I_{xx}-I_{zz})^2\sin^2\chi(2\cos^2\chi-\sin^2\chi)^2$$
$$\left\{\cos^2\chi\sin^2 i(1+\cos^2 i)+4\sin^2\chi(1+6\cos^2 i+\cos^4 i)\right\}. \quad (5.1)$$

It is clearly observed from the above formulae that the misalignment between the rotation and magnetic field axes is necessary to emit gravitational radiation, whereas this is not compulsory for dipolar radiation. A comparison between the values of $L_{\rm GW}$ and $L_{\rm EM}$ is given in Table 1. Since $L_{\rm GW}$ is larger than $L_{\rm EM}$, it will be another unique way of separating B-WDs from regular WDs.

6. Conclusion

Although SNeIa have been used as standard candles, some of them are powered by super-Chandrasekhar WDs which have not been directly detected so far because they are very dim. It is already known that magnetic fields and rotation can increase the mass of WDs. In this article, we have argued that continuous GW is one of the best ways to detect some of these B-WDs directly, provided they posses tri-axial systems. Therefore, if there are some super-Chandrasekhar WDs in the sky, the future space-based GW detectors should be able to detect them.

Acknowledgement

S. K. would like to thank the IAU for providing a travel grant to attend the symposium. The authors further thank Lilia Ferrario of the ANU for thoroughly reading the manuscript with encouraging amendments.

References

Bonazzola, S. & Gourgoulhon, E. 1996, *A&A*, 312, 675
Braithwaite, J. 2009, *MNRAS*, 397, 763
Das, U. & Mukhopadhyay, B. 2013, *Phys. Rev. Lett.*, 110, 071102
Ferrario, L., De Martino, D., & Gaensicke, B. 2015, *Space Sci. Rev.*, 191, 111
Heyl, J. S. 2000, *MNRAS*, 317, 310
Howell, D. A. *et al.* 2006, *Nature*, 443, 308
Kalita, S. & Mukhopadhyay, B. 2018, *JCAP*, 09, 007
Kalita, S. & Mukhopadhyay, B. 2019, *MNRAS*, 490, 2692
Ong, Y. C. 2018, *JCAP*, 09, 015

Pili, A. G., Bucciantini, N., & Del Zanna, L. 2014, *MNRAS*, 439, 3541
Ruiter, A. J. *et al.* 2010, *ApJ*, 717, 203
Scalzo, R. A. *et al.* 2010, *ApJ*, 713, 1073
Subramanian, S. & Mukhopadhyay, B. 2015, *MNRAS*, 454, 752
Zimmermann, M. & Szedenits, E. Jr. 1979, *Phys. Rev. D*, 20, 351

White Dwarfs as Probes of Fundamental Physics: Tracers of Planetary, Stellar and Galactic Evolution
Proceedings IAU Symposium No. 357, 2019
M. A. Barstow, S. J. Kleinman, J. L. Provencal & L. Ferrario, eds.
doi:10.1017/S1743921320000307

Laboratory studies of Vacuum Ultra-Violet (VUV) emission spectra of heavy element ions

W.-Ü Lydia Tchang-Brillet[1], Ali Meftah[1,2], Djamel Deghiche[2], Jean-François Wyart[1,3], Christian Balança[1], Norbert Champion[1] and Christophe Blaess[1]

[1]Laboratoire d'Etude du Rayonnement et de la Matière en Astrophysique et Atmosphèère (LERMA), Observatoire de Paris-Meudon, PSL Research University; Sorbonne Universités, CNRS, F-92190 Meudon, France
email: lydia.tchang-brillet@obspm.fr

[2]Laboratoire de Physique et Chimie Quantique, University Mouloud Mammeri Tizi-Ouzou, Algeria

[3]Laboratoire Aimé Cotton, CNRS UMR9188, Univ Paris-Sud, Univ Paris-Saclay, bâtiment 505, 91405 Orsay CEDEX, France

Abstract. Reliable spectroscopic data are needed for interpretation and modeling of observed astrophysical plasmas. For heavy element ions, which have complex spectra, experimental data are rather incomplete. To provide valuable fundamental quantities, such as precise wavelengths, level energies and semi-empirical transition probabilities, we are carrying out laboratory studies of high-resolution VUV emission spectra of moderately charged ions of transition metals and rare earth elements. Experimental and theoretical methods are summarized. Examples of studies are described.

Keywords. heavy elements, spectra, energy levels, transition probabilities, vacuum ultraviolet

---

1. Introduction

A huge amount of spectroscopic and collisional data is needed for collisional-radiative modeling of astrophysical plasmas in non-local thermal equilibrium (NLTE) conditions. In particular, heavy elements like transition metals and lanthanides of different ionization stages have attracted attention since their detection in high resolution observational spectra collected in the ultraviolet (UV) region by Hubble Space Telescope (HST/GHRS and STIS). For instance, singly and doubly charged transition metals and lanthanides are present in atmospheres of chemically peculiar stars (See Ryabchikova *et al.* 2006). Spectral lines from Fe V and Ni V observed in white dwarf spectra have been used to test the variation of the fine structure constant α versus gravitation (Prevals *et al.* 2013). Lanthanide and actinide ions up to three times charged are predicted to be formed through the r-process in neutron star mergers, and to strongly contribute to the opacity of the ejected matter (Karsen*et al.* 2013).

The present contribution aims to summarize experimental and parametric studies of triply and quadruply charged lanthanide ions, carried on within our collaborating team. After Nd IV (Wyart *et al.* 2007) and Nd V (Meftah *et al.* 2008; Deghiche *et al.* 2015), followed by Tm IV (Mefta *et al.* 2007), Yb V (Meftah *et al.* 2013) and Er IV (Meftah *et al.* 2016), extensions to other Thulium ions and to higher configurations of Nd IV (Arab *et al.* 2019) are underway. The electronic configurations of these ions have $5p^6 4f^N$

and $4f^{N-1}nl$, $5p^5 4f^N nl$ open subshells, which result in dense and complex spectra in the vacuum ultraviolet (VUV) wavelength range (300–3000Å). To disentangle these spectra in the laboratory and to derive energy levels and fundamental properties of these ions, high spectral resolution is needed and a systematic isoelectronic or isoionic approach is helpful.

Although *ab initio* theoretical calculations provide a great amount of atomic data needed, experimental studies remain essential. For instance, in Nd IV, a recent full relativistic *ab initio* calculation (Gaigalas *et al.* 2019) led to calculated wavelengths with deviations within 20% from the experimental ones (Wyart *et al.* 2007). Indeed, a precision of 10^{-6} on VUV wavelengths can only be achieved by experimental measurements. The knowledge of experimental energy levels is a starting point for validation of *ab initio* calculations. Furthermore, reliable semi-empirical radiative transition probabilities and Landé factors can be derived from parametric calculations based on experimental energies (see below). In Nd IV, a comparison of gA values (weighted probability for spontaneous emission) from Cowan's code parametric calculations (Arab *et al.* 2019) and from a more sophisticated semi-empirical calculation (Enzonga Yoca & Quinet 2014) fitting the same experimental energies agree within 10%, whereas the full relativistic *ab initio* values of gA may deviate by a factor 2 (Gaigalas *et al.* 2019).

2. Experimental methods

We record emission spectra on the 10.7-meter high-resolution normal incidence VUV spectrograph of the Meudon Observatory. This instrument is equipped with a 3600 lines/mm holographic concave grating, leading to a linear dispersion of 0.25 Å/mm in the focal plane in the first order. The resulting resolution is about 150 000 with a slit width of 30 μm. The plate-holder was designed to accommodate two photographic plates of 18″ x 2″, allowing simultaneous recording of a wavelength range of 240 Å between 300 and 3000 Å. Photographic plates provide precise wavelength measurements but they are no longer available. Instead, we use photo-stimulable image plates, which show enough sensitivity in the VUV range (Reader *et al.* 2000). They have linear response in intensity over five orders of magnitude. Indeed, a succession of CCDs paving the focal plane would have been too expensive and yet unsatisfactory with respect to resolution and focusing onto the Rowland circle. Emission spectra of ionized lanthanides are produced using high-voltage low inductance vacuum spark sources with an anode made of the element of interest. By varying the peak electric current, it is possible to discriminate lines from different ionization stages, generally three of them, according to their intensity behaviors. Impurity lines with well known wavelengths are used as internal standards for wavelength calibration of the spectrograms. The uncertainties of the measured wavelengths are estimated to be $\pm 0.001 - 0.005$Å. Figure 1 displays two wavelength sections of experimental thulium spectra recorded on image plates.

3. Analysis and parametric calculations

The analysis of a complex spectrum aims for determination of the energy level scheme of the ion from the observed lines by applying the Ritz combination principle taking into account selection rules on quantum numbers for electric dipole transitions. For each transition between one upper level and one lower level of opposite parities, $\Delta E = E_u - E_l = \sigma$, where σ is the wavenumber of the line and E, the energy value of a level in cm^{-1}. Furthermore, the observed intensities must be consistent with theoretical predictions of transition probabilities. Consequently, the analysis must be supported by theoretical studies. We apply the Racah-Slater semi-empirical method, by means of Cowan's codes (RCN/RCN2/RCG/RCE) (Cowan 1981).

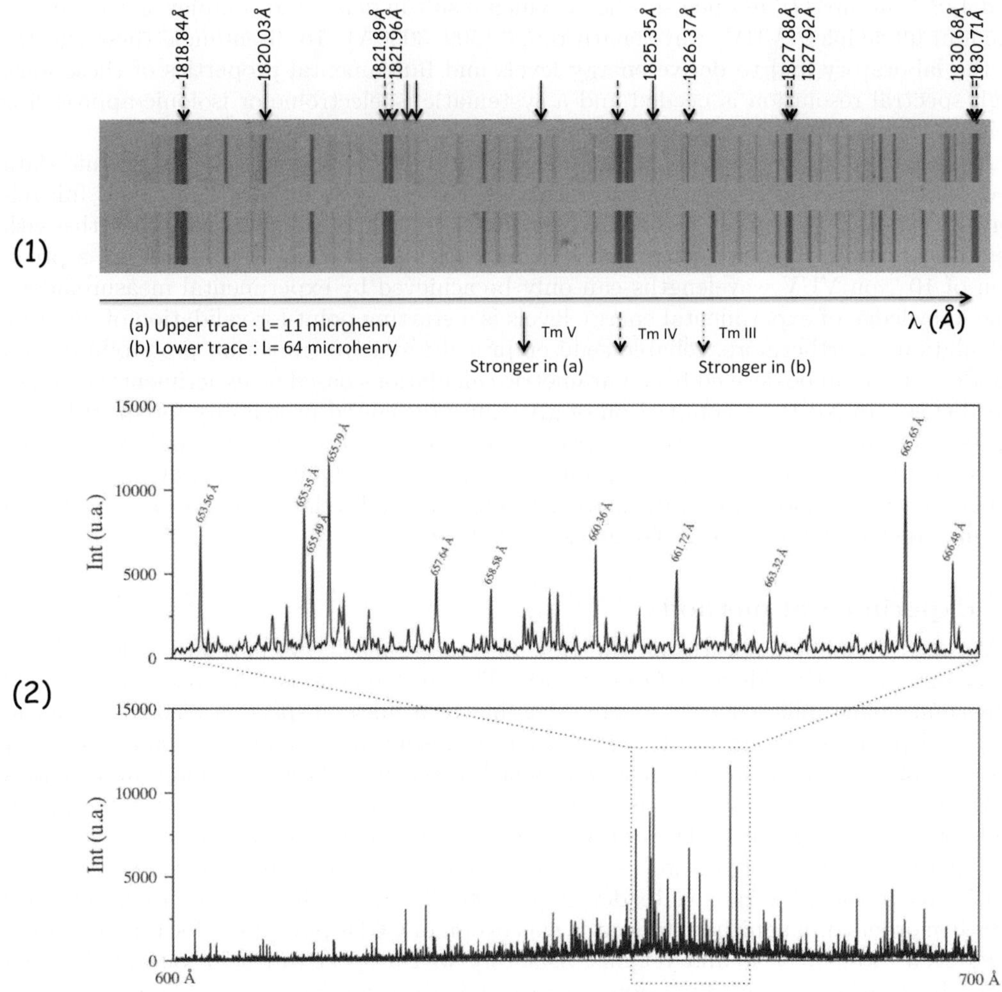

Figure 1. Sections of the thulium vacuum spark emission spectrum. (1) Spectra recorded on image plates in two different discharge conditions, showing different behaviors of Tm III – V lines (1815 – 1830 Å); (2) Digitized intensity profiles in a shorter wavelength region (600 – 700 Å) showing mainly Tm V lines in the enlarged section.

In the first step, *ab initio* predictions of level energies from unknown configurations and transition probabilities are derived by Hartree-Fock calculations including relativistic corrections (HFR). The Hamiltonian is diagonalized in the basis of one or several configurations, to explicitly take into account configuration interactions. Matrix elements are written as a sum of products of an angular part c_{ij}^{α} given by Racah algebra and a radial integral P_{α}: $H_{ij} = \sum_i c_{ij}^{\alpha} P_{\alpha}$.

A second semi-empirical step is performed as soon as some experimental values of energies become available. The radial integrals P_{α} are treated as adjustable energy parameters in a least square fit (LSQ) minimizing the mean standard deviation ΔE defined as : $\Delta E = \sqrt{(N_l - N_P)^{-1} \sum_s i(E_i^{exp} - E_i^{calc})^2}$ where N_l is the number of fitted levels and N_P the number of parameters. An iteration of the cycle LSQ/diagonalization provides improved predictions for unknown levels and line intensities. One may mention a few refinements in the present approach. Most of the time, the HFR values of average

Table 1. Identified transitions in Tm V used for the determination of three $4f^{10}6p$ levels.

$E_u(cm^{-1})$	J_u	$4f^{10}6p$ term	$\lambda(\AA)$	$\sigma(cm^{-1})$	Int^1_{exp}	$gA(s^{-1})$	$E_l(cm^{-1})$	J_l	$4f^{10}(5d+6s)$ term
249871.29	7.5	^6H	896.49	111546.14	150	1.585E+10	138325.83	7.5	^6H
			900.59	111038.31	157 p	8.208E+09	138833.94	8.5	^6I
			906.20	110350.91	141	1.989E+10	139521.49	6.5	^6G
			1837.74	54414.58	181	9.472E+09	195456.43	8.5	^6I
			1903.77	52527.23	22	2.197E+09	197343.07	7.5	^4I
250401.53	8.5	^4K	892.25	112076.21	45	7.533E+09	138325.83	7.5	^6H
			896.32	111567.29	169	1.536E+10	138833.94	8.5	^6I
			918.04	108927.71	180	2.331E+10	141474.16	9.5	^6K
			959.40	104231.81	38	2.887E+09	146171.21	9.5	^6K
			1820.03	54943.94	156	6.833E+09	195456.43	8.5	^6I
			1884.74	53057.54	121	5.137E+09	197343.07	7.5	^4I
257574.44	7.51	^6K	893.63	111903.42	51	3.305E+09	145671.02	8.5	^4K
			893.52	111916.91	103	1.031E+10	145657.94	7.5	^6H
			922.68	108379.89	223	1.989E+10	149195.71	8.5	^6L
			931.53	107349.37	47	3.860E+09	150226.91	7.5	^4I

Notes: [1]Experimental intensities in arbitrary units, estimated from the area under the line profile.
p: perturbed

energies of configurations, E_{av}, need to be scaled according to the observed positions of transition arrays on the wavenumber scale, which indicate the relative distances between configurations. Furthermore, the HFR values of radial integrals (electrostatic and spin-orbit) are overestimated and need to be scaled by an appropriate scaling factor (SF) derived from neighboring ions. Indeed, isoelectronic or isoionic ions show rather consistent values of scaling factors and provide reliable choices for initial parameter values of the LSQ fit (Wyart 2011). This allows reliable correspondences between experimental and calculated energies in the fit, in case of very close levels of the same J.

Analyses have gained efficiency using the interactive IDEN code (Azarov 1993; Azarov *et al.* 2018) based on pattern recognition with visual display of the experimental data together with the predictions of the Cowan codes on level energies, wavenumbers, and line intensities. The final experimental energy values are optimized by using the LOPT code (Kramida 2011), which minimizes the differences between the whole set of wavenumbers from wavelength measurements and those calculated from the experimental energies by the Ritz principle (Ritz wavelengths). These latter generally have reduced uncertainties compared to the observed wavelengths.

4. Example results

As an example of typical results derived from spectral analyses, Table 1 shows preliminary results on the Tm V spectrum analysis. For three previously unknown levels of $4f^{10}6p$, the experimental energy, the quantum number J, the wavelengths and wavenumbers of transitions involved in their determinations are reported, together with their intensities compared to calculated transition probabilities gA, the energy and designation of the lower level. Complete results on energy levels, line list and parameter values will be published later elsewhere.

Our previously published data are available on the https://molat.obspm.fr webpage.

References

Arab, K., Deghiche, D., Meftah, A., Wyart, J.-F., Tchang-Brillet, W.-Ü. L., Champion, N., Blaess, C., & Lamrous, O. 2019, *JQSRT*, 229, 145

Azarov, V. I. 1993, *Physica Scripta*, 48, 656

Azarov, V. I., Kramida, A., & Vokhmentsev, M. Ya 2018, *Comput. Phys. Comm.*, 225, 149

Cowan, R. D. 1981 *The Theory of Atomic Structure and Spectra*, University of California Press
Deghiche, D., Meftah, A. Wyart, J.-F., Champion, N., Blaess, C., & Tchang-Brillet, W.-Ü. L. 2015, *Physica Scripta*, 90, 095402
Enzonga Yoca, S. & Quinet, P. 2014, *J. Phys. B Atomic Molecular Physics*, 47, 035002
Gaigalas, G., Kato, D., Rynkun, P., Radziute, L., & Tanaka, M. 2019, *ApJS*, 240, 29
Kasen, D., Badnell, N. R., & Barnes, J. 2013, *ApJ*, 774, 25
Kramida, A. E. 2011, *Computer Physics Communications*, 182, 419
Meftah, A., Wyart, J.-F., Champion, N., & Tchang-Brillet, W.-Ü. L. 2007, *European Physical Journal D*, 44, 35
Meftah, A., Wyart, J.-F., Sinzelle, J., Tchang-Brillet, W.-Ü. L., Champion, N., Spector, N., & Sugar, J. 2008, *Physica Scripta*, 77, 055302
Meftah, A., Wyart, J.-F., Tchang-Brillet, W.-Ü. L., Blaess, C., & Champion, N. 2013, *Physica Scripta*, 88, 045305
Meftah, A. *et al.* 2016, *J. Phys. B*, 49, 5002
Preval, S. P., Barstow, M. A., Holberg, J. B., & Dickinson, N. J. 2013, *MNRAS*, 436, 659
Reader, J., Sansonetti, C. J., & Deslattes, R. D. 2000, *Applied Optics*, 39, 637
Ryabchikova, T., Ryabtsev, A., Kochukhov, O., & Bagnulo, S. 2006, *A&A*, 456, 329
Wyart, J.-F., Meftah, A., Tchang-Brillet, W.-Ü. L., Champion, N., Lamrous, O., Spector, N., & Sugar, J. 2007, *J. Phys. B Atomic Molecular Physics*, 40, 3957
Wyart, J.-F. 2011, *Canadian Journal of Physics*, 89, 451

White Dwarfs as Probes of Fundamental Physics: Tracers of Planetary, Stellar and Galactic Evolution
Proceedings IAU Symposium No. 357, 2019
M. A. Barstow, S. J. Kleinman, J. L. Provencal & L. Ferrario, eds.
doi:10.1017/S1743921320000538

Neutral helium line profiles through the simulation of local interactions

Patrick Tremblay, Alain Beauchamp and Pierre Bergeron

Département de Physique, Université de Montréal, C. P. 6128, Succ. Centre-Ville, Montréal, QC H3C 3J7, Canada
email: patrick@astro.umontreal.ca

Abstract. For the past 25 years, we have been considering the Stark effect for neutral helium lines in DB white dwarfs using the standard Stark broadening theory in both the impact regime (in the center of the lines) and the quasi-static regime (in the wings) for the electrons, while neglecting the effect of ions in motion. Although this is probably a good approximation based on previous theoretical work, the transition between the two regimes for the electrons and the contribution of the ions very near the core might be poorly represented. To better represent these particularities, we report here the results of a new series of simulations that treat the local dynamics and interactions of both electrons and ions around a neutral helium atom. From these simulations, we produce new improved line profiles, which we compare with our previous analytical results.

Keywords. line: profiles, atomic processes, methods: n-body simulations, methods: statistical, techniques: spectroscopic

1. Introduction

There are two general methods for measuring the physical parameters of white dwarf stars. First, the effective temperature ($T_{\rm eff}$) and surface gravity ($\log g$) can be measured by fitting optical spectra with the predictions of detailed model atmospheres, the so-called spectroscopic technique. Alternatively, the photometric energy distributions can be fitted with synthetic photometry, yielding in this case a measurement of $T_{\rm eff}$ as well as the stellar radius if the distance to the star is known. The mass-radius relation for white dwarfs can then be used to infer all other quantities, and most importantly the stellar mass. In principle, both the spectroscopic and photometric techniques should yield similar results.

Genest-Beaulieu *et al.* (2019) compared the mass distributions of DB stars as a function of effective temperature using both techniques (see their Figure 10). While the photometric mass distribution is well centered around the average mass of 0.6 $M_\odot$ at all temperatures, the spectroscopic mass distribution shows significant departures from this average mass, which also depend on the range of temperature considered. Genest-Beaulieu *et al.* have invoked several possibilities to account for these discrepancies, including flux calibration issues, line broadening theory, and convective energy transport. In this paper, we explore the possibility that Stark broadening may be responsible for some of the problems, at least at high effective temperatures where the Stark effect dominates.

For instance, we show in Figure 1 our best fit to a very high signal-to-noise spectrum of BPM 17088 (WD 0308–565) obtained with the X-shooter spectrograph at the ESO Very Large Telescope, and provided to us by S. Moehler (private communication). While the overall fit is excellent, the high quality of this optical spectrum reveals some small, but significant discrepancies, in particular in the line cores and in the far wings of some

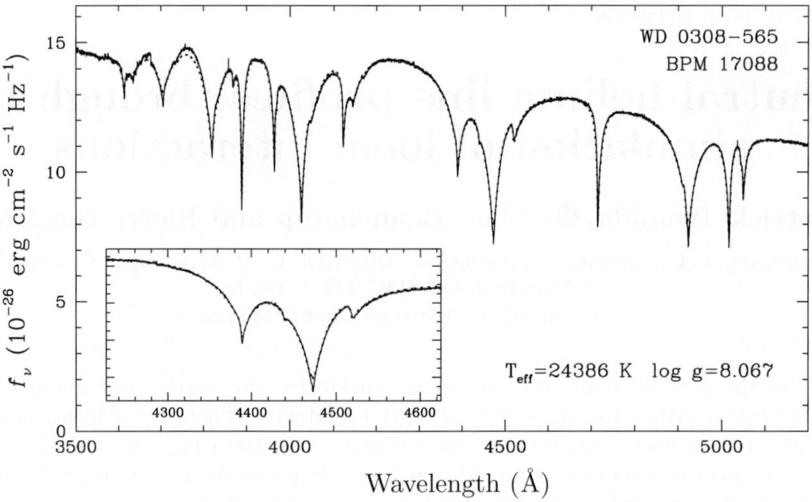

Figure 1. Our best spectroscopic fit (dotted line) to the high-resolution spectrum of the DB white dwarf BPM 17088 (WD 0308–565). The inset is a zoomed-in view near He I λ4471.

absorption lines. At this temperature ($T_{\rm eff} \sim 24,000$ K), however, we cannot completely rule out that the observed discrepancies might be related to some inaccurate treatment of convective energy transport. To disentangle both problems, we feel it is thus necessary to revisit our treatment of the Stark broadening of neutral helium lines.

2. Stark broadening

Stark broadening takes place in the presence of electric fields produced by charged particles in the emitter environment. Strong electric fields present in the photospheric regions of DB white dwarfs produce two effects: the atomic energy levels of the helium atom become a function of the electric field intensity, and the mixing of the different neighboring atomic levels leads to transitions that are normally not allowed by the selection rules for electric dipole transitions. In a few cases, these transitions may be observed as forbidden components. The $2P^3 - 4D^3$ λ4471 transition is the classic example often used to describe Stark broadening of a neutral helium line. Figure 2 displays the behavior of the upper levels involved in the He I λ4471 Stark-broadened profiles (m=0 only, assuming a field along the z direction). For the sake of simplicity, we assume here that the lower levels are unaffected by the electric field, and we investigate the behavior of the upper levels only. As observed in the figure, the energy of the perturbed upper levels is highly dependent on the field intensity, in a non-linear way. The final profile is an average over every possible temporal sequence of local electric fields that is consistent with the distribution of dynamic perturbers in the vicinity of the emitting helium atom.

Our former modeling of Stark broadening was based on the standard theory, as developed by Barnard *et al.* (1969), and applied by Beauchamp (1995, 1997) in the context of DB white dwarfs. Within this framework, the ions are considered at rest but in various static configurations — the so-called quasi-static regime. In contrast, the broadening due to the much faster electrons is described by the impact approximation in the line cores, and by the one-perturber approximation in the line wings (Baranger 1962), the latter of which converges to the quasi-static regime. Effects due to perturbing ions and electrons are treated differently within this theoretical framework.

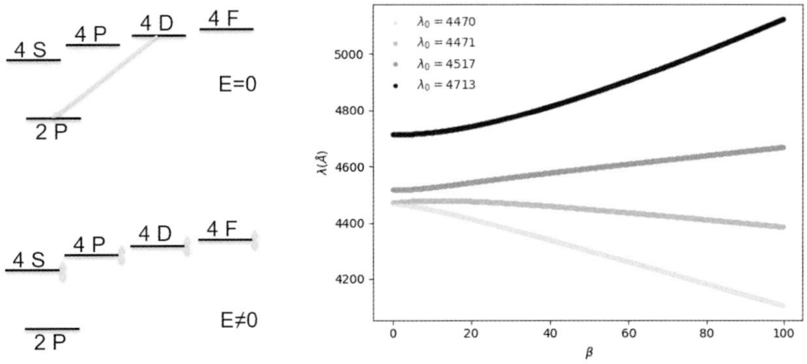

Figure 2. Left: Energy level diagrams in the absence (top) and presence (bottom) of an electric field. Right: Wavelengths of the transitions between perturbed $4S^3$ to $4F^3$ upper levels and the unperturbed $2P^3$ lower level as a function of the normalized electric field strength $\beta = F/F_0$, where F_0 is the normal field strength for an electron density of $N_e = 10^{16}$ cm^{-3}.

3. An improved approach

Our goal here is to improve these line profile calculations using well-established simulation techniques developed by Gigosos *et al.* (1996, 2014), which treat the effects of dynamic ions and electrons in a unified way. The core of the method consists in generating spherical simulation spaces that reproduce the dynamical environment of the emitter. Each simulation space is defined in the emitter referential, that is, a static emitter is placed at the center of each simulation space, and the perturber velocity follows a Maxwell distribution parameterized by the temperature and the reduced mass of the emitter-perturber pair. For a given temperature and electron number density, the radius of the simulation space is defined as the distance at which the perturbers contribution to the electric field at the emitter position becomes negligible, as described by a Debye-screened potential. Ions and electrons are randomly distributed assuming an electrically neutral plasma, and moving along classical trajectories, which are straight lines for a neutral emitter. The number of perturbers in the simulation volume is allowed to vary, as proposed by Hegerfeldt *et al.* (1988). The set of electric field intensities generated by this process approximately follows the statistical distributions from Hooper (1968).

The physical environment described above allows the electric field to vary both in amplitude and orientation depending on the location of the perturbers. This variation is different for each simulation space, and can thus have different behaviors. For instance, if one or several perturbers pass near the emitter, the field will vary rapidly, as opposed to the case when they are relatively far from the emitter, in a region where the field vector will remain mostly constant. From these simulations, a given line profile can be computed by following these steps:

- A time sequence of 50,000 electric field vectors is extracted for each simulation space.
- The energy of the perturbed levels is computed for each electric field.
- The autocorrelation function, as defined by Griem (1974), is calculated for each simulation space.
- An arithmetic average of the autocorrelation functions obtained amongst 25,000 simulation volumes is performed. The end result is a representation of the line profile in the time domain.
- A Fourier transform is applied to the averaged autocorrelation function to obtain the line profile, in the frequency domain.

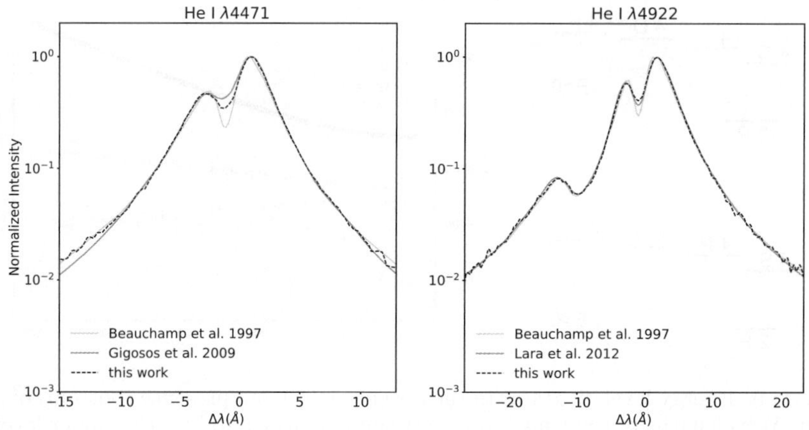

Figure 3. He I λ4471 (left panel) and λ4922 (right panel) line profiles obtained from our simulations compared to the work of Beauchamp *et al.* (1997), Gigosos *et al.* (2009), and Lara *et al.* (2012, for λ4922).

4. Test results

Our simulation code has been tested for the He I λ4471 and λ4922 transitions for a temperature of 20,000 K and an electron density of 10^{16} cm^{-3}. The results displayed in Figure 3 reveal that our simulations can reproduce the overall shape of the line profiles rather well. Also, the fact that we can reproduce both line profiles implies that our theoretical framework is adaptive and can most likely be applied to any transition in the optical. One obvious problem in our simulations is the noise observed in the line wings. This has been traced back to the numerical Fourier transform of the autocorrelation function, resulting from the noise present over long simulation times in the autocorrelation function. We are currently exploring several correction procedures to obtain smoother line profiles.

Our calculations represent the first step towards providing a complete grid of improved line profiles for He I. These will need to be compared in detail with the earlier work of Beauchamp and Gigosos, and eventually included in the calculations of model spectra for DB white dwarfs. Our simulation framework will hopefully permit, in the future, to treat various emitters (such as hydrogen), perturbers, and broadening mechanisms.

References

Baranger, M. 1962, *Pure and Applied Physics*, 13, 493
Barnard, A.J., Cooper, J., & Shamey, L.J. 1969, *Astronomy and Astrophysics*, 1, 28
Beauchamp, A. 1995, *PhD thesis, Université de Montréal*
Beauchamp, A., Wesemael, F., & Bergeron, P. 1997, *Astrophysical Journal Supplement Series*, 108, 559
Genest-Beaulieu, C., & Bergeron, P. 2019, *Astrophysical Journal*, 882, 106
Gigosos, M. A. 2014, *Journal of physics D: Applied Physics*, 47
Gigosos, M. A. & González, M. Á. 1996, *J. Phys. B: At. Mol. Opt. Phys.*, 29
Gigosos, M. A. & González, M. Á. 2009, *Astronomy and Astrophysics*, 503, 1
Griem, H. R. 1974, *Spectral Line Broadening by Plasmas*, (New York: Academic)
Hegerfeldt, G. C. & Kesting, V. 1988, *APS*, 37, 5
Hooper, C. F. 1968, *Physical Review*, 165, 1
Lara, N., González, M. Á., & Gigosos, M. A. 2012, *Astronomy and Astrophysics*, 542, A75

White Dwarfs as Probes of Fundamental Physics: Tracers of Planetary, Stellar and Galactic Evolution
Proceedings IAU Symposium No. 357, 2019
M. A. Barstow, S. J. Kleinman, J. L. Provencal & L. Ferrario, eds.
doi:10.1017/S1743921320000253

White-dwarf asteroseismology: An update

Alejandro H. Córsico[1,2]

[1]Facultad de Ciencias Astronómicas y Geofísicas, Universidad Nacional de La Plata, Paseo del Bosque s/n, (1900) La Plata, Argentina

[2]Instituto de Astrofísica de La Plata, IALP (CCT La Plata), CONICET-UNLP
email: acorsico@fcaglp.unlp.edu.ar

Abstract. The vast majority of stars that populate the Universe will end their evolution as white-dwarf stars. Applications of white dwarfs include cosmochronology, evolution of planetary systems, and also as laboratories to study non-standard physics and crystallization. In addition to the knowledge of their surface properties from spectroscopy combined with model atmospheres, the global pulsations that they exhibit during several phases of their evolution allow spying on the deep interior of these stars. Indeed, by means of asteroseismology, an approach based on the comparison between the observed pulsation periods of variable white dwarfs and the periods predicted by representative theoretical models, we can infer details of the internal chemical stratification, the total mass, and even the stellar rotation profile and strength of magnetic fields. In this article, we review the current state of the area, emphasizing the latest findings provided by space-mission data.

Keywords. stars: oscillations, stars: evolution, white dwarfs

1. Introduction

White-dwarf stars constitute the end fate of ~97 % of all the stars that populate the Universe, including our Sun. Indeed, all the stars with a mass lower than $\sim 8M_{\odot}$ in the Main Sequence (MS) will passively end their lives as white dwarfs. The reader interested in details about the formation and evolution of white dwarfs can consult the review article by Althaus *et al.* (2010). Here, we will only give a brief review of the main characteristics of these stars. White dwarfs are compact stars, characterized by mean densities of the order of $\overline{\rho} \sim 10^6$ gr/cm^3 (in comparison, $\overline{\rho_{\odot}} = 1.41$ gr/cm^3) and radii of roughly $R_{\star} \sim 0.01 R_{\odot}$. They cover a wide interval of surface temperatures ($4000 \lesssim T_{\rm eff} \lesssim 200\,000$ K) and, consequently, a vast range of luminosities ($0.0001 \lesssim L_{\star}/L_{\odot} \lesssim 1000$). The mass distribution of white dwarfs is characterized by a very pronounced peak at $M_{\star} \sim 0.6 M_{\odot}$ (the average mass), although the mass range is very wide ($0.15 \lesssim M_{\star}/M_{\odot} \lesssim 1.3$). White dwarfs with masses in the range $0.45 \lesssim M_{\star}/M_{\odot} \lesssim 1.0$ likely have cores made of ^{12}C and ^{16}O, while the less massive ones ($M_{\star}/M_{\odot} \lesssim 0.45$) probably harbor ^{4}He cores, and the most massive ones ($M_{\star}/M_{\odot} \gtrsim 1.0$) could have cores made of ^{16}O, ^{20}Ne and ^{24}Mg. Due to the high densities characterizing white dwarfs, the equation of state that governs most of the structure of a white dwarf is that corresponding to a Fermi gas of degenerate electrons, which provide most of the pressure. In turn, the non-degenerate ions contribute to the mass of the star and the heat content accumulated during the previous evolutionary stages. Loosely speaking, the evolution of a white dwarf consists in a gradual cooling, during which the energy sources of nuclear reactions are irrelevant, although there are some exceptions, such as very hot pre-white dwarfs, extremely low-mass (ELM) white dwarfs, and average-mass white dwarfs coming from low-metallicity progenitors.

Table 1. The different parts of the chemical structure (first row), their origin (second row), and their uncertainties (third row), corresponding to the DA white dwarf model shown in Fig. 1.

	Core	**Intershell**	**He buffer**	**Envelope**
Origin	The result of He-core burning and the following steady He burning	Built up during the TP-AGB phase	Shaped by gravitational settling. Prior H burning	Shaped by gravitational settling. Primordial H
Uncertainties	$^{12}C(\alpha,\gamma)^{16}O$ reaction rate, extra-mixing	Number of TPs, extra-mixing rotation (?)	Number of TPs, IFM relationship	Metallicity, C dredge-up, late He flashes

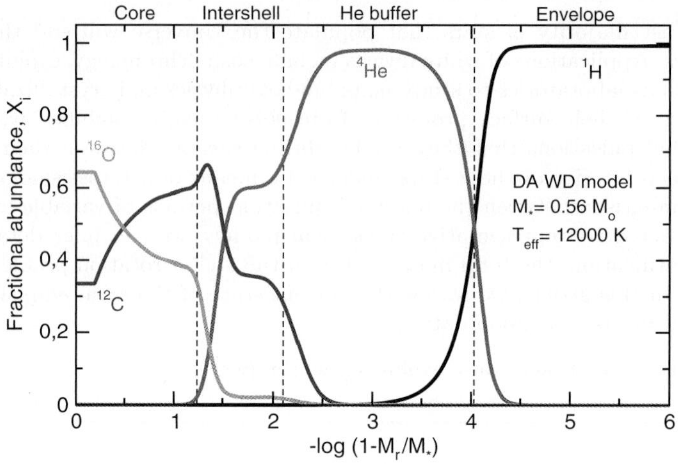

Figure 1. Internal chemical structure of a typical DA white dwarf model with $M_\star = 0.56 M_\odot$, $T_{\rm eff} \sim 12\,000$ K, and H envelope thickness of $\log(M_{\rm H}/M_\star) \sim -4$, resulting from the complete evolution of a single $1 M_\odot$ progenitor from the ZAMS to the white dwarf stage. Plotted is the mass fraction ($X_{\rm i}$) of ^{16}O (green), ^{12}C (blue), ^{4}He (red), and ^{1}H (black), in terms of the outer mass fraction coordinate. The centre of the star is at $-\log(1 - M_r/M_\star) = 0$. We separate with vertical dashed lines four different parts of the internal chemical structure of the model, the origin and the uncertainties of which are described in the text.

By virtue of the high surface gravities ($\log g \sim 8$ [cgs]), the different chemical species in these stars are well separated due to the effect of gravitational settling. According to the dominant chemical species on the surface of white dwarfs, they are classified into two main groups, those of the spectral type DA ($\sim 80\%$ of the total, atmospheres rich in H) and those of spectral type DB ($\sim 15\%$ of the total, atmospheres rich in He). Figure 1 shows the stratified chemical structure of a DA white dwarf model as a function of the outer mass coordinate. This coordinate strongly amplifies the outer regions of the star. As can be seen, 99% of the total mass of the star $[-\log(1 - M_r/M_\star) \lesssim 2]$ is composed mostly of C and O, the mass of the He layer being $M_{\rm He} \lesssim 0.01 M_\star$ and the mass of the H envelope being $M_{\rm H} \lesssim 0.0001 M_\star$. For illustrative purposes, we have divided the chemical structure of the star into four parts: the core, the intershell region, the He-buffer, and the H envelope. In Table 1, we include a short account of the origin and the uncertainties playing a role at each part of the chemical structure (see, for details, Córsico *et al.* 2019a).

Since white dwarfs are very old objects, with ages of the order of $\tau \sim 10^9 - 10^{10}$ years, they convey extremely valuable information about the evolution of stars "from the cradle to the grave", and about the rate of stellar formation along the complete history of our Galaxy. The study of white dwarfs has many applications to several fields. One one hand, they can provide an estimate of the age of stellar populations like open and globular clusters, and the halo and the galactic disk, by using the luminosity function of white dwarfs (see, for instance, García-Berro *et al.* 2010). On the other hand, due to the existence of a mass limit for white dwarfs (the Chandrasekhar mass), they are of utmost relevance as progenitors of supernovae of type Ia, and also cataclysmic variables like novae, that involve very energetic events ($E \sim 10^{44} - 10^{51}$ erg) of mass transfer onto the white dwarf surface from its companion star. Finally, by virtue of the extreme conditions of pressure and density prevailing inside white dwarfs, they allow the investigation of phenomena such as crystallization due to Coulomb interactions, the assessment of constraints on fundamental particles like axions and neutrinos, the variation of fundamental constants, etc. Classically, the approaches employed to study white dwarfs have been spectroscopy, photometry and astrometry, that allow us to infer the effective temperature, surface gravity and composition, parallaxes and distances, etc. Of particular interest in this article is the technique named "asteroseismology", that allows us to extract information of the *interior* of stars by studying how they pulsate. In the next section we provide a brief description of this technique applied to pulsating white-dwarf stars.

2. Stellar pulsations and white-dwarf asteroseismology

Asteroseismology is based on the well-known physical principle establishing that *"studying how a system vibrates in its normal modes allows us to infer its mechanical properties"*. In the case of stars, the vibrations are global pulsations that allow us to "see" the sub-photospheric layers, otherwise inaccessible by means of classical techniques†. Basically, asteroseismology consists in the comparison of theoretical periods of stellar models with the periods observed in real pulsating stars. In the case of white-dwarf asteroseismology, there are other tools to extract information for several quantities — in particular the stellar mass — on the basis of the asymptotic behavior of g (gravity) modes of high radial order. These tools will be described in Section 2.3. White-dwarf asteroseismology allows us to derive information about the stellar mass, the chemical stratification, the core chemical composition, the existence and strength of magnetic fields, the properties pf stellar rotation, the physics of convection, etc. Very detailed reports about white-dwarf asteroseismology can be found in the review articles by Fontaine & Brassard (2008); Winget & Kepler (2008); Althaus *et al.* (2010) and Córsico *et al.* (2019a).

2.1. *A little bit on nonradial stellar pulsations*

Stellar pulsations are eigenmodes of the stars that can be thought of as standing waves in 3 dimensions. Each star has a unique spectrum of discrete eigenfrequencies, the natural frequencies of the star, that is set by its internal structure, mass, effective temperature, radius, etc. In a sense, these natural frequencies are like the "fingerprints" of the star. The eigenfrequencies are associated with eigenfunctions that provide the spatial variation of the different physical parameters of the star when it pulsates. In white-dwarf stars, the frequency spectrum of pulsations is extremely sensitive to the details of the internal chemical stratification. Pulsations in stars can be radial, which retain the spherical symmetry (Cepheids, RR Lyrae, Miras, etc). Radial pulsations are a particular case of a very general class of oscillatory movements, called nonradial pulsations. The latter do not preserve the spherical symmetry. Nonradial pulsations are routinely detected in the Sun, and also in variable solar-type stars, red giants, δ Scuti, γ Doradus, β Cephei, SPB, WR, sdB, white dwarfs and pre-white dwarf stars. The detection of nonradial pulsations has experienced an unprecedented flourishing in the last years thanks to the

† The only alternative way to obtain direct information from the stellar interior is through neutrinos, which escape from the depths of the stars without interacting with matter.

advent of the space missions, such as MOST (Walker *et al.* 2003), CoRoT (Baglin *et al.* 2009), *Kepler* (Borucki *et al.* 2010) and K2 (Howell *et al.* 2014) — which have already ended — the TESS (Transiting Exoplanet Survey Satellite) mission (Ricker *et al.* 2015) — that is currently working — and the future missions Cheops (Moya *et al.* 2018) and Plato (Piotto 2018), among others.

In the frame of the linear theory, that assumes small amplitudes of pulsation, the deformations of a star when it pulsates in spheroidal modes are specified by the Lagrangian displacement vector (Unno *et al.* 1989):

$$\vec{\xi}_{k\ell m} = \left(\vec{\xi}_r, \vec{\xi}_\theta, \vec{\xi}_\phi\right)_{k\ell m} \tag{2.1}$$

where the components in spherical coordinates are:

$$\vec{\xi}_r = \xi_r(r) Y_\ell^m(\theta, \phi) e^{i\sigma t} \vec{e}_r \tag{2.2}$$

$$\vec{\xi}_\theta = \xi_h(r) \frac{\partial Y_\ell^m}{\partial \theta} e^{i\sigma t} \vec{e}_\theta \tag{2.3}$$

$$\vec{\xi}_\phi = \xi_h(r) \frac{1}{\sin\theta} \frac{\partial Y_\ell^m}{\partial \phi} e^{i\sigma t} \vec{e}_\phi \tag{2.4}$$

Here, $Y_\ell^m(\theta, \phi)$ are the spherical harmonics, σ is the pulsation frequency, and $\xi_r(r)$ and $\xi_h(r)$ are the radial and horizontal eigenfunctions, respectively. Each eigenmode has a sinusoidal temporal dependence, an angular dependence by means of spherical harmonics, and a radial dependence given through the radial and horizontal eigenfunctions, which must inevitably be obtained (for realistic star models) through the numerical resolution of the differential equations of stellar pulsations (see Unno *et al.* 1989). Pulsation modes are characterized by three "quantum numbers": *(i)* harmonic degree $\ell = 0, 1, 2, 3, \cdots, \infty$, that represents $(\ell - m)$ nodal lines (parallels) on the stellar surface, *(ii)* azimuthal order $m = -\ell, \cdots, -2, -1, 0, +1, +2, \cdots, +\ell$, that represents m nodal lines (meridians) on the stellar surface, and *(iii)* the radial order $k = 0, 1, 2, 3, \cdots, \infty$, that represents concentric spherical nodal surfaces on which the fluid displacement is null. Within the set of spheroidal modes†, there are two families of modes that differ according to the dominant restoring force. The p (pressure) modes, on one hand, involve large variations of pressure and displacements mostly radial, compressibility being the dominant restoring force. They are characterized by short periods (high frequencies). The g (gravity) modes, on the other hand, are associated with small variations of pressure and large tangential displacements. In this case, the restoring force is gravity through buoyancy, and the modes are characterized by long periods (small frequencies). These are the modes usually detected in white dwarfs. In the case in which $\ell > 1$, there exist a third sub-class of modes, the f modes. They have intermediate characteristics between those of the p and g modes, and do not have radial nodes ($k = 0$), except for stellar models with very high central densities.

2.2. *White-dwarf pulsations*

2.2.1. *A brief historical account*

The first theoretical hints of pulsations in white dwarfs were revealed by the pioneering work of Sauvenier-Goffin (1949) and Ledoux & Sauvenier-Goffin (1950), who found instability in radial ($\ell = 0$) modes with periods of ~ 10 s due to hydrogen nuclear burning.

† Spheroidal modes are characterized by $(\vec{\nabla} \times \vec{\xi})_r = 0$ and $\sigma^2 = 0$, where $\vec{\xi}$ is the Lagrangian displacement vector and σ the pulsation frequency (Unno *et al.* 1989).

These periods were not detected at that time, and this led to the conclusion that H burning is not the source of energy in a white dwarf, and allowed Mestel (1952) to elaborate his theory of cooling. The first pulsating white dwarf, HL Tau 76, was discovered by Landolt (1968), with a detected period of ~740 s, too long to be due to a radial mode (Faulkner & Gribbin 1968; Ostriker & Tassoul 1968). Two additional pulsating white dwarfs, G44–32 (600-800 s) and R548 (200-300 s) (Lasker & Hesser 1969, 1971) were discovered shortly after, but again, the periods measured were too long to be associated with radial modes. The first demonstration that the periods detected in these white dwarfs were due to nonradial g modes came from Warner & Robinson (1972) and Chanmugam (1972), in agreement with the theoretical work of Baglin (1969) and Harper & Rose (1970). An additional hint that these periods were due to nonradial modes came from the detection of rotational splitting in the observed frequencies (Robinson *et al.* 1976), since radial-mode frequencies do not exhibit splitting due to rotation. A definitive proof of the nonradial g-mode nature of the periods of pulsating white dwarfs was provided by the important work of McGraw (1979), who confirmed that the variability is due to changes in surface temperature and not to changes of the radius, which is typical of g modes. Finally, Robinson *et al.* (1982) demonstrated that the variations in the stellar radius are quite small ($\Delta R_\star \sim 10^{-5} R_\star$), and that the changes in effective temperature ($\Delta T_{\rm eff} \sim 200$ K) is what matters most in the pulsations of white dwarfs.

2.2.2. The present situation

Currently, it is a well established fact that, along their evolution, white dwarfs go through at least one stage of pulsation instability in which they become pulsating variable stars, showing light curves with variations in the optical and in the far UV parts of the electromagnetic spectrum (Winget & Kepler 2008; Fontaine & Brassard 2008; Althaus *et al.* 2010; Córsico *et al.* 2019a). The changes in brightness, with amplitudes between 0.001 mmag and 0.4 magnitudes, are due to g modes with harmonic degree $\ell = 1, 2$. In general terms, g modes in white dwarfs probe the envelope regions, due to the very low values of the Brunt-Väisälä frequency (the critical characteristic frequency for the g-mode spectrum) in the core regions. This is contrary to what happens in normal (non-degenerate) pulsating stars. At present, more than 350 pulsating white dwarfs and pre-white dwarfs have been discovered through observations from Earth, in their great majority extracted from the Sloan Digital Sky Survey (SDSS), and in the last years also through space missions such as the already finished *Kepler*/K2 Mission, and currently by TESS (Córsico *et al.* 2019a). Pulsating white dwarfs exhibit a wide variety of light curves, some sinusoidal and with small amplitudes, others nonlinear and with large amplitudes. They are multimodal pulsating stars (they pulsate in more than one period), and frequently exhibit harmonics and linear combinations of eigenfrequencies that are not related to genuine modes of pulsation, but instead to nonlinear effects. At present, there are six classes of confirmed pulsating white dwarfs and pre-white dwarfs known (ZZ Ceti or DAV stars, GW Lib stars, V777 Her or DBV stars, GW Vir stars, ELMV stars and pre-ELMV stars) and two tentative classes of white-dwarf pulsators, that need confirmation (hot DAV stars and DQV stars). In Table 2 we show in a compact way the main characteristics of these families of pulsating white dwarfs and pre-white dwarfs, while in Figure 2 we show their location in the $\log T_{\rm eff} - \log g$ diagram (details can be found in Córsico *et al.* 2019a). In Table 2, the second column indicates the discovery year of the first object of each class and number of known objects at the date of writing this article (November 2019), the third column shows the range of effective temperatures at which they are detected (instability strip), the fourth column provides the range of

Table 2. Properties of the different sub-classes of variable pulsating white dwarfs and pre-white dwarfs, sorted by decreasing effective temperature. The tentative classes of pulsators are labeled with a question mark in parentheses.

Class	Year of disc. (#)	$T_{\rm eff}$ [$\times$1000 K]	$\log g$ [C.G.S.]	Periods [s]	Amplitudes [mag]	Main surface composition
GW Vir (PNNV)	1984 (10)	100 – 180	5.5 – 7	420 – 6000	0.01 – 0.15	He, C, O
GW Vir (DOV)	1979 (9)	80 – 100	7.3 – 7.7	300 – 2600	0.02 – 0.1	He, C, O
Hot DAV (?)	2013 (3)	30 – 32.6	7.3 – 7.8	160 – 705	0.001 – 0.015	H
V777 Her (DBV)	1982 (27)	22.4 – 32	7.5 – 8.3	120 – 1080	0.05 – 0.3	He (H)
DQV (?)	2008 (6)	19 – 22	8 – 9	240 – 1100	0.005 – 0.015	He, C
GW Lib	1998 (20)	10.5 – 16	8.35 – 8.7	100 – 1900	0.007 – 0.07	H, He
ZZ Cet (DAV)	1968 (260)	10.4 – 12.4	7.5 – 9.1	100 – 1400	0.01 – 0.3	H
pre-ELMV	2013 (5)	8 – 13	4 – 5	300 – 5000	0.001 – 0.05	He, H
ELMV	2012 (11)	7.8 – 10	6 – 6.8	100 – 6300	0.002 – 0.044	H

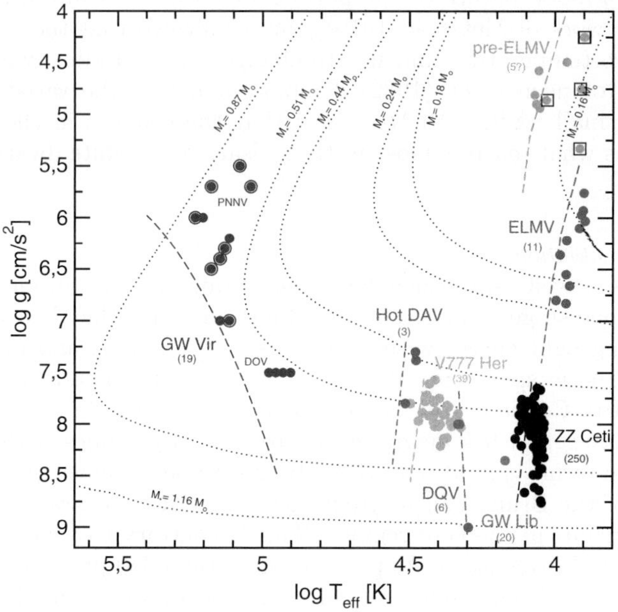

Figure 2. Confirmed and tentative sub-classes of pulsating white-dwarf and pre-white-dwarf stars (circles of different colors) in the $\log T_{\rm eff} - \log g$ diagram (adapted from Córsico *et al.* 2019a). Stars emphasized with squares surrounding the light green circles can be identified as pre-ELMV stars as well as SX Phe and/or δ Scuti stars. GW Vir stars indicated with blue circles surrounded by blue circumferences are PNNVs. In the case of GW Lib stars, only the location of the prototypical object, GW Librae, has been included (magenta dot). Two post-VLTP (Very Late Thermal Pulse) evolutionary tracks for H-deficient white dwarfs (0.51 and $0.87 M_\odot$; Miller Bertolami & Althaus 2006), four evolutionary tracks of low-mass He-core H-rich white dwarfs (0.16, 0.18, 0.24, and $0.44 M_\odot$; Althaus *et al.* 2013), and one evolutionary track for ultra-massive H-rich white dwarfs (Camisassa *et al.* 2019) are plotted for reference. Dashed lines indicate the theoretical blue edge of the instability domains.

surface gravity, the fifth column indicates the range of periods detected, the sixth column contains the range of amplitudes of the variations in the light curves, and the seventh column shows the surface composition. Pulsation periods are usually between ~ 100 s and ~ 1400s, although PNNVs and ELMVs exhibit much longer periods, up to ~ 6300 s. Interestingly, the periods of g modes of white dwarfs are of the same order of magnitude as the periods of p modes in non-degenerate pulsating stars.

2.3. *The tools of white-dwarf asteroseismology*

The tools employed in asteroseismology of pulsating white dwarf stars have been described in several review articles (Winget & Kepler 2008; Fontaine & Brassard 2008; Althaus *et al.* 2010). Here, we will focus on the information that the existence of a constant (or nearly constant) period separation among the periods observed for a given star can give about the stellar mass, in particular in the case of GW Vir stars and DBVs. Also, we will describe the search for asteroseismological models that replicate the individual observed periods in real stars, that is, the period-to-period fits. Finally, we will revise the information about the stellar rotation that is possible to extract from the frequency splittings.

2.3.1. *Clues to the stellar mass from the period spacing*

For g-modes with high radial order k (long periods), the separation of consecutive periods ($|\Delta k| = 1$) becomes nearly constant at a value given by the asymptotic theory of nonradial stellar pulsations. Specifically, the asymptotic period spacing (Tassoul *et al.* 1990) is given by:

$$\Delta\Pi_\ell^{\mathrm{a}} = \Pi_0 / \sqrt{\ell(\ell+1)}, \tag{2.5}$$

where

$$\Pi_0 = 2\pi^2 \left[\int_{r_1}^{r_2} \frac{N}{r} dr\right]^{-1}, \tag{2.6}$$

N being the Brunt-Väisälä frequency. This expression is rigorously valid for chemically homogeneous stars. In principle, one can compare the asymptotic period spacing computed from a grid of models with different masses and effective temperatures with the mean period spacing exhibited by the star, and then infer the value of the stellar mass. This method has been applied in numerous studies of pulsating PG 1159 stars (see, for instance, Córsico *et al.* 2007a,b, 2008, 2009, and references therein). For the method to be valid, the periods exhibited by the pulsating star must be associated with high order g modes, that is, the star must be within the asymptotic regime of pulsations. Furthermore, the interior of white-dwarf stars are supposed to be chemically stratified and characterized by strong chemical gradients built up during the progenitor star life. So, the direct application of the asymptotic period spacing to infer the stellar mass of white dwarfs is somewhat questionable. A better way to compare the models to the period spacing of the observed pulsation spectrum, is to calculate the average of the computed period spacings, using:

$$\overline{\Delta\Pi}(M_\star, T_{\mathrm{eff}}) = \frac{1}{(n-1)} \sum_k^{n-1} \Delta\Pi_k, \tag{2.7}$$

where $\Delta\Pi_k$ is the "forward" period spacing defined as $\Delta\Pi_k = \Pi_{k+1} - \Pi_k$, and n is the number of computed periods laying in the range of the observed periods. The theoretical period spacing of the models as computed through Eq. (2.5) and (2.7) share the same general trends (that is, the same dependence on $M_\star$ and T_{eff}), although $\Delta\Pi_\ell^{\mathrm{a}}$ is usually somewhat higher than $\overline{\Delta\Pi}$, particularly for the case of low radial-order g modes. We can compare the measured mean period spacing of a given pulsating white dwarf with the average of the computed period spacings for several sequences with different stellar masses, by fixing the effective temperature and its uncertainties for the target star derived from spectroscopy. In this way, an estimate of the stellar mass can be obtained. As an example of application, in Fig. 3 we show the run of the average of the computed period

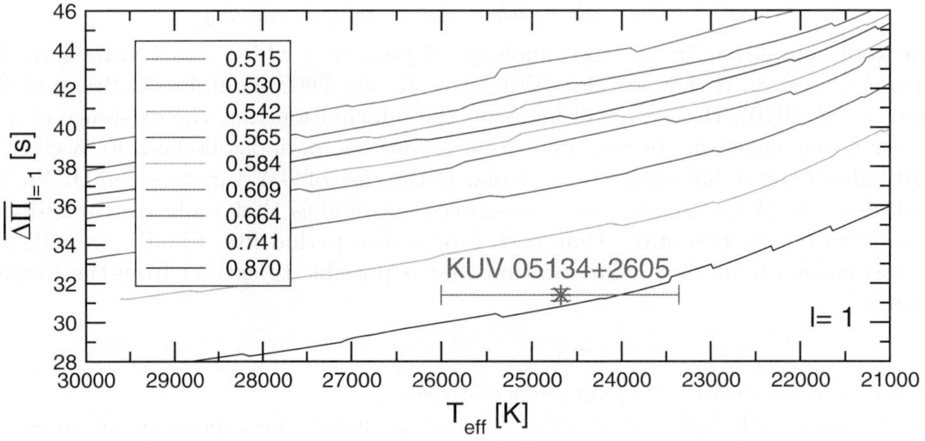

Figure 3. Average of the computed period spacings for $\ell = 1$ corresponding to DB white dwarf sequences with different stellar masses. The location of the target star KUV 05134+2605 is shown with a red star symbol ($T_{\rm eff} = 24\,700 \pm 1300$ K) for each solution of observed mean period spacing, $\Delta\Pi_{\ell=1} = 31.4 \pm 0.3$ s. By simple linear interpolation, we found that the mass of the star according to its period spacing is $M_\star = 0.85 \pm 0.05 M_\odot$.

spacings [Eq. (2.7)] with $\ell = 1$, in terms of the effective temperature for a set of DB white dwarf evolutionary sequences. In the plot, we have included the measured period spacing for a DBV star, KUV 05134+2605 ($T_{\rm eff} = 24\,700 \pm 1300$ K and $\log g = 8.21 \pm 0.06$; Bergeron *et al.* 2011), whose period spectrum exhibits $\Delta\Pi_{\ell=1} = 31.4 \pm 0.3$ s. By means of a linear interpolation, we obtain the mass of the star according to its period spacing, $M_\star = 0.85 \pm 0.05 M_\odot$.

A cautionary word is necessary here. In the case of GW Vir stars and DBV stars, the period spacing is sensitive mostly to the stellar mass and effective temperature and very weakly to the thickness of the He envelope in the case of DBVs (see Tassoul *et al.* 1990) and the thickness of the C/O/He envelope in the case of the GW Vir stars (Kawaler & Bradley 1994). In the case of ZZ Ceti stars, however, the period spacing depends on the stellar mass, the effective temperature, and the thickness of the H envelope with a comparable sensitivity. Consequently, the method is not — in principle — directly applicable to ZZ Ceti stars due to the intrinsic degeneracy of the dependence of $\Delta\Pi$ with the three parameters $M_\star$, $T_{\rm eff}$, and $M_{\rm H}$ (Fontaine & Brassard 2008). An illustration of this has been recently given in the asteroseismological study of ultra-massive ZZ Ceti stars by Córsico *et al.* (2019b).

2.3.2. Constraints from period-to-period fits

Another way to infer the stellar mass, along with the effective temperature and also many details of the internal structure of pulsating white dwarfs, is through their individual pulsation periods. In this approach, we seek a pulsation white dwarf model that best matches the pulsation periods of the target star. The goodness of the match between the theoretical pulsation periods (Π_k) and the observed individual periods ($\Pi_{{\rm obs},i}$) is measured by means of a quality function defined as:

$$\chi^2(M_\star, T_{\rm eff}) = \frac{1}{N} \sum_{i=1}^{N} \min[(\Pi_{{\rm obs},i} - \Pi_k)^2], \tag{2.8}$$

where N is the number of observed periods. The white dwarf model that shows the lowest value of χ^2 is adopted as the "best-fit model" (see Córsico *et al.* 2007a,b, 2008, 2009,

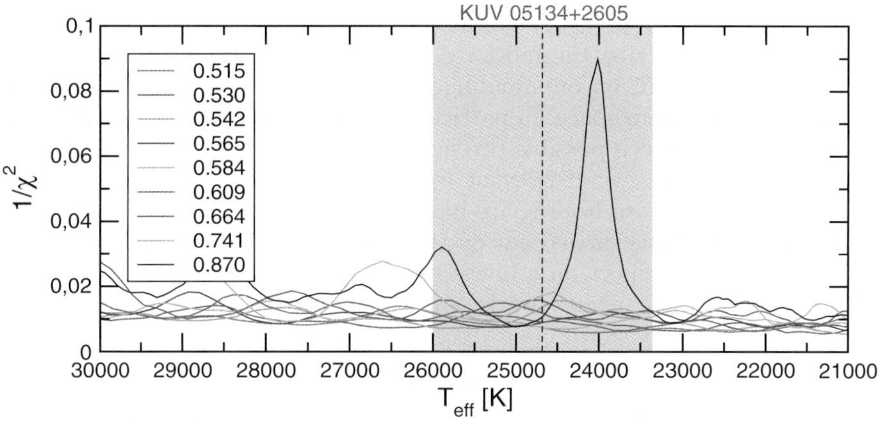

Figure 4. Inverse of the quality function of the period fit in terms of the effective temperature. The vertical gray strip indicates the spectroscopic $T_{\rm eff}$ of KUV 05134+2605 and its uncertainties ($T_{\rm eff} = 24\,680 \pm 1322$ K). The strong maximum in $(\chi^2)^{-1}$ corresponds to the best-fit model, with $M_\star = 0.870 M_\odot$ and $T_{\rm eff} = 23\,976$ K.

2012; Bell *et al.* 2019). The quality of the period fits is assessed by means of the average of the absolute period differences, $\overline{\delta} = (\sum_{i=1}^{N} |\delta_i|)/N$, where $\delta_i = \Pi_{{\rm obs},i} - \Pi_k$, and by the root-mean-square residual, $\sigma = \sqrt{(\sum |\delta_i|^2)/N} = \sqrt{\chi^2}$. Also, in order to have an indicator of the quality of the period fit, we can compute the Bayes Information Criterion (BIC; Koen, & Laney 2000):

$$\text{BIC} = N_{\rm p} \left(\frac{\log N}{N} \right) + \log \sigma^2, \qquad (2.9)$$

where $N_{\rm p}$ is the number of free parameters, and N the number of observed periods. The smaller the value of BIC, the better the quality of the fit. Following the same example of the previous section, in Fig. 4 we show the inverse of the quality function of the period fit for the DBV star KUV 05134+2605 in terms of the effective temperature. The vertical gray strip indicates the spectroscopic $T_{\rm eff}$ and its uncertainties. Clearly, the model that best reproduces the observed periods has a stellar mass of $0.87 M_\odot$ and an effective temperature $T_{\rm eff} \sim 24\,000$ K.

2.3.3. Rotational splittings

When a pulsating star is rotating, the frequencies of nonradial pulsation modes split into $2\ell + 1$ components, and their separation is proportional to the velocity of rotation of the star. In this way, when a pulsating white dwarf exhibits rotational splitting, it is possible to constrain its rotation period. If the rotation is slow and rigid, the rotation frequency Ω of the white dwarf is connected with the frequency splitting $\delta\nu$ through the coefficients $C_{k,\ell}$ — that depend on the details of the stellar structure — and the values of m ($-\ell, \cdots, -1, 0, +1, \cdots, +\ell$), by means of the expression $\delta\nu = m(1 - C_{k,\ell})\,\Omega$ (Unno *et al.* 1989). Note that, when present, frequency splitting ($\ell = 1$: triplets, $\ell = 2$: quintuplets, etc) allows us to identify the harmonic degree ℓ and the azimuthal order m of the modes. Thanks to this effect, the rotation rate of many pulsating white dwarf stars has been measured (see next Section).

3. Recent findings

In this section, we describe the most recent results of white-dwarf asteroseismology. In particular, we will focus on outstanding findings that have been the result of uninterrupted observations from space, in particular, through the *Kepler*/K2 Mission. These findings would not have been possible through observations from the ground. We mainly describe the results from several relevant works that study the pulsation properties of V777 Her and ZZ Ceti stars from space-based observations. We also describe the first asteroseismological analysis carried out on the DBV star EC01585–1600 with the TESS data. Finally, we briefly report on a investigation of the possible existence of pulsating warm ($T_{\rm eff} \sim 19\,000$ K) DA white dwarfs, and the null results from TESS.

3.1. *Outbursting ZZ Ceti stars*

The *Kepler* spacecraft observations of ZZ Ceti stars revealed a new type of phenomenon never observed before from the ground: outburst-like events in DAV stars (Bell *et al.* 2017). The first one, WD J1916+3938 (KIC4552982), was observed during 1.5 years. 20 pulsation modes with periods typical of ZZ Ceti stars were detected, along with 178 enhancements of brightness typical of outburst phenomena, with peaks of up to 17 % above the quiescent level, involving very energetic events ($E \sim 10^{33}$ erg), with a mean recurrence period of about 2.7 days (Bell *et al.* 2015). The second outbursting ZZ Ceti star, PG1149+057, was studied by Hermes *et al.* (2015). This star shows flux enhancements of up to 45% above the quiescent level! It was revealed that for this star, the outbursts actually affect the normal pulsations (in amplitude and frequency). This leads to the important conclusion that outbursts are an intrinsic phenomenon of the (otherwise isolated) star. At present, a total of 8 outbursting isolated ZZ Ceti stars have been discovered (Bell *et al.* 2016, 2017), close to the red edge of the ZZ Ceti instability strip. A possible explanation is connected with parametric instability via mode coupling of white dwarf pulsations (Dziembowski 1982; Wu & Goldreich 2001; Luan & Goldreich 2018).

3.2. *Dichotomy of mode line widths in ZZ Ceti stars*

By observing 27 DAVs through the K2 Campaign of the *Kepler* space telescope, Hermes *et al.* (2017) discovered a dichotomy of mode line widths in the power spectrum of pulsating DA white dwarfs. Specifically, low-frequency modes with periods longer than roughly 800 s are generally incoherent over the length of observations, while higher-frequency modes (< 800 s) are observed to be much more stable in phase and amplitude. This phenomenon is incompatible with stochastic excitation, that for white dwarfs would be able to excite pulsations with periods of the order of ~ 1 sec. The dichotomy of model line width can be related to the oscillation of the outer convection zone of a DA white dwarfs during pulsations. The oscillation of the base of the outer convection zone would affect g-mode eigenfunctions having the outer turning point located at the base of the convection zone (Montgomery *et al.* 2019). Deepening the study of this phenomenon can lead to a better understanding of the properties of the outer convection zone of DA white dwarfs.

3.3. *White-dwarf rotation rates from asteroseismology*

The rotation rates of 27 isolated DA white dwarfs have been determined via asteroseismology using rotational splittings by Hermes *et al.* (2017) with observations from *Kepler*/K2 Mission. In this way, the number of pulsating white dwarfs with measured rotation rates has been doubled (see Table 10 of Córsico *et al.* 2019a). The results indicate that the mean rotation period is of $\sim 30-40$ hours, although fast rotating stars have also been found, with periods of $\sim 1-2$ hours. Evidence has been found for a link between high mass and fast rotation, although additional massive white dwarfs are required to confirm this trend. If true, this trend would favor the hypothesis that the high-mass white dwarfs are the result of mergers, which are supposed to rotate very fast. Hermes *et al.* (2017) find that the majority of isolated descendants of $1.7-3.0M_\odot$ ZAMS progenitors rotate at ~ 1.5 days, instead of minutes! This indicates that most internal angular momentum must be lost on the first-ascent giant branch.

3.4. *Amplitude and frequency variations of g modes in a DBV white dwarf*

By using observations from the *Kepler* mission, Zong *et al.* (2016a) have detected amplitude and frequency variations of the components of the triplets of frequencies caused by rotation in the DBV star KIC 08626021. The timescale for the quasi-periodic modulations of the variations is of about 600 days, undetectable from ground-based observations. A similar phenomenon has been detected with *Kepler* in the pulsating sdB star KIC 10139564 (Zong *et al.* 2016b). It is thought that the modulations of frequencies and amplitudes are not related to any evolutionary effect (e.g., neutrino cooling), since the timescales involved are several orders of magnitude shorter than the cooling rate of DB white dwarfs. These modulations cannot be attributed to signatures of orbiting companions around the star, because different timescales for different triplets are detected. Therefore, a possible explanation is nonlinear resonant mode coupling in rotationally split triplets (Buchler *et al.* 1997; Goupil *et al.* 1998). If so, this star could constitute a new window to study nonlinear pulsations. On the other hand, the detected frequency modulations can potentially prevent a measurement of the evolutionary (cooling) rate of period change of the star.

3.5. *The DBV pulsator WD0158–160: TESS observations*

The first pulsating white dwarf analyzed with TESS is the DBV pulsator WD0158–160 (also called EC01585–1600, G272–B2A, TIC257459955; Bell *et al.* 2019). TESS performs extensive time-series photometry that allows to discover pulsating stars, and, in particular, pulsating white dwarfs. The TESS Asteroseismic Science Consortium (TASC) Working Group 8 (WG8) focuses on TESS observations of evolved compact stars that exhibit photometric variability, including hot subdwarfs, white dwarfs, and pre-white dwarfs with mag < 16, with short (120 sec) cadence. TIC257459955 is a known DBV white dwarf with $T_{\rm eff} = 24\,100$ K and $\log g = 7.88$ (Rolland *et al.* 2018), or alternatively, $T_{\rm eff} = 25\,500$ K and $\log g = 7.94$ (Voss *et al.* 2007). Bell *et al.* (2019) find 9 independent frequencies suitable for asteroseismology plus frequency combinations. The periods of genuine eigenmodes are in the range $[245-866]$ sec, with a period spacing of $\Delta\Pi = 38.1 \pm 1.0$ s associated to $\ell = 1$. The comparison with the average of the computed period spacings (see Section 2.3.1) gives an estimate of the stellar mass. Here, it is assumed that the period spacing is primarily dependent on $T_{\rm eff}$ and $M_\star$ (and very weakly on $M_{\rm He}$). This method indicates a stellar mass of $M_\star = 0.621 \pm 0.06M_\odot$, or alternatively, $M_\star = 0.658 \pm 0.10M_\odot$ (according to the two different spetroscopic determinations of $T_{\rm eff}$), larger than the spectroscopic estimates ($M_\star = 0.542-0.557M_\odot$). The star also has been analyzed by performing period-to-period fits (Section 2.3.2) by employing a fully evolutionary model approach (La Plata Group) and a parametric approach (Texas Group). There exist families of asteroseismic solutions, but the solution that satisfies the spectroscopic parameters and the astrometric constraints from Gaia, is characterized by a DB white dwarf model with $M_\star \sim 0.60M_\odot$, $T_{\rm eff} \sim 25\,600$ K, $M_{\rm He} \sim 3 \times 10^{-2}M_\star$, $d \sim 67$ pc, and a rotation period of ~ 7 or ~ 14 hours. It is worth emphasizing that the asteroseismological solutions from the La Plata Group and the Texas Group are similar regarding the location of the chemical transition zones, which are mainly responsible for setting the period spectrum of a white dwarf model.

3.6. *The possible existence of pulsating warm DA white dwarfs*

In their theoretical pioneering work, Winget *et al.* (1982) (see also Winget 1982) predicted the possible existence of warm ($T_{\rm eff} \sim 19\,000$ K) pulsating DA white-dwarf stars, hotter than ZZ Ceti stars. However, to date, no pulsating warm DA white dwarf has been detected. Althaus *et al.* (2019) re-examined the pulsational predictions for such white dwarfs on the basis of new full evolutionary sequences and also analyzed a sample of warm DA white dwarfs observed by the TESS satellite in order to search for the possible pulsational signals. Althaus *et al.* (2019) computed white-dwarf evolutionary sequences with very small H content, appropriate for the study of warm DA white dwarfs, employing a new full-implicit treatment of time-dependent element diffusion. Also, they computed non-adiabatic pulsations in the effective temperature range of $30\,000 - 10\,000$ K, focusing on $\ell = 1$ g modes with periods in the range $50 - 1500$ s. These authors found that extended and smooth He/H transition zones inhibit the excitation of g modes due to partial ionization of He below the H envelope, and only in the case that the H/He transition is assumed to be much more abrupt, do the models exhibit pulsational instability. In this case, instabilities are found only in white dwarf models with H envelopes in the range of $-14.5 \lesssim \log(M_{\rm H}/M_\star) \lesssim -10$ and at effective temperatures higher than those typical of ZZ Ceti stars, in agreement with the previous study by Winget *et al.* (1982). Althaus *et al.* (2019) found that none of the warm DAs observed by the TESS satellite are pulsating. This study suggests that the non-detection of pulsating warm DA white dwarfs, if white dwarfs with very thin H envelopes do exist, could be attributed to the presence of a smooth and extended H/He transition zone. This could be considered as an indirect proof that element diffusion indeed operates in the interior of white dwarfs.

4. Conclusions

From the beginning to the present, white-dwarf asteroseismology has undergone impressive progress, in recent years by the availability of space missions that provide unprecedented high-quality data, but also from ground-based observations, mainly with the spectral observations of the Sloan Digital Sky Survey (SDSS, York *et al.* 2000). This progress from the observational front has been accompanied by the development of new detailed models of white dwarf stars and novel implementations of asteroseismological techniques. The combination of these factors is leading the white-dwarf asteroseismologists to realize the desire to know the details of the inner structure and evolutionary origins of these stars. These studies will soon be driven by the new observations from space (TESS, PLATO, Cheops). Therefore, the future of this area is, at the very least, very promising.

Acknowledgements

I would warmly thanks the Local Organising Committee of the IAU Symposium 357, in particular to Professor Martin Barstow, for support that allowed me to attend this conference.

References

Althaus, L. G., Córsico, A. H., Isern, J., García-Berro, E. *et al.* 2010, *A&ARv*, 18, 471
Althaus, L. G., Miller Bertolami, M. M., & Córsico, A. H. 2013, *A&A*, 557, A19

Althaus, L. G., Córsico, A. H., Uzundag, M., *et al.* 2019, arXiv e-prints, arXiv:1911.02442

Baglin, A. 1969, *Astrophysical Letters*, 3, 119

Baglin, A., Auvergne, M., Barge, P., *et al.* 2009, *Transiting Planets*, 71

Bell, K. J., Córsico, A. H., Bischoff-Kim, A., *et al.* 2019, *A&A*, 632, A42

Bell, K. J., Hermes, J. J., Montgomery, M. H., *et al.* 2017, 20th European White Dwarf Workshop, 303

Bell, K. J., Hermes, J. J., Montgomery, M. H., *et al.* 2016, *ApJ*, 829, 82

Bell, K. J., Hermes, J. J., Bischoff-Kim, A., *et al.* 2015, *ApJ*, 809, 14

Bergeron, P., Wesemael, F., Dufour, P., *et al.* 2011, *ApJ*, 737, 28

Borucki, W. J., Koch, D., Basri, G., *et al.* 2010, *Science*, 327, 977

Buchler, J. R., Goupil, M.-J., & Hansen, C. J. 1997, *A&A*, 321, 159

Camisassa, M. E., Althaus, L. G., Córsico, A. H., *et al.* 2019, *A&A*, 625, A87

Chanmugam, G. 1972, *Nature Physical Science*, 236, 83

Moya, A., Barceló Forteza, S., Bonfanti, A., *et al.* 2018, *A&A*, 620, A203

Córsico, A. H., De Gerónimo, F. C., Camisassa, M. E., *et al.* 2019b, arXiv e-prints, arXiv:1910.07385

Córsico, A. H., Althaus, L. G., Miller Bertolami, M. M., *et al.* 2019a, *A&ARv*, 27, 7

Córsico, A. H., Althaus, L. G., Miller Bertolami, M. M., *et al.* 2012, *A&A*, 541, A42

Córsico, A. H., Althaus, L. G., Miller Bertolami, M. M., *et al.* 2009, *A&A*, 499, 257

Córsico, A. H., Althaus, L. G., Kepler, S. O., *et al.* 2008, *A&A*, 478, 869

Córsico, A. H., Althaus, L. G., Miller Bertolami, M. M., *et al.* 2007, *A&A*, 461, 1095

Córsico, A. H., Miller Bertolami, M. M., Althaus, L. G., *et al.* 2007, *A&A*, 475, 619

Dziembowski, W. 1982, *Acta. Astron.*, 32, 147

Faulkner, J. & Gribbin, J. R. 1968, *Nature*, 218, 734

Fontaine, G. & Brassard, P. 2008 *PASP*, 120, 1043

García-Berro, E., Torres, S., Althaus, L. G., *et al.* 2010, *Nature*, 465, 194

Goupil, M. J., Dziembowski, W. A., & Fontaine, G. 1998, *Baltic Astronomy*, 7, 21

Harper, R. V. R. & Rose, W. K. 1970, *ApJ*, 162, 963

Hermes, J. J., Gänsicke, B. T., Kawaler, S. D., *et al.* 2017, *ApJS*, 232, 23

Hermes, J. J., Montgomery, M. H., Bell, K. J., *et al.* 2015, *ApJL*, 810, L5

Howell, S. B., Sobeck, C., Haas, M., *et al.* 2014, *PASP*, 126, 398

Kawaler, S. D., & Bradley, P. A. 1994, *ApJ*, 427, 415

Koen, C. & Laney, D. 2000, *MNRAS*, 311, 636

Landolt, A. U. 1968, *ApJ*, 153, 151

Lasker, B. M. & Hesser, J. E. 1969, *ApJL*, 158, L171

Lasker, B. M. & Hesser, J. E. 1971, *ApJL*, 163, L89

Ledoux, P. J. & Sauvenier-Goffin, E. 1950, *ApJ*, 111, 611

Luan, J. & Goldreich, P. 2018, *ApJ*, 863, 82

McGraw, J. T. 1979, *ApJ*, 229, 203

Mestel, L. 1952, *MNRAS*, 112, 583

Miller Bertolami, M. M. & Althaus, L. G. 2006, *A&A*, 454, 845

Montgomery, M. H., Hermes, J. J., & Winget, D. E. 2019, arXiv e-prints, arXiv:1902.05615

Ostriker, J. P. & Tassoul, J.-L. 1968, *Nature*, 219, 577

Piotto, G. 2018, *European Planetary Science Congress*, EPSC2018-969

Ricker, G. R., Winn, J. N., Vanderspek, R., *et al.* 2015, *Journal of Astronomical Telescopes, Instruments, and Systems*, 1, 014003

Robinson, E. L., Nather, R. E., & McGraw, J. T. 1976, *ApJ*, 210, 211

Robinson, E. L., Kepler, S. O., & Nather, R. E. 1982, *ApJ*, 259, 219

Rolland, B., Bergeron, P., & Fontaine, G. 2018, *ApJ*, 857, 56

Sauvenier-Goffin, E. 1949, *Annales d'Astrophysique*, 12, 39

Tassoul, M., Fontaine, G., & Winget, D. E. 1990, *ApJS*, 72, 335

Unno, W., Osaki, Y., Ando, H., *et al.* 1989, Nonradial oscillations of stars

Voss, B., Koester, D., Napiwotzki, R., *et al.* 2007, *A&A*, 470, 1079

Walker, G., Matthews, J., Kuschnig, R., *et al.* 2003, *PASP*, 115, 1023

Warner, B. & Robinson, E. L. 1972, *Nature Physical Science*, 239, 2
Winget, D. E. & Kepler, S. O. 2008, *ARAA* 46,157
Winget, D. E., van Horn, H. M., Tassoul, M., *et al.* 1982, *ApJL*, 252, L65
Winget, D. E. 1982, *Ph.D. Thesis*
Wu, Y. & Goldreich, P. 2001, *ApJ*, 546, 469
York, D. G., Adelman, J., Anderson, J. E., *et al.* 2000, *AJ*, 120, 1579
Zong, W., Charpinet, S., Vauclair, G., *et al.* 2016, *A&A*, 585, A22
Zong, W., Charpinet, S., & Vauclair, G. 2016, *A&A*, 594, A46

White Dwarfs as Probes of Fundamental Physics: Tracers of Planetary, Stellar and Galactic Evolution
Proceedings IAU Symposium No. 357, 2019
M. A. Barstow, S. J. Kleinman, J. L. Provencal & L. Ferrario, eds.
doi:10.1017/S1743921320001386

Variable white dwarfs

H. L. Shipman

University of Delaware, Dept. of Physics and Astronomy, Newark, DE 19716
email: harrys@udel.edu

Abstract. Asteroseismology of white dwarf stars has led to a number of interesting results pertaining to the long term evolution and present state of white dwarf interiors. I will review recent results and will give a not necessarily comprehensive view of the prospects for further progress in this area. Two – but only two white dwarf stars — have shown the expected cooling as they age. Careful observations of a few white dwarfs with rich pulsational properties reveal interior compositions as well as the thickness of their surface layers. A few very well observed stars have revealed changes in their pulsational spectra which we don't understand yet.

Keywords. white dwarfs, oscillations

1. Introduction

Variability in white dwarf stars was discovered roughly 50 years ago by Landolt (1968). Since then, pulsating white dwarf stars have made the transition from astronomical curiosities into one powerful tool for understanding the late stages of stellar evolution. I highlight some important transitions but admit I have not identified the people or papers whom I should credit. When the pulsation spectrum of PG 1159-035 was analyzed to give the star's mass and cooling (or shrinking) rate in the late 1970s, it began the use of variable white dwarf stars to determine stellar properties. Other milestones include the determination of the thickness of hydrogen layers in DA whites, the use of white dwarf variables to determine white dwarf rotation rates, and the observational testing of convection theories in white dwarf stars.

In this brief contribution I wish to make three main points. The use of a single mode to draw conclusions of broader astrophysical interest is fraught with difficulties and should be avoided. Variable white dwarf stars can provide much information about the composition of white dwarf stars, including their deep interiors all the way to their centers. As we look to the future, there are many more stars to be analyzed, and there is much that can be done even with small (1-2 meters) telescopes.

2. The Danger of Analyzing a Star with only one mode

One person who articulated the dangers of drawing too much insight from a single mode was Kepler *et al.* (2003), While this story is quite familiar to the small group of astronomers who analyze variable white dwarfs, I think it is worth retelling to a broader audience. Some years ago the Whole Earth Telescope project (hereafter WET) became involved in the analysis of the DB white dwarf EC 20058-053. Preliminary work with one easily observable mode indicated that there was a long term variation with an approximately 13 years period which would indicate the presence of a planet around this object. The data for this mode seemed convincing, When a second mode was added to the data set it seemed even more convincing.

Rather than the embarrassment of adding to the already large pile of discoveries of extra-solar planets whose discovery had to be retracted (for a popular history of this work see Bartusiak (2018)). James Dalessio and Judi Provencal were persistent, for a third independent mode was also present in the data and we needed to get more data to be sure. Fortunately, the more senior members of the team, including me, accepted their arguments. The period was there, but the variations were completely out of phase with the variations produced by the two dominant modes. If the variations were produced by a planetary companion, all the pulsations would have the same phase. The team knew that the variations were not produced by a planet. The cause is not yet firmly established, but we can exclude the presence of an extra-solar planet as the cause (Dalessio *et al.* 2013).

A talk I heard in the summer of 2019 at an astronomy teaching conference by Stephanie Slater gave me some support for the possible answer. Her talk was on the techniques of Polynesian wayfinders. In the talk she told stories of wayfinders who could feel the paths of ocean waves which were in several separate modes, sometimes in opposite directions. I tried this myself by observing ocean waves from aboard a ship. I could not observe the claimed seven different modes which could be detected by experienced navigators, but I could detect two different modes in some circumstances. This suggestion was also made in the paper on EC 20058 (Dalessio *et al.* 2013). (Disclosure statement: I was one of the "et al" authors on both this paper and the GD 358 paper, obtaining some of the observational data).

There was extensive discussion at the Hilo conference on the presence of debris disks around white dwarf stars as indicated by the presence of lines of heavy elements in the spectra of some white dwarf stars. After the conference yet before this article was written (Dec 3 2019), close analysis of these lines in the white dwarf star WD J0914+1914 by a group led by Boris Gaensicke has indicated the presence of a massive exoplanet around this object (Gaensicke *et al.* 2019; Garlick 2019). I am including this since I have confidence in Gaensicke's work based on previous papers. I hope this result will pass the test of time. (I am respecting the embargo on the news release with the timing of the present paper.)

3. Recent Results on the Deep Interior of Two White Dwarf Stars

Two recently published papers indicate the power of applying time-variable astrophysics to white dwarf stars (Giammichele 2018; Kim *et al.* 2019). Both papers illustrate the power of making extensive observations over long periods of time: one with Kepler spacecraft observations which provide continuous data over a period of two months, and one with 36 years of data including a few WET campaigns.

I believe it will not be necessary in the future to obtain several decades of data in order to obtain enough information to determine the detailed run of chemical composition of other white dwarf stars. In Giammichele's wonderful words, "the star must cooperate" by having enough observable modes to permit this sort of analysis (Smart 2018). What remains to be determined is whether future ground-based observations, which can only achieve millimagnitude accuracy but can, with effort, obtain adequate cadence over long periods of time, will suffice. Another possible route to obtaining data is to piggyback on spacecraft like TESS which may produce long-term light curves as they search for earth-sized exoplanets.

Both papers required a fair amount of effort in developing models to interpret the data. I am most familiar with the analysis in Kim *et al.* (2019) where she reports that over a million models were used to interpret the data. This amount of effort is in great contrast to the work several decades ago described by Van Horn (2015), where several hours of work with the computers available at the time were required to produce a single model. I can remember the later stages of this era, where months were required to develop a program to calculate single profiles of neutral helium lines in B stars.

4. Prospects for the Future

These two recent results indicate that the power of time-variable astrophysics will continue to produce interesting results which will impact many other areas of astronomy. The long series of observations of GD 358 have uncovered many changes in the pulsation behavior over periods of years which remain to be understood. The results described above for EC 20058 remain to be fully understood, though it is unclear to me that increased understanding in this particular case will lead to dramatic insights. My own choices for additional targets for observation are G 29-38, GD 514 (and other high mass white dwarf stars), PG 1351+48, and GD 358. These four stars are within reach of current telescopes in the 1-2 m class. Several telescopes of this type have survived their abandonment by national observatories. One of the cooperating observatories in WET has continued to operate in spite of being located in a rather unstable part of the world. We have been able to obtain good data from an amateur group in Texas.

Acknowledgments

I thank the Whole Earth Telescope group for helpful discussions. Particular thanks go to Judi Provencal for decades of productive collaboration. I thank the NSF, NASA, and the Crystal Trust for supporting my astrophysics work over the years.

References

Bartusiak, M. 2018, *Dispatches from Planet 3*, New Haven: Yale University Press, chapter 3

Bischoff-Kim, A. *et al.* 2018, *ApJ*, 871, 13

Dalessio, J., Provencal, J. L., Shipman, H. L., *et al.* 2013, *ApJ*, 765,5

Gaensicke, B. *et al.* 2019, *Nature* 576, 61

Garlick, P. 2019, https://warwick.ac.uk/newsandevents/pressreleases/

Giammichele, N. 2018, *Nature* 571, 639

Kepler, S. O. *et al.* 2003, *A&A*, 401, 639

Landolt, A. 1968, *ApJ*, 153, 151

Smart, A. 2018, *PhT*, (March 2018), 71, 16

Van Horn, H. M 2015, Unlocking the Secrets of White Dwarf Stars, Heidelberg: Springer, chapter 9

White Dwarfs as Probes of Fundamental Physics: Tracers of Planetary, Stellar and Galactic Evolution
Proceedings IAU Symposium No. 357, 2019
M. A. Barstow, S. J. Kleinman, J. L. Provencal & L. Ferrario, eds.
doi:10.1017/S1743921320001362

Evolution and asteroseismology of ultra-massive DA white dwarfs

F. C. De Gerónimo[1,2], A. H. Córsico[1,2], M. E. Camisassa[1,2] and L. G. Althaus[1,2]

[1]Facultad de Ciencias Astronómicas y Geofísicas, Universidad Nacional de La Plata, Paseo del Bosque s/n, (1900) La Plata, Argentina
email: fdegeronimo@fcaglp.unlp.edu.ar

[2]Instituto de Astrofísica La Plata, CONICET-UNLP

Abstract. Ultra-massive ($\gtrsim 1 M_\odot$) oxygen/neon (ONe) core white dwarfs (WDs) are the result of the evolution of isolated progenitor stars with masses above $6-9 M_\odot$. It is expected that hydrogen-rich (DA) ultra-massive WDs harbor crystallized cores at the typical temperatures of the ZZ Ceti instability strip. These stars offer a unique opportunity to study the processes of crystallization and to infer their core chemical composition. We present a study of the evolution and asteroseismology of ultra-massive DA WDs. We found that all pulsating WDs known to date with $M_\star \gtrsim 1.1 M_\odot$ should have more than 80% of their mass crystallized, if a ONe-core is assumed. Finally, we present a complete asteroseismological analysis to the well known ZZ Ceti BPM 37093 and a preliminary analysis to GD 518 and SDSS J0840+5222.

Keywords. White Dwarfs, Oscillations

---

1. Introduction

Ultra-massive ($\gtrsim 1 M_\odot$) ONe-core WDs stars are the final evolutionary stage of progenitor stars with masses above $6-9 M_\odot$. These progenitors are expected to reach the necessary conditions in their interiors to ignite carbon in semi-degenerate conditions during the evolution through the super asymptotic giant branch (SAGB) phase, leaving as a result a very dense ONe core. In particular, ultra-massive DA WDs are expected to harbor highly crystallized ONe cores by the time that they reach the ZZ Ceti instability strip.

Through detailed asteroseismological analysis it is possible to obtain details of the chemical composition, in particular, the degree of crystallization.

We present a study of the evolution and asteroseismology of ultra-massive DA WDs based on the development of updated and realistic chemical profiles, reliable to perform detailed asteroseismological analysis.

2. Input physics

Our WD evolutionary sequences were computed by adopting the initial realistic chemical profiles from Siess (2010), resulting from the complete evolution of the progenitor stars through the SAGB phase. Our evolutionary sequences are characterized by stellar masses $M_\star = 1.10, 1.16, 1.22$, and $1.29 M_\odot$.

All the evolutionary models were computed with the `LPCODE` evolutionary code (Althaus *et al.* 2005). This numerical code considers all the physical ingredients necessary for the description of WD evolution. Energy release and chemical redistribution induced by the process of phase separation during crystallization are included by adopting the oxygen/neon phase diagram of Medin & Cumming (2010).

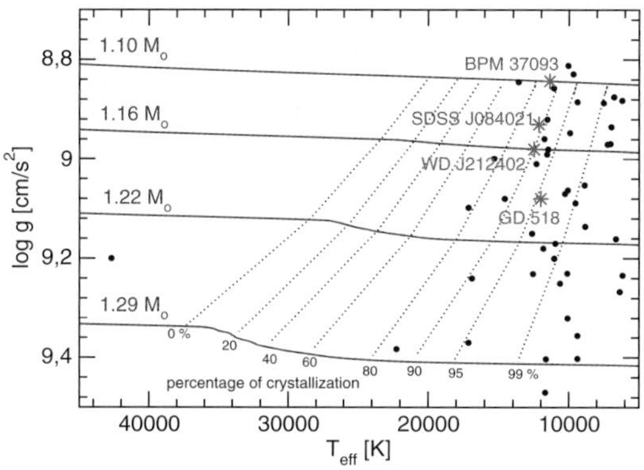

Figure 1. Evolutionary sequences for the ultra-massive DA WDs.

The pulsational properties of the nonradial g(gravity)-modes were computed by using the adiabatic version of the `LP-PUL` pulsation code (Córsico *et al.* 2005). Of particular interest, for the treatment of crystallization and its effect on the pulsation spectrum of g modes, we adopted the "hard sphere" boundary conditions which assume that the amplitude of the eigenfunctions of g modes is reduced substantially below the solid/liquid interface (see Montgomery & Winget 1999).

3. Results

The main properties of our evolutionary sequences are shown on Fig. 1. There we show the spectroscopic Hertzsprung-Russell diagram (the $\log g - T_{\rm eff}$ plane) together with the crystallized mass fraction. The position of the ultra-massive DA WDs known at date are plotted with black star symbols while the ZZ Ceti stars are plotted with black dots. As can be seen from the figure, ZZ Ceti stars with $M_\star \geqslant 1.1 M_\odot$ should have more than 80% of their mass crystallized.

Diffusion processes and gravitational settling play a key role in the development of a pure H envelope and eroding the chemical profiles all along the WD cooling. Due that the crystallization temperature of ^{20}Ne is larger that the crystallization temperature of ^{16}O, a phase separation is induced by the crystallization process. These changes in the chemical profiles are translated into no-negligible changes on the period spectrum. The signatures of the phase separation and the induced chemical redistribution are seen as peaks in the Brunt-Väisälä frequency. These features are completely irrelevant for the pulsation properties of the more massive models due to they are within the crystallized region.

4. Asteroseismological analyses

4.1. *BPM 37093*

This white dwarf was the first ultra-massive ZZ Ceti star discovered (Kanaan *et al.* 1992). Recently, Nitta *et al.* (2016) determined $T_{\rm eff} = 11370$ K $\log g = 8.843$. While the asteroseismological study performed for this star by Metcalfe *et al.* (2004) indicates a crystallized mass fraction $\sim 90\%$, Brassard & Fontaine (2005) questioned this result and argued that the crystallized region should be between 32% and 82%. In the following we present our detailed asteroseismological analysis based on an updated and realistic set of ONe-core ultra-massive white dwarf models, by considering the set of eight pulsation modes (511.7, 531.1, 548.4, 564.1, 582.0, 600.7, 613.5, 635.1) from Metcalfe *et al.* (2004).

Table 1. References: (a) Nitta *et al.* (2016). (b) Camisassa *et al.* (2019)

Quantity	Spectroscopy	Asteroseismology
$T_{\rm eff}$ [K]	$11\,370 \pm 500^{(a)}$	$11\,650 \pm 40$
$M_\star/M_\odot$	$1.098 \pm 0.1^{(b)}$	1.16 ± 0.014
$\log g$ [cm/s^2]	$8.843 \pm 0.05^{(a)}$	8.970 ± 0.025
$M_{\rm cr}/M_\star$	$0.935^{(b)}$	0.923

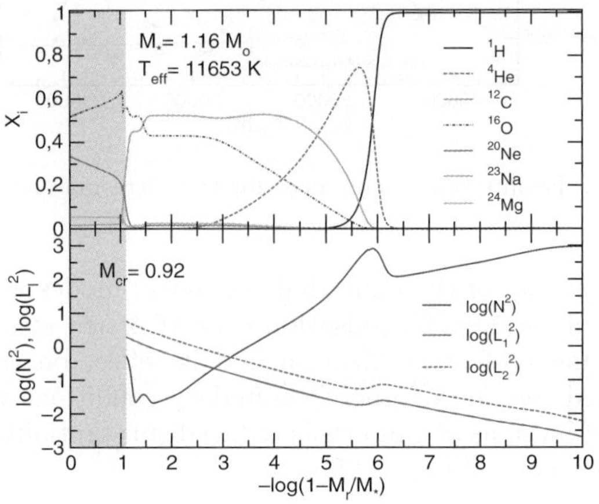

Figure 2. Chemical structure (upper panel), and the squared Brunt-Vaïsälä and Lamb frequencies (lower panel) of our best-fit model for BPM 37093.

4.1.1. Period spacing analysis

The large number of modes detected allow us to search for a constant period spacing. We found, by performing three significance tests (Kolmogorov-Smirnov, inverse variance and the Fourier Transform) a constant period spacing of ~ 17 s. This indicates that BPM 37093 is dominated by $\ell = 2$ modes and the stellar mass should be 1.16 or $1.29 M_\odot$.

4.1.2. Matching the observed periods

We searched for the model that best fit the observed periods by employing the function:

$$\chi^2(M_\star, M_{\rm H}, T_{\rm eff}) = \frac{1}{N}\sum_{i=1}^{N} \min[(\Pi_i^{\rm O} - \Pi_k^{\rm T})^2], \qquad (4.1)$$

where Π_k^T refers to the theoretical periods and Π_k^O the observed ones.

Table 1 shows the stellar parameters derived for our best-fit model ($\sigma^2 = 1.6$) together with the spectroscopic determinations. In Fig. 2 we show the chemical structure (upper panel) and the run of the Brunt-Väisälä and Lamb frequencies (bottom panel) derived for BPM 37093. A crystallized mass of 92% is obtained.

According to our mode identification of the pulsation periods, we derived a rotation period of ~ 55 h, consistent with the values derived from asteroseismology for WD stars.

4.2. *GD 518*

Hermes *et al.* (2013) discovered pulsations in this star (modes of 440.2 s, 513.2 s and 583.7 s). They determinations indicates $T_{\rm eff} \sim 12\,030$ K and log $g \sim 9.08$. According to the ONe-core WD evolutionary tracks of Camisassa *et al.* (2019), this star should have a stellar mass of 1.198 $M_\odot$.

Due to the few modes exhibited we only attempted to perform period-to-period fits. We obtained a best-fit model ($\sigma^2 = 0.56$) characterized by $T_{\rm eff} \sim 12\,060$ K, $M_\star = 1.22\ M_\odot$ and log $g = 9.15$. Our model reveals a crystallized mass of 95.5%.

4.3. *SDSS J084021.23+522217.4*

This star was discovered by Curd *et al.* (2017) and their determinations indicate $T_{\rm eff} = 12\,160$ K, log $g = 8.93$ and $M_\star = 1.16 M_\odot$. Their preliminary asteroseismological analysis, by considering CO-core WD models, reveals $M_\star = 1.14 M_\odot$, $M_{\rm H} = 5.8 \times 10^{-7} M_\star$, $M_{\rm He} = 4.5 \times 10^{-4} M_\star$, $0.50 \leqslant M_{\rm cr}/M_\star \leqslant 0.70$ and $11\,850 \leqslant T_{\rm eff} \leqslant 12\,350$ K.

By contrast our analysis is based on ONe-core WD models. Our best-fit model ($\sigma^2 = 0.14$) is characterized by $T_{\rm eff} = 12550$ K, $M_\star = 1.10 M_\odot$, $M_{\rm H}/M_\star = 1.02 \times 10^{-7}$, $M_{\rm He}/M_\star = 3.0 \times 10^{-4}$ and 81% of crystallized mass.

5. Conclusions

We presented the results of the evolutionary and asteroseismological study of ultra-massive ZZ Ceti stars by employing an expanded set of grid of realistic ONe-core WD models presented in Camisassa *et al.* (2019). We found that all pulsating white dwarfs known to date in the literature with masses above $1.10 M_\odot$ should have more than 80% of their mass crystallized.

Detailed asteroseismological analysis of BPM 37093 indicates $T_{\rm eff} = 11\,650$ K, $M_\star = 1.16 M_\odot$, $\log(M_{\rm H}/M_\star) = -6$ and $M_{\rm cr}/M_\star = 0.92$ A rotation period of 55 h has been inferred. Preliminary analysis to GD 518 and SDSS J084021 indicates good agreement with the spectroscopic determinations, finding $M_{\rm cr}/M_\star = 0.97$ and 0.81 respectively.

Our results were obtained by assuming that ultra-massive WDs come from single-star evolution and must harbor ONe cores but, however, it cannot be discarded that these objects are the result of mergers of two WDs and harbor CO cores. New photometric observations are needed in order to perform reliable asteroseismological analyses that could help us to derive the core chemical composition and infer their evolutionary origin.

Acknowledgments

F.C.D.G and A.H.C. warmly thank the Local Organizing Committee of the IAU Symposium No. 357 for support that allowed him to attend this conference.

References

Althaus, L. G., Serenelli, A. M., Panei, J. A., *et al.* 2005, *A&A*, 435, 631
Brassard, P. & Fontaine, G. 2005, *ApJ*, 622, 572
Camisassa, M. E., Althaus, L. G., Córsico, A. H., *et al.* 2019, *A&A*, 625, A87
Córsico, A. H., Althaus, L. G., Montgomery, M. H., *et al.* 2005, *A&A*, 429, 277
Curd, B., Gianninas, A., Bell, K. J., *et al.* 2017, *MNRAS*, 468, 239
Hermes, J. J., Kepler, S. O., Castanheira, B. G., *et al.* 2013, *ApJL*, 771, L2
Kanaan, A., Kepler, S. O., Giovannini, O., *et al.* 1992, *ApJL*, 390, L89
Medin, Z. & Cumming, A. 2010, *Phys. Rev. E*, 81, 036107
Metcalfe, T. S., Montgomery, M. H., & Kanaan, A. 2004, ApJL, 605, L133
Montgomery, M. H. & Winget, D. E. 1999, *ApJ*, 526, 976
Nitta, A., Kepler, S. O., Chené, A.-N., *et al.* 2016, *IAU Focus Meeting*, 29B, 493
Siess, L. 2010, *A&A*, 512, A10

White Dwarfs as Probes of Fundamental Physics: Tracers of Planetary, Stellar and Galactic Evolution
Proceedings IAU Symposium No. 357, 2019
M. A. Barstow, S. J. Kleinman, J. L. Provencal & L. Ferrario, eds.
doi:10.1017/S1743921320001027

Validation of asteroseismic fitting with the new white dwarf evolution code

Agnès Kim

Penn State Scranton, 120 Ridgeview Drive, Scranton, PA 18447, USA
email: axk55@psu.edu

Abstract. The new version of the White Dwarf Evolution Code (Bischoff-Kim & Montgomery 2018) overcomes limitations of earlier versions by utilizing MESA modules for the equations of state and opacities, now allowing regions of the model with a mix of helium, carbon, and oxygen. This single improvement allows us to almost exactly replicate models output by other stellar evolution codes. Armed with this new capability, we use as a star to fit, a hydrogen atmosphere white dwarf model from the La Plata group (using the LPCODE). We present results of fitting different subsets of periods for that model. This allows us some validation of our fitting methods, knowing exactly what properties we should be recovering in our best fit model.

Keywords. stars: interiors, stars: oscillations (including pulsations), stars: white dwarfs

1. Introduction

The goal of asteroseismic fitting is to produce a model that is a good representation of the star. We go from the premise that if the pulsation periods of our model match the observed pulsation spectrum, then our model matches the interior conditions and stellar parameters of the star. While we can never gain absolute knowledge of the interior structure of a star, we can check whether our tools and methods can recover the properties of a "known star", by attempting to fit a model produced with a code other than the one used to do the fitting.

The code used to compute the models involved in the fitting process is the WDEC (Bischoff-Kim & Montgomery 2018) and the synthetic star we sought to recover was produced using the LPCODE (Romero *et al.* 2012). We also began to address the question of period sampling. While to our knowledge, all normal modes of vibrations should be excited in a white dwarf interior, there are damping mechanisms and observational constraints that lead to the detection of a subset of all possible modes. In some cases, we observe only a few modes (one to three), while in others we get up to a dozen mode (for DAVs and DBVs). The mix of observed frequencies also determines how well the fits are constrained. In what follows, we detail the methodology further and present some results of this exercise. We learn valuable lessons.

2. Method

2.1. *Models*

The fitting techniques we are testing involve the computation of model grids with WDEC and a search for the best fit model on that grid. The best fit parameters are sometimes refined via a simplex search, but we did not do that in this work. We used a DAV model from the La Plata group that had a temperature of 11,288 K, a mass of 0.593 $M_\odot$, and a hydrogen layer mass of $10^{-4.285}$. The model was chosen to represent a typical DAV. The composition profiles of this model (the synthetic star) are shown

Table 1. Parameters used in the fits. For a description of each, see Bischoff-Kim (2018) and Bischoff-Kim & Montgomery (2018). For each parameter varied, we list the range followed by the step size.

Oxygen Profile	Envelope Profiles	Other
$h_1 = 0, 1; 0.1$	$M_{\rm env} = 1.5, 10; 0.5$	$T_{\rm eff} = 10,000 - 13,000; 500$ K
$h_2 = 0, 1; 0.1$	$M_{\rm He} = M_{\rm env}, 10; 0.5$	$M = 0.4, 1.0; 0.05 M_\odot$
$h_3 = 0.68$; fixed	xhe_bar $= 0.68$; fixed	MLT $\alpha = 0.65$; fixed
$w_1 = 0.44$; fixed	$\alpha 1 = 16$; fixed	
$w_2 = 0.12$; fixed	$\alpha 2 = 8$; fixed	
$w_3 = 0.41$; fixed	$M_{\rm H} = 4.0, 10; 0.5$	
	parameters of the WDEC target model	
$h_1 = 0.71$	$M_{\rm env} = 1.6$	$T_{\rm eff} = 11288$ K
$h_2 = 0.77$	$M_{\rm He} = 2.2$	$M = 0.593 M_\odot$
	$M_{\rm H} = 4.24$	

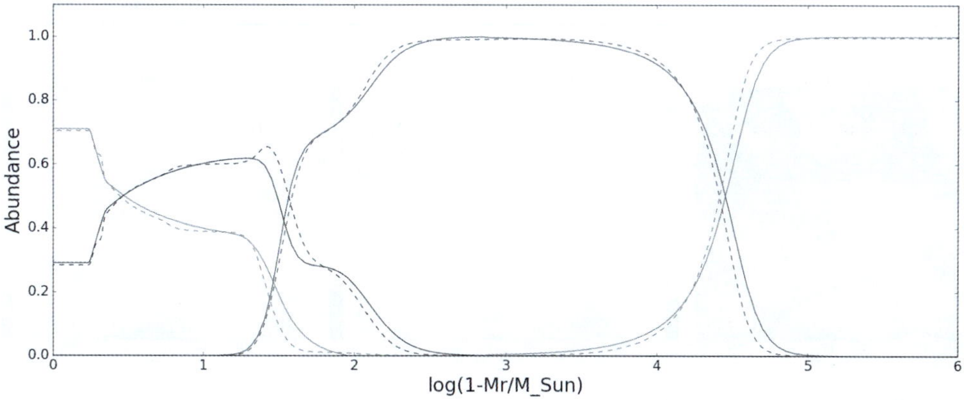

Figure 1. Composition profiles for the LPCODE model (dashed lines) and the matching WDEC model (solid lines). The center of the model is to the left and the surface to the right. The inner most profile (red) is oxygen, followed by carbon (black),helium (blue) and hydrogen (green).

in Fig. 1 with the dashed lines. We produced a WDEC model as close as possible to the LPCODE model. The composition profile of that model is shown with solid lines in Fig. 1. The effective temperature and mass of the model are the same as that of the LPCODE model. We call this model the "WDEC target model". Its parameters are listed in Table 1. In the grids used here, we varied 7 parameters. We also include the values we set for the other parameters that can be varied in the fitting process. The 6 oxygen profile parameters of Table 1 define Oxygen composition profiles by providing 5 points with vertical (composition) coordinates defined by h_i and horizontal (mass coordinates) defined by w_i. The envelope profile parameters define the helium and hydrogen composition profiles by specifying the location of transition zones ($M_{\rm env}$, $M_{\rm He}$, $M_{\rm H}$) and how smooth the transitions are ($\alpha 1$ and $\alpha 2$). xhe_bar denotes the helium abundance in the region where it is mixed with carbon.

2.2. *Period sets and fitting*

To simulate fitting a star, we must pick a set of periods to fit from the complete list of possible modes for the synthetic star. We work with three sets of periods. The first set is similar to the modes we observe in G117-B15A, where we have a consecutive sequence of three, low k modes. The second set is similar to the pulsation spectrum of EC14012, with

Table 2. The lists of periods (in seconds) used in our fitting exercises. For each period sampling (superstar, G117-B15A like, EC14012), we list the periods of the grid point model, the WDEC target model, and the LPCODE model, respectively (see text for more details). All periods listed are $\ell = 1$ modes. Radial overtones (k) are listed.

k	The superstar			G117-B15A like			EC14012 like		
1	129.087	140.051	135.388						
2	192.787	201.948	197.131	192.787	201.948	197.131			
3	256.689	269.803	279.542	256.689	269.803	279.541			
4	286.012	289.583	301.235	286.012	289.583	301.235			
5	304.199	315.773	325.593						
6	362.257	379.909	385.368				362.257	379.909	385.368
7	408.054	429.740	432.723						
8	462.873	484.940	491.634						
9	511.129	534.928	544.735				511.129	534.928	544.735
10	548.731	560.193	566.079				548.731	560.193	566.079
11							576.757	603.513	612.191
15							737.931	817.847	827.041
19							922.234	993.246	997.359
21							992.598	1076.844	1096.776

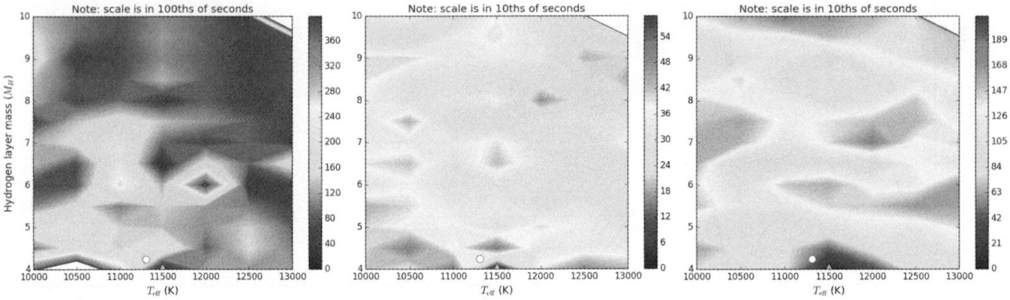

Figure 2. Results of the first level of fitting (fitting the periods of the model on the nearest grid point). Shown are quality of fit maps for the hydrogen layer mass. Blue indicates good fits, while red is worse. The scale shown are in the units labeled at the top of each figure and are essentially a standard deviation between the expected list of period and the periods of the models in the grid. The triangle at the very bottom of each map at $T_{\rm eff} = 11500$ K is the location of the best fit, while the circle is the expected location of the best fit. Left: G117-B15A like period spectrum; center: same for EC14012, right: consecutive $k = 1-10$ periods.

7 higher k modes, most of them non-consecutive. The last set includes an $\ell = 1$ sequence, complete from $k = 1$ to 10. To our knowledge, no observed period spectrum offers such a complete sequence though some pulsating white dwarfs come close, for instance TIC 257459955 (Bell *et al.* 2019). We call this spectrum the "superstar" (2).

We performed three levels of fitting. The first one was to find the model at the grid point nearest the WDEC target model and fit the periods of that model. The best fit model in this case is a perfect fit and we expect the optimization process to find it easily. The second was to fit the WDEC target model. Lastly, we fit the periods of the LPCODE model. This last test brings us closest to the challenge of fitting an observed period spectrum. All sets of periods are listed in Table 2.

3. Results

In Fig. 2 we present the results of the first level of fitting (fitting the periods of the model on the nearest grid point). We focus on the fitness map for the thickness of the hydrogen layer. We show the results for all three sets of periods discussed above and

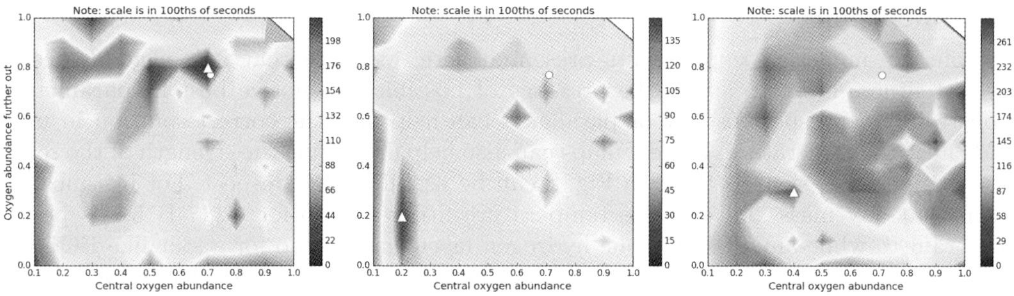

Figure 3. Result of the higher level fitting described in section 2.2. Shown are quality of fit maps for the core parameters for the G117-B15A like period spectrum. See caption of Figure 2 for more information. Left: fitting the grid model periods; center: fitting the target WDEC model periods, right: fitting the LPCODE model periods.

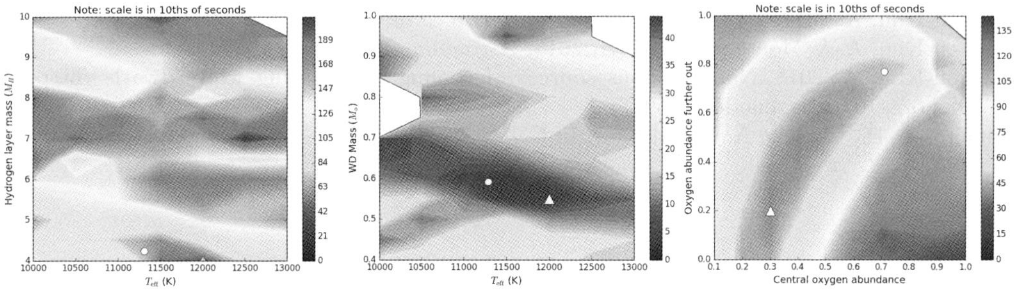

Figure 4. Quality of fit maps for the core parameters for the periods spectrum with the consecutive $k = 1 - 10$ $\ell = 1$ sequence of modes. See caption of Figure 2 for more information. Left: hydrogen layer mass; center: mass and effective temperature, right: core parameters.

listed in Table 2. In rather simplistic terms, we expect the presence of higher k modes to assist in constraining the hydrogen layer mass. The transition occurs in the outer parts of the model, where higher k modes tend to be trapped. We find that indeed, we get stronger constraints with the EC14012 like spectrum and the "superstar spectrum" (where all modes from $k = 1$ to 10 are present).

In Fig. 3, we focus on the G117-B15A like spectrum and attempt the higher level fitting described in section 2.2. We look at the fitness map for the core parameters. While we do recover the correct core parameters when fitting the grid point model (a sanity check), the method fails when we move to fitting periods of models that are off the grid, or not computed with WDEC. We do not recover the correct best fit parameters. In the case of the target WDEC model, we do find a best fit near where we expect it, but there is a much stronger minimum elsewhere in parameter space. The same is true when fitting the periods of the LPCODE model.

In Fig. 4, we examine various results of fitting the LPCODE model when using the consecutive the $k = 1 - 10$ mode sequence. Due to the larger range of consecutive modes, we expect better results. We do get good results, with the exception again of the core parameters, even though the region of best fits does contain the correct parameters. Effective temperature is harder to constrain. It should be noted that the two core parameters have to do with the oxygen abundance in different regions of the model. We know from experience that we are more sensitive to the location of transition zones, so the results may be a little on the pessimistic side.

4. Conclusion

From the numerical experiments presented here, we learn that it is a good idea to look at fitness maps to assess the presence of possible alternative best fit parameters. Constraints from spectroscopy or parallaxes can help pick the correct solution in mass and effective temperature. Fitness maps can also help determine the strength of the result of a fitting. Plots such as those in Fig. 2 can be useful in that respect. For instance, we learn that any mass and effective temperature fitting we do for G117-B15A is poorly constrained, while any fitting of the hydrogen layer mass we do for a star like EC14012 should give us a strong result. This statement comes with a caveat. Each observed period spectrum is unique and offers a different set of constraints in the asteroseismic fitting. The presence or absence of a single mode can dramatically affect the nature of the fitting (Bischoff-Kim 2017).

References

Bell, K. J., Córsico, A. H., Bischoff-Kim, A., *et al.* 2019, arXiv e-prints, arXiv:1910.04180

Bischoff-Kim, A. 2017, *European Physical Journal Web of Conferences*, 06011

Bischoff-Kim, A. & Montgomery, M. H. 2018, *Astronomical Journal*, 155, 187

Bischoff-Kim, A. 2018, Non-luminous sources of cooling in pulsating white dwarfs (https://doi.org/10.5281/zenodo.1715917).

Romero, A. D., Córsico, A. H., Althaus, L. G., *et al.* 2012, arXiv e-prints, arXiv:1204.6101

White Dwarfs as Probes of Fundamental Physics: Tracers of Planetary, Stellar and Galactic Evolution
Proceedings IAU Symposium No. 357, 2019
M. A. Barstow, S. J. Kleinman, J. L. Provencal & L. Ferrario, eds.
doi:10.1017/S1743921320000241

The chemical structure of the hot pulsating DB white dwarf KIC 08626021 from asteroseismology

S. Charpinet[1], **P. Brassard**[2], **N. Giammichele**[1] **and Gilles Fontaine**[2]

[1]Institut de Recherche en Astrophysique et Planétologie, CNRS, Université de Toulouse, CNES, 14 avenue Edouard Belin, F-31400 Toulouse, France
email: stephane.charpinet@irap.omp.eu

[2]Dèpartement de Physique, Université de Montrèal, Québec H3C 3J7, Canada

Abstract. Giammichele *et al.* (2018) proposed a full determination, largely independent of evolution calculations, of the chemical composition and stratification inside the hot pulsating DB white dwarf KIC 08626021. However, Timmes *et al.* (2018) pointed out that neglecting the effects of neutrino cooling, such as in the static models used in Giammichele *et al.* study, could impact significantly the derived seismic solution and compromise conclusions drawn upon it. Here we present a reanalysis of KIC 08626021, using improved static models which now incorporate more realistic luminosity profiles that reflect the still significant energy losses induced by neutrino emission mechanisms in hot DB white dwarfs. We show that this effect has only a limited impact on the derived seismic model properties and, more importantly, that all the conclusions brought by Giammichele *et al.* (2018) remain entirely valid.

Keywords. white dwarfs, stars: interiors, stars: oscillations

1. Context

Giammichele *et al.* (2018) established from asteroseismology, using a new flexible parameterization of the chemical stratification in white dwarf cores (Giammichele *et al.* 2017a,b), that the pulsating DB white dwarf KIC 08626021 has a central homogeneous core $\sim 15\%$ richer in oxygen (in mass) and $\sim 40\%$ more massive than predicted by standard evolutionary models. This finding provides strong and challenging constraints to the modeling of stellar cores during the helium-burning phase that precedes (e.g., Fields *et al.* 2016; De Gerónimo *et al.* 2017, 2018). However, this result has been questioned by Timmes *et al.* (2018), who pointed out that the impact of neutrino cooling on the thermal structure of the star – an effect neglected in the static equilibrium models employed by Giammichele *et al.* (2018) – is still significant in a hot DB white dwarf like KIC 08626021. They found that low-order g-mode frequencies could differ by as much as ~ 70 μHz (~ 30 μHz on average) between hot DB white dwarf models neglecting and including neutrino losses. Such differences could induce significant shifts in the derived parameters of KIC 08626021, possibly changing the conclusions drawn upon its seismically determined structure. We have therefore undertaken a reanalysis of this star, now accounting properly for the effects of neutrino emission. Details on this new analysis are published in Charpinet *et al.* (2019) and briefly described below.

2. New static models with neutrinos

The models used in Giammichele *et al.* (2018) are static structures constructed on the assumption that the luminosity profile is proportional to the mass distribution ($L_r \propto M_r$;

hereafter referred to as the "0th-order" approximation). This assumption is well verified in white dwarfs that have cooled sufficiently, such as DA pulsators found in the $T_{\rm eff} =$ 13,000–11,000 K range. However, for hot white dwarfs, such as the helium-atmosphere DB pulsators found in the $T_{\rm eff} =$ 32,000–22,000 K range, cooling from neutrino emission is still important, with a significant impact on the thermal structure and pulsation frequencies, as illustrated by Timmes *et al.* (2018). The above assumption on the luminosity profile is therefore no longer appropriate in that case.

To address this problem, we introduce improved static models that are constructed from the relation $L_r = M_r \cdot (L_*/M_* - k_\nu)$, where k_ν accounts for all effects (mainly neutrino cooling, but also residual contraction) that would cause the luminosity profile to differ from the $L_r \propto M_r$ relation. The quantity k_ν is interpolated from a grid of luminosity profiles obtained from "generic" white-dwarf cooling sequences (evolutionary models) covering a wide range of values for $T_{\rm eff}$, mass, He-envelope thickness, and core composition (C/O relative mass fraction). These models are generic in the sense that their structure is a simple C/O core of homogeneous composition surrounded by a pure helium envelope, providing the main, "1st-order" correction to the luminosity profile. A "2nd-order" correction accounting for a particular chemical stratification, e.g., as inferred from a seismic model solution, can be obtained if necessary, by calibrating the interpolation of k_ν relative to a specific cooling sequence. This approach produces improved static white-dwarf models that satisfactorily account for the impact of neutrino cooling (and other less important effects) on the thermal structure (see Charpinet *et al.* 2019).

When comparing adiabatic frequencies for relevant g-modes computed from two identical models representative of KIC 08626021, one including the effects of neutrino emission as described above, and the other neglecting them, we find frequency differences that are indeed comparable to the variations reported by Timmes *et al.* (2018).

3. Updated seismic solution for KIC 08626021

KIC 08626021 was reanalysed with our updated static models, following an iterative approach as described in Charpinet *et al.* (2019), in order to achieve a "2nd-order" solution. Compared to Giammichele *et al.* (2018), we find that using the new improved models has no negative impact on the achieved quality of the optimal fit. They lead to the identification of a very well defined seismic solution that also reproduces the frequencies of KIC 08626021 to the precision of the observations ($\sim$0.6 nHz) and that does not induce changes in the mode identification. The new solution is however shifted relative to the former model uncovered by Giammichele *et al.* (2018). The most relevant parameters of this updated solution are given in Table 1 and Fig. 1 shows the new chemical stratification uncovered compared to the former solution of Giammichele *et al.* (2018).

We first find that the new improved solution for KIC 08626021 remains close to the former model, as far as global stellar parameters ($T_{\rm eff}$, $\log g$, M_*, and R_*) are concerned. Shifts in these parameters remain close to the 1σ-uncertainties quoted by Giammichele *et al.* (2018). Therefore, systematic errors introduced by neglecting neutrino cooling do not propagate evenly among the optimized model parameters. Larger variations appear when comparing the internal stratifications of the new optimal model with the former one, but also impacting unevenly the characteristics of the profiles. The chemical composition in the intermediate C/O and C/He layers are the most affected. We find that the helium mass fraction in the mixed C/He layer, $X({\rm He})_{\rm env}$, is now 0.38 ± 0.07 instead of $0.18 \pm$ 0.04, while the amount of oxygen (in mass fraction) in the intermediate C/O layer reaches ~ 0.6 instead of ~ 0.4. In contrast, the other important characteristics of the chemical profiles remain nearly unchanged. In particular, all chemical transitions are found to be located nearly at the same fractional mass depths, except for a slight outward shift of the core-to-envelope transition ($\log q_2$). The inner core, for its part, remain nearly the

Table 1. Updated properties of KIC 08626021 when accounting for neutrino cooling in the seismic models (this work; Charpinet *et al.* 2019) compared to the former solution from Giammichele *et al.* (2018).

Quantity	Giammichele *et al.* (2018)	Charpinet *et al.* (2019) (This work)
$\log g$ (cm s^{-2})	7.92 ± 0.01	7.905 ± 0.014
$T_{\rm eff}$ (K)	$29,968 \pm 198$	$30,114 \pm 210$
$X(\rm He)_{env}$	0.18 ± 0.04	0.38 ± 0.07
$\log(q_1)$	-7.63 ± 0.20	-7.83 ± 0.36
$\log(q_2)$	-3.23 ± 0.10	-3.48 ± 0.07
$X(\rm O)_{centre}$	0.86 ± 0.04	0.84 ± 0.04
$M(\rm He)/M_*$ (%)	0.0113 ± 0.006	0.0133 ± 0.0063
$M(\rm C)/M_*$ (%)	21.96 ± 4.2	20.76 ± 1.1
$M(\rm O)/M_*$ (%)	78.03 ± 4.2	79.23 ± 1.1
$M_*/M_\odot$	0.570 ± 0.005	0.562 ± 0.006
$R_*/R_\odot$	0.0138 ± 0.0001	0.0139 ± 0.0001

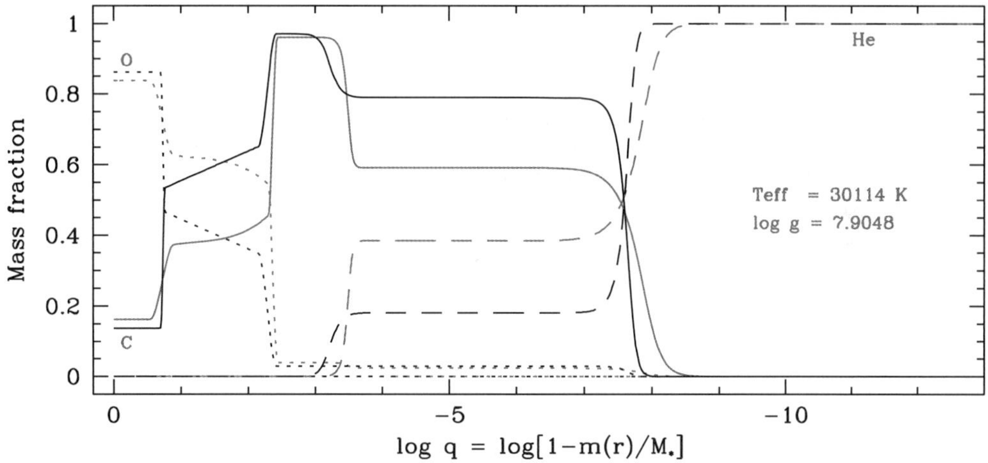

Figure 1. Stratification in helium, carbon, and oxygen derived from the new optimal seismic solution obtained with models incorporating neutrino cooling (red curves) and compared with the former solution from Giammichele *et al.* (2018, black curves). This plot is taken from Figure 2 of Charpinet *et al.* (2019).

same with the central oxygen mass fraction, $X(\rm O)_{centre}$, estimated at 0.84 ± 0.04 (i.e., very close to the former, 0.86 ± 0.04, value) and the homogeneous central part extending up to a fractional mass depth of $\log q \sim -0.71$ (i.e., encompassing a mass of $\sim 0.45\ M_\odot$), which is identical to the value of $\log q \sim -0.7$ ($\sim 0.45\ M_\odot$) found from the former solution. Consequently, all the conclusions regarding the inner core size and its C/O composition ratio expressed in Giammichele *et al.* (2018) are unchanged.

4. Conclusion

Despite some changes in the determined internal structure, we find that including the effects of neutrino cooling in our static DB white dwarf models has little impact on the main characteristics of the optimal seismic solution for KIC 08626021. The only significantly affected quantities are the C/O and C/He mass fractions in the intermediate layers of the star. Other quantities, such as the location of helium transitions in the envelope, or the central oxygen mass fraction and extent of the homogeneous central part of the core are either unchanged or only slightly displaced. *Therefore, despite significant effects on calculated frequencies induced by omitting neutrinos and their impact on the thermal structure of the star, we find that all conclusions and implications drawn from the analysis of Giammichele et al. (2018) remain completely valid.*

Acknowledgments

We dedicate this work to our inspiring friend and colleague Gilles Fontaine, gone too early. Stéphane Charpinet and Noemi Giammichele acknowledge financial support from the Centre National d'/'Etudes Spatiales (CNES, France) and from the Agence Nationale de la Recherche (ANR, France) under grant ANR-17-CE31-0018, funding the INSIDE project. This work was granted access to the high-performance computing resources of the CALMIP computing centre under allocation number 2019-p0205. P.B. and G.F. acknowledge the contribution of the Canada Research Chair Program.

References

Charpinet, S., Brassard, P., Giammichele, N., & Fontaine, G. 2019, *A&A* (Letters), 628, L2

De Gerónimo, F. C., Althaus, L. G., Córsico, A. H., Romero, A. D., & Kepler, S. O. 2017, *A&A*, 599, A21

De Gerónimo, F. C., Althaus, L. G., Córsico, A. H., Romero, A. D., & Kepler, S. O. 2018, *A&A*, 613, A46

Fields, C. E., Farmer, R., Petermann, I., Iliadis, C., & Timmes, F. X. 2016, *ApJ*, 823, 46

Giammichele, N., Charpinet, S., Fontaine, G., & Brassard, P. 2017a, *ApJ*, 834, 136

Giammichele, N., Charpinet, S., Brassard, P., & Fontaine, G. 2017b, *A&A*, 598, A109

Giammichele, N., Charpinet, S., Fontaine, G., *et al.* 2018, *Nature*, 554, 73

Timmes, F. X., Townsend, R. H. D., Bauer, E. B., *et al.* 2018, *ApJ* (Letters), 867, L30

White Dwarfs as Probes of Fundamental Physics: Tracers of Planetary, Stellar and Galactic Evolution
Proceedings IAU Symposium No. 357, 2019
M. A. Barstow, S. J. Kleinman, J. L. Provencal & L. Ferrario, eds.
doi:10.1017/S1743921320000265

Searching for ZZ Ceti white dwarfs in the Gaia survey

Olivier Vincent, Pierre Bergeron and David Lafrenière

Département de Physique, Université de Montréal, Montréal, Québec, H3T 1J4
emails: ovincent@astro.umontreal.ca, bergeron@astro.umontreal.ca,
david@astro.umontreal.ca

Abstract. The Gaia satellite recently released parallax measurements for nearly 400,000 white dwarf stars, allowing for precise measurements of their physical parameters. By combining these parallaxes with Pan-STARRS and CFIS-u photometry, we measured the effective temperatures and surface gravities for all white dwarfs within 100 pc and identified a sample of ZZ Ceti white dwarf candidates within the instability strip. We report the results of a photometric follow-up, currently under way, aimed at identifying new ZZ Ceti stars among this sample using the PESTO camera attached to the 1.6-m telescope at the Mont Mégantic Observatory. Our goal is to verify that ZZ Ceti stars occupy a region in the $\log g - T_{\rm eff}$ plane where no nonvariable stars are found, supporting the idea that ZZ Ceti pulsators represent a phase through which all hydrogen-line (DA) white dwarfs must evolve.

Keywords. white dwarfs, stars: variable, surveys, techniques: photometric

1. Introduction

In recent years, the number of known hydrogen-line pulsating white dwarfs, known as DAV or ZZ Ceti stars, has been growing rapidly. The traditional approach to identify candidates consists in measuring the white dwarf atmospheric parameters, $T_{\rm eff}$ and $\log g$, by fitting the hydrogen Balmer lines, and then to determine whether the object falls within the ZZ Ceti instability strip. The boundaries of this strip have yet to be determined accurately, but they have been empirically shown to be located between 11,000 K $< T_{\rm eff} <$ 13,000 K with a strong dependency on $\log g$ (Green *et al.* 2015). Once a candidate has been identified, variability can be confirmed by measuring its light curve.

The Sloan Digital Sky Survey (SDSS) has shown to be an invaluable source of white dwarf spectra for this purpose, and roughly half of the currently known DAV stars have been found using this survey. Mukadam *et al.* (2004), for instance, have nearly doubled the number of known DAVs at that time, with an impressive discovery of 35 new pulsators. While the spectroscopic approach to finding new ZZ Ceti stars has been very efficient, it is limited by the fact that wide-scale spectroscopic surveys do not focus on white dwarfs, and so the tedious task of finding new candidates falls into the hands of smaller ground-based telescopes. While that may change in the future, a new approach has been provided by the Gaia DR2 parallax measurements along with Pan-STARRS photometry.

The photometric technique relies on observed fluxes averaged over various bandpasses, such as the Pan-STARRS *grizy* magnitudes. These photometric measurements are fitted with synthetic photometry obtained from model atmospheres, with the effective temperature and solid angle $\pi R^2/D^2$ considered free parameters, where R is the stellar radius and D the distance from Earth. Since the distance is known from the parallax measurement, the stellar radius is obtained directly, while the stellar mass and surface gravity can be derived from the mass-radius relation for white dwarfs.

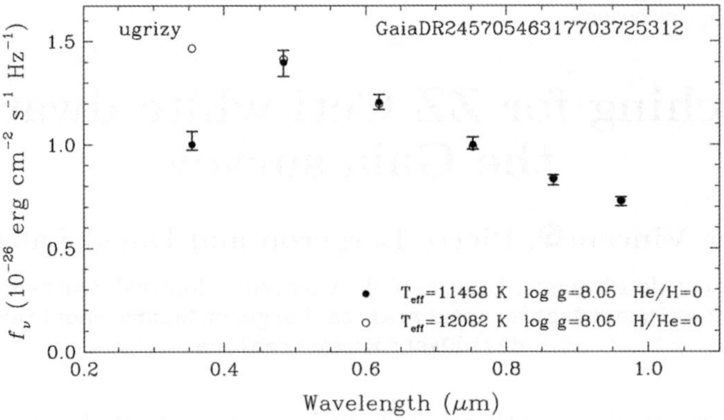

Figure 1. Sample photometric fit to a ZZ Ceti white dwarf candidate using Pan-STARRS *grizy* and CFIS-u photometry (error bars), combined with the Gaia parallax. Filled (open) circles correspond to our best fit under the assumption of a pure hydrogen (pure helium) atmospheric composition. Note that the CFIS-u data point was not used in these fits. The results clearly indicate that this object is a hydrogen-atmosphere white dwarf.

2. Sample Selection

Our initial sample consists of all white dwarf candidates from the Gaia DR2 with $D < 100$ pc and a parallax error less than 10%. This limit on the distance was chosen as to minimize the effects of interstellar reddening. These candidates were then cross-matched with the Pan-STARRS catalog, and all objects were fitted with the photometric technique assuming pure hydrogen atmospheres. We then used the instability strip empirically determined by Green *et al.* (2015) to select our ZZ Ceti candidates in the $\log g$ – $T_{\rm eff}$ plane.

To further reduce the number of false candidates, the u bandpass comes in handy: its wavelength coverage includes the Balmer jump, which is a very distinctive feature between hydrogen- and helium-atmosphere white dwarfs. Indeed, hydrogen-atmosphere white dwarfs show a significant drop in the u-band flux, whereas their helium-atmosphere counterparts show a more continuous flux distribution. Figure 1 shows an example of a photometric fit to one of our ZZ Ceti white dwarf candidates using Pan-STARRS *grizy* combined with the u photometry from the Canada-France Imaging Survey (CFIS; Ibata *et al.* 2017). Here the u magnitude is not used in the fitting but helps to discriminate between the pure hydrogen and pure helium solutions. In the example displayed in Figure 1, we can clearly see the drop in the u-flux caused by the Balmer jump, as correctly predicted by the pure hydrogen model.

It is also worth noting that without the u band, the *grizy* photometry tends to give similar solutions for both hydrogen- and helium-atmosphere models. There are likely many helium-atmosphere white dwarfs polluting our sample, but unfortunately, u magnitudes are only available for a fraction of our sample. On one hand, the SDSS does not cover nearly as much sky as the Gaia survey, and furthermore, the SDSS white dwarf sample has already been investigated extensively for white dwarf pulsators, leaving us with very few new ZZ Ceti candidates. Luckily, the CFIS survey is currently under way, and will eventually provide u-band photometry for many additional targets in our sample. While CFIS will provide data solely for the northern hemisphere, it will cover more sky area than SDSS, and with a much better precision. For a given measurement error, CFIS-u is about 3 magnitudes deeper than SDSS. We have early access to this new survey, and CFIS-u data are being included in our analysis as the data are being released.

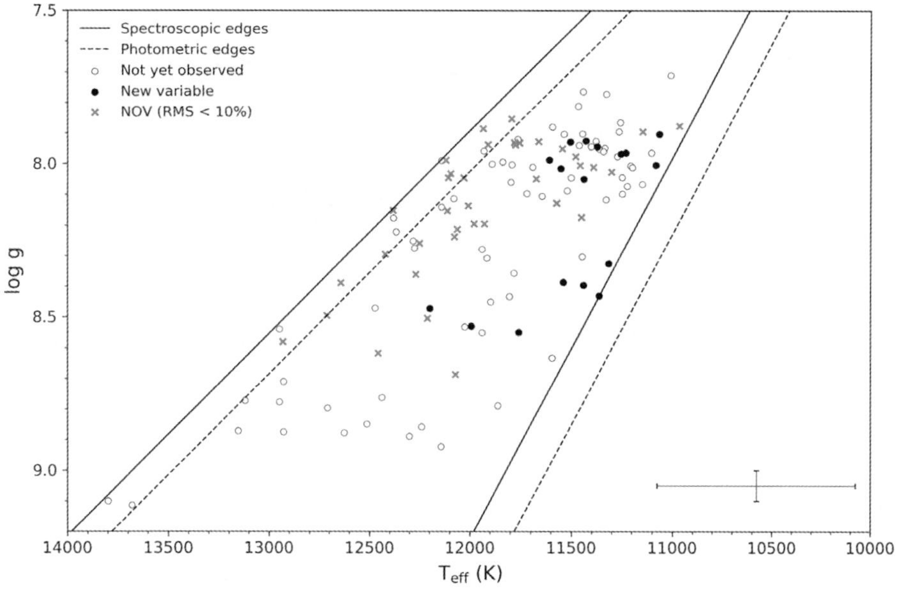

Figure 2. Overview of our photometric survey of ZZ Ceti candidates; the various symbols are described in the legend. The solid lines represent the empirical instability strip inferred from spectroscopy (Green *et al.* 2015), while the dashed lines indicate the same strip shifted by 200 K (see text).

3. Survey Overview

Our survey strategy is to observe as many candidates as possible for two hours using a 10-second exposure time, with a higher priority given to candidates confirmed to be hydrogen-rich through photometric fits including the u bandpass. Our ongoing survey began in late summer 2018 with the Mont Mégantic Observatory 1.6-m telescope equipped with the PESTO instrument, which uses a frame transfer EMCDD to achieve an observing efficiency near 100%. Our final sample contains 134 ZZ Ceti candidates, excluding known variables and objects in the southern hemisphere. Figure 2 shows an overview of the current state of our survey. Each object displayed is a ZZ Ceti candidate, where black filled dots represent new pulsators discovered in our survey, while crosses are objects not observed to vary (NOV), with a photometric error less than 10%.

As of September 2019, 17 new pulsators had been identified. A few selected light curves of these new ZZ Ceti variables are displayed in Figure 3. A wide variety of pulsators has been detected so far. For example, GaiaDR2647899806626643200 is a long-period (~2000 s) and large amplitude (~20%) variable, whereas GaiaDR24491980748701631616 has a shorter period of 350 seconds and smaller amplitude of ~4%. There are also a few pulsators showing multiple pulsation modes, such as GaiaDR24217793816094052480.

Many objects in Figure 2, not observed to vary, are located well within the ZZ Ceti instability strip. While this may appear worrisome at first glance, there are several possible explanations for this result:
(1) The first possibility is the case where the photometric error is larger than the amplitude of the pulsation, in particular at the blue edge of the instability strip. Small amplitude modes can easily be lost within the photometric noise, and it is important to keep this in mind when assessing the variability of an object. Our average photometric error is about 3%, which implies that all pulsators with relative amplitudes close to this value are less likely to be detected in our survey.

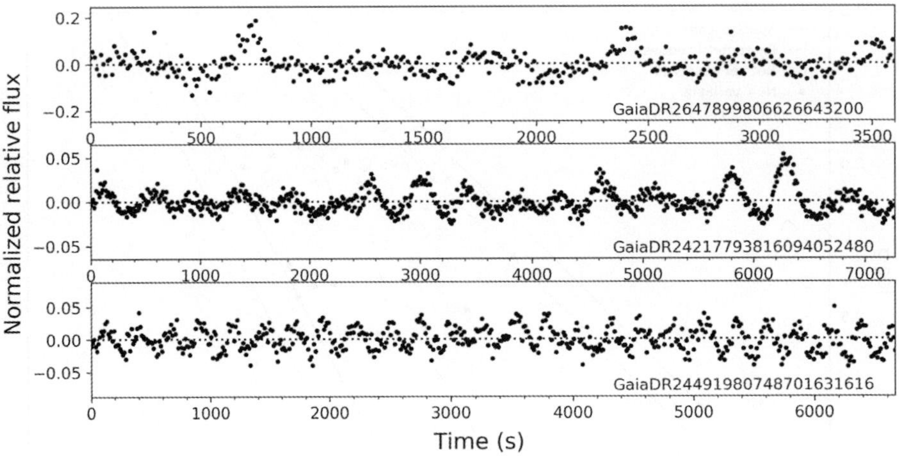

Figure 3. Selected light curves for newly discovered ZZ Ceti white dwarfs.

(2) As mentioned above, we lack u-band photometry for many candidates, and consequently, our sample is most likely polluted by helium-atmosphere white dwarfs. This situation should improve as more CFIS-u photometry becomes available.
(3) There are many cases where the Pan-STARRS magnitudes are contaminated by the presence of an unresolved companion, such as a bright neighbouring star or an M-dwarf. In the latter case, this can be confirmed with infrared photometry. These contaminations may throw off our photometric solutions and misplace the object inside (or outside) the instability strip. We could also be dealing with unresolved double degenerate binaries, which are difficult to identify without spectroscopic measurements.
(4) Systematic differences between the spectroscopic and photometric techniques have been recently reviewed by Bergeron *et al.* (2019). In particular, they found that the photometric technique tends to yield cooler effective temperatures compared to the spectroscopic values when relying on *grizy* photometry alone, and that the inclusion of the SDSS-u magnitude could reduce these differences significantly. Cloutier (2019) found that the photometric temperatures of ZZ Ceti stars based on *grizy* Pan-STARRS magnitudes were 200 K cooler, on average, than the spectroscopic values. Given that our initial selection of ZZ Ceti candidates was based on the empirical strip determined from spectroscopy, we reproduced in Figure 2 the expected photometric instability strip, shifted by 200 K with respect to the spectroscopic values. Many NOV objects near the blue edge are now falling outside of the strip, indicating that hotter ZZ Ceti candidates are more likely to be false positives. Another interesting feature is how the shift puts all the new variables much closer to the middle of the strip, rather than close to the red edge.

References

Bergeron, P., Dufour, P., Fontaine, G. *et al.* 2019, *ApJ*, 876, 67

Cloutier, S. 2018, M.Sc. thesis, Université de Montréal

Green, E. M., Limoges M.-M., Gianninas, A. *et al.* 2015, in Astronomical Society of the Pacific Conference Series, Vol. 493, 19th European Workshop on White Dwarfs, ed. P. Dufour, P. Bergeron & G. Fontaine, 237

Ibata, R. A., McConnachie, A., Cuillandre, J.-C. *et al.* 2017, *ApJ*, 848, 128

Mukadam, A. S., Mullally, F., Nather, R. E. *et al.* 2017, *ApJ*, 607, 982

White Dwarfs as Probes of Fundamental Physics: Tracers of Planetary, Stellar and Galactic Evolution
Proceedings IAU Symposium No. 357, 2019
M. A. Barstow, S. J. Kleinman, J. L. Provencal & L. Ferrario, eds.
doi:10.1017/S1743921320000320

Pulsating white dwarfs and convection

J. L. Provencal[1], M. H. Montgomery[2], H. L. Shipman[1] and The WET Team[3]

[1]University of Delaware, Dept. of Physics and Astronomy, Newark, DE 19716
email: `jlp@udel.edu`

[2]Dept. of Astronomy & University of Texas, Austin, TX
email: `mikemon@astro.as.utexas.edu`

[3]See www.physics.udel.edu/darc for a list of WET collaborators

Abstract. Convection is a highly turbulent, three dimensional process that is traditionally treated using a simple, local, time independent description. Convection is one of the largest sources of theoretical uncertainty in stellar modeling. We outline recent progress in studies using pulsating white dwarfs to constrain convection and calibrate mixing length theory.

Keywords. convection, white dwarfs, oscillations

1. Introduction

The dynamics of convection influence many key aspects of stellar modeling, ranging from the effects of convective overshoot to supernovae to stellar ages. We know that convection is a highly turbulent, three dimensional process that involves nonlocal motions. Yet convective energy transport in stellar environments is typically modeled using Mixing Length Theory (MLT). MLT is a simple one dimensional theory in which convective cells of the same size travel a given distance (the "mixing length") before dissipating and releasing or absorbing energy. MLT, with its free mixing length parameter α, lacks predictive value. More physically self-consistent approaches based on hydrodynamic simulations are beginning to supplant it or at the very least calibrate it (i.e. Tremblay *et al.* 2015), but such models need to be rigorously validated. An empirical determination of convection dynamics in different stellar environments is thus an important goal with implications for a wide range of astrophysical fields.

2. Convective Light Curve Fitting

Pulsating WDs are invaluable asteroseismic laboratories to probe stellar convection. The g-mode pulsations ($P \approx 100$–1500 s) are linked to the inevitable development of surface hydrogen or helium CZs as every WD cools. The CZ is very sensitive to temperature changes. Pulsating hydrogen atmosphere WDs (DAVs) experience local $\Delta T = \pm 200$K, while helium pulsators (DBVs) swing through variations as large as $\Delta T = 3000$ K. These changes produce significant variations in the local heat capacity of the CZ. All of this means that, at any given time, the CZ of a pulsating WD cannot be considered as a structure of uniform depth over the entire star. The direct result is an observed nonlinear light curve with narrow peaks and wider valleys. Fig. 1 illustrates the increased nonlinearities in observed light curves as a function of temperature.

The complete methodology of convective light curve fitting can be found in Montgomery (2005) and Montgomery *et al.* (2010). In very broad brush strokes, convective light curve fitting uses observed nonlinearities to recover the parameter τ_0, which

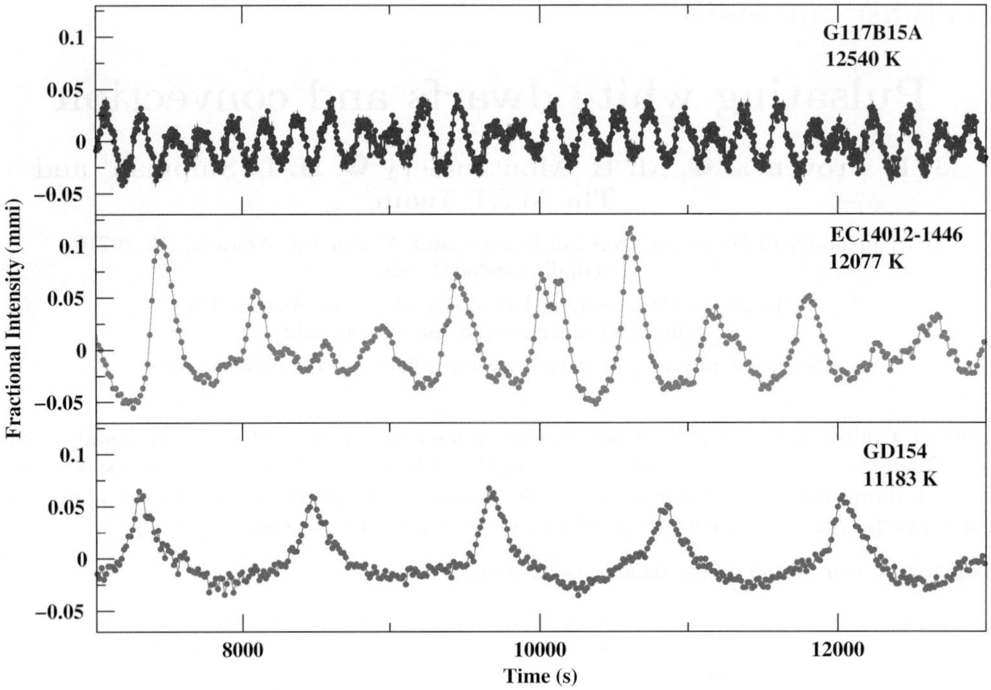

Figure 1. The general evolution of pulsating white dwarf light curves as a function of T_{eff}. Nonlinearities increase as the convection zone deepens and evolves in cooler stars.

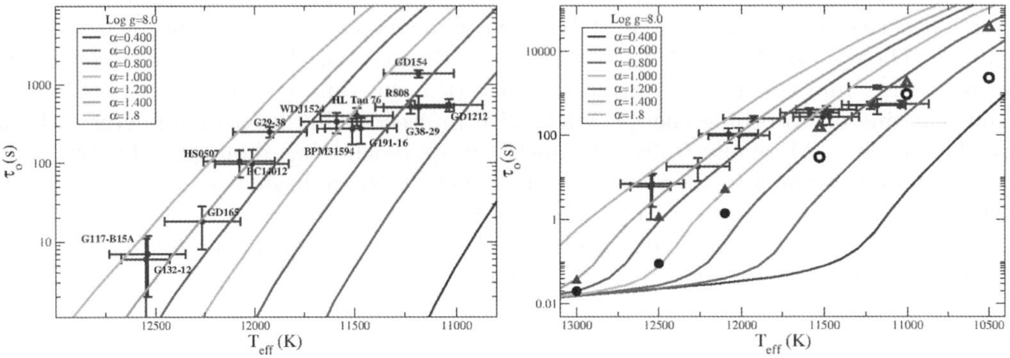

Figure 2. For DA pulsating white dwarfs. Left: Comparison of convective light curve fitting results (blue points) with predictions of MLT (colored lines). Right: Similar comparison, but including the recent 3D radiation hydrodynamical convection simulations of Tremblay *et al.* (2015). The triangular symbols use the flux boundary to define the CZ, while the circular symbols use the Schwarzchild criterion. Note changes in the x and y axes.

represents the time-averaged heat capacity of the CZ in each star. The Whole Earth Telescope (WET) is in the midst of a long term project to produce an empirical description of τ_o as a function of effective temperature across the DAV and DBV instability strips. Our goal is to provide insight into convection dynamics in WD environments. Fig. 2 presents the comparison of current convective light curve fitting results for DAVs with the expectations of MLT (left) and 3D radiation hydrodynamical convection simulations of Tremblay *et al.* (2015) (right). Each τ_0 determination in each panel is based on

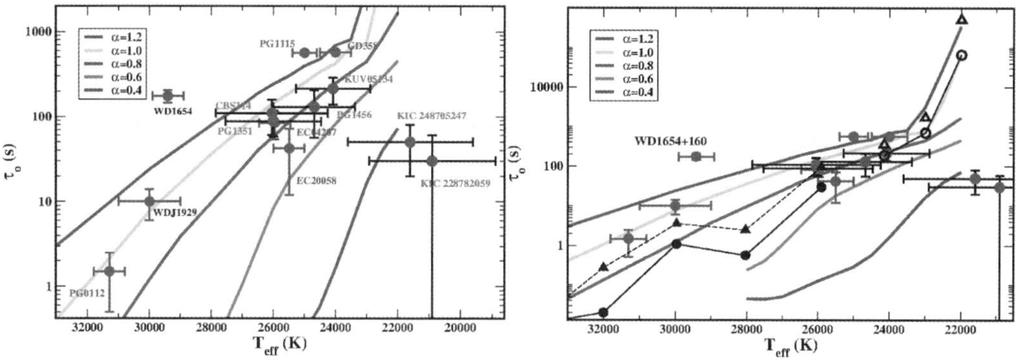

Figure 3. For DB white dwarfs. Left: Comparison of convective light curve fitting results (blue points) with predictions of MLT (colored lines). Blue error bars represent UV temperature determinations, while black error bars are optical determinations. Black object labels contain no photospheric metals. Purple labels contain traces of photospheric hydrogen, carbon, and other metals. Right: Similar comparison, but including the recent 3D radiation hydrodynamical convection simulations of Cukanovaite *et al.* (2019). The triangular symbols use the flux boundary, while the circular symbols use the Schwarzchild criterion. Note changes in the x and y axes.

extensive Whole Earth Telescope (WET) ground based photometric campaigns Provencal *et al.* (2012). Once we have determined τ_o for an object, we require an accurate effective temperature to place the WD at its location in the T_{eff} plane. For this, we rely on published spectroscopic T_{eff} measurements.

Fig. 2 (left) clearly shows that τ_o does not follow the predictions of MLT for any single value of the mixing parameter α. The value of τ_0 reaches a plateau at $\approx$800 s near $T_{\rm eff}$=12000 K. Our results argue for an increasing time-averaged heat capacity and a high convective efficiency for stars hotter than 12,000 K, but a nearly constant time-averaged heat capacity (and a changing convective efficiency) for stars cooler than $\approx$12,000 K. This translates into a much slower deepening of the CZ for cooler pulsators than predicted by theoretical calculations. The 3D models of Tremblay *et al.* (2015) (right) offer a theoretical improvement, predicting a change in alpha, or convective efficiency, as a function of $T_{\rm eff}$. However, there remains some differences, such as the observed plateau, with the 3D predictions.

Fig. 3 presents a similar comparison for the DBVs. To date, we have determined τ_o for 11 DBVs. The situation here is more complex than for the DAVs. Our greatest limitation in interpreting the behavior of τ_o across the DB instability strip lies in the errors associated with $T_{\rm eff}$ determinations. This well known issue stems from the high temperatures of the DBVs and resulting difficulties of determining temperatures from optical spectroscopy, as well as the effects of trace amounts of hydrogen and other elements. In Fig. 3 left, objects with optical temperature determinations have black error bars, while UV determinations have blue error bars. Objects containing elements such as carbon or hydrogen in their photospheres have purple name labels. In Fig. 3 right, we present comparisons with the recent 3D convective simulations of Cukanovaite *et al.* (2019). 3D helium convection, in addition to the issues with overshooting, etc., is further complicated by the presence of a second convection zone (HeII) that may or may not be merged with the He I convection zone. In short, much work remains to be completed before we can understand the behavior of the DB pulsators.

An additional probe of convection may be given by a comparison of the predicted maximum pulsation periods for given values of $\tau_{\rm o}$ with the observed maximum pulsation periods for stars in our sample (Fig. 4). Brickhill 1991 shows that g-mode pulsations

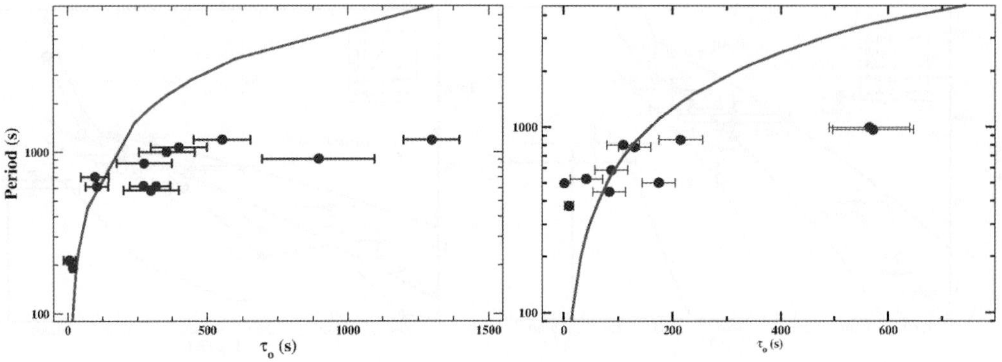

Figure 4. Left: Comparison of observed pulsation periods (blue points) with predicted maximum periods (red line) for DAVs. Right: The same comparison, but for DBVs. In both cases, some mechanism is limiting observed pulsation periods to $\approx$ 1000 s.

should occur when convective driving exceeds radiative damping. This condition is met when the pulsation period P is less than P_{max}, where $P_{max} = 2\pi\tau_o$. In Fig. 4, we compare observed maximum pulsation periods (blue points) with the predicted maximum period (red line). Agreement is fairly good for hotter stars (low values of τ_o, but observed pulsation periods do not exceed $\approx$1000 s for cooler stars. Combined with our results from convective light curve fitting, an unknown mechanism is functioning to limit tau_o and keep observed pulsation periods from increasing as as predicted.

3. Summary and Implications

Convective light curve fitting uses the observed nonlinearities in pulsating white dwarf light curves to probe white dwarf convection in both hydrogen and helium atmospheres. Our results for hydrogen atmospheres show a plateau in the parameter tau_o, which is related to the thermal response timescale at the base of the convection zone. Our results show that no single value of the MLT parameter α is appropriate for all white dwarfs. Our results compare more favorably with recent 3D simulations, but some differences remain that require additional investigation. The situation is even more complex for the DBVs, where difficulties in temperature determinations, the existence of trace metals, and the possible presence of two convection zones complicate our analysis. A comparison of observed pulsation frequencies with theoretical predictions further argues that an unknown mechanism, acting to limit tau_o and keep observed pulsation periods from increasing as as predicted, is missing from our analysis.

References

Brickhill, A. J. 1991, *MNRAS*, 252, 334
Cukanovaite, E., Tremblay, P. E., Freytag, B. *et al.* 2019, *MNRAS*, 490, 1010
Montgomery, M. H. 2005, *ApJ*, 633, 1142
Montgomery, M. H., Provencal, J. L., *et al.* 2010, *ApJ*, 716, 84
Provencal, J. L. *et al.* 2012, *ApJ*, 751, 91
Tremblay, P. E., Ludwig, H. G., *et al.* 2015, *ApJ*, 799, 142

White Dwarfs as Probes of Fundamental Physics: Tracers of Planetary, Stellar and Galactic Evolution
Proceedings IAU Symposium No. 357, 2019
M. A. Barstow, S. J. Kleinman, J. L. Provencal & L. Ferrario, eds.
doi:10.1017/S1743921320000228

Accreting pulsating white dwarfs: Probing heating and rotation

Paula Szkody[1], Boris Gänsicke[2], Odette Toloza[2], Patrick Godon[3,4], Edward Sion[3], Stella Kafka[5], Keaton Bell[1], Zachary Vanderbosch[6] and AAVSO observers[4]

[1]Astronomy Department, University of Washington, Seattle, WA 98195
email: szkody@astro.washington.edu

[2]Department of Physics, University of Warwick, Coventry, CV4 7AL, UK

[3]Astrophysics and Planetary Science, Villanova University, Villanova, PA 19085

[4]Rowland Department of Physics & Astronomy, JHU, Baltimore, MD 21218

[5]AAVSO, 48 Bay State Road, Cambridge, MA 02138

[6]Department of Astronomy, University of Texas, Austin TX 78712

Abstract. The white dwarfs in close, interacting binaries provide a natural laboratory for exploring the effects of heating and angular momentum from the accreting material arriving on the surface from the companion. This study is even more fruitful when it involves a pulsating white dwarf, which allows an exploration of the effects of the accretion on the interior as well as in the atmosphere. The last decade has seen the accomplishment of UV (HST) and optical (ground) studies of several accreting white dwarfs that have undergone a dwarf nova outburst that heated the white dwarf and subsequently returned to its quiescent temperature. The most recent study involves V386 Ser, which underwent its first known outburst in January 2019, after 19 years at quiescence. V386 Ser is unique in that its quiescent pulsation shows a triplet, with spacing indicating a rotation period of 4.8 days, extremely slow for accreting white dwarfs. This paper presents the result of HST ultraviolet spectra obtained 7 months after its outburst that shows the first clear confirmation of shorter period modes being driven following the heating from a dwarf nova outburst.

Keywords. accretion, dwarf novae, white dwarfs

1. Introduction

Dwarf novae are close binaries consisting of a late main sequence star that is transferring material through the Roche lobe to a white dwarf via an accretion disk. In 18 known cases, the mass transfer is low enough that the accretion disk does not totally swamp the light of the white dwarf and non-radial pulsations are viewed. These accreting pulsators differ from single ZZ Ceti stars in several ways. They are generally hotter due to the accretion (12–16,000K) compared to the instability strip for ZZ Cetis (Gianninas, Bergeron & Fontaine 2007), they have mixed compositions due to the mass transfer (0.1 solar) compared to pure hydrogen for DAs or pure helium for DBs, they rotate faster ($\sim$200 sec) compared to the rotation of days for single white dwarfs, and they are more massive ($0.8M_{\odot}$) compared to single white dwarfs. The temperatures determined from UV spectroscopy with HST show that the instability strip for the pulsators is wide and contains dwarf novae with comparable temperatures that are not known to pulsate (Szkody *et al.* 2010). The width of the strip was explained by Arras, Townsley & Bildsten (2006) as due to helium enrichment of the atmosphere due to the mass transfer.

Table 1. Known Accreting Pulsators.

Name	Year	P_{orb} [min]	T[° K]	P_{pulse} [sec]
GW Lib	1998	76.8	15,000	236,376,648
V455 And	2004	81.1	10,500	320–370
GY Cnc	2004	81.5	14,500	335,581,595
V386 Ser	2004	80.5	14,500	221,346,609
SDSS2205+11	2004	82.8	15,000	330,475,575
EQ Lyn	2005	77.8	15,100	1192–1230
PQ And	2005	80.6	12,000	1358,1967,1988
LV Cnc	2005	81.3	13,500	214,260
MT Com	2005	119.5	12,000	668,1236,1344
V355 UMa	2006	82.5	12,500	641,1065
PP Boo	2006	88.8	10,000	559
OV Boo	2008	66.6	14,200	500,660,1140
EZ Lyn	2008	85.0	13,000	256,756
DY CMi	2011	85.6		238,684
BW Scl	2012	78.2	14,800	618,1242
SDSS1457+51	2012	77.9		582–642,1200
SDSS0755+14	2017	84.8	15,900	257–262
RXJ0232-37	2017	95.3	13,200	267

While it remains unclear why some of the pulsators appear to stop pulsating at random times, it is known that the increased accretion during a dwarf nova outburst heats the white dwarf (Sion 1995; Godon *et al.* 2006) so that it moves out of the instability strip. Subsequent cooling on a human observable timescale of months to a few years allows a unique probe of how efficiently the convection zone drives pulsations. The 18 known accreting pulsators are listed in Table 1, along with their discovery years, their orbital periods, white dwarf temperatures and observed pulsation periods at quiescence. Of these, four have been observed after a dwarf nova outburst, using HST and ground observations prior to 2019. Diverse results were obtained. EQ Lyn had one observation a year after outburst, but showed no pulsations then and returned to its quiescent pulsation period within 3 yrs (Mukadam *et al.* 2011). V455 And showed a 1 min shorter period at 2 yrs past its outburst but has complications as it contains a magnetic white dwarf (Szkody *et al.* 2013a). GW Lib is still cooling after its 2007 outburst, shows a sporadic changing mix of periods between 5 min, 19 min and 4 hrs (Toloza *et al.* 2016) and EZ Lyn only showed pulsations after its outburst (Szkody *et al.* 2013b). In 2019, V386 Ser underwent its first known dwarf nova outburst.

2. V386 Ser

V386 Ser was first discovered in the SDSS by Szkody *et al.* (2002) and pulsations were found by Woudt & Warner (2004). An international campaign over 11 nights showed the dominant pulsations period at 609 sec was an evenly spaced triplet, implying a very slow rotation of 4.8 days (Mukadam *et al.* 2011). A low resolution UV spectrum obtained with the Solar Blind Channel on HST could be fit with a quiescent white dwarf of about 14,000K, while simultaneous optical data showed the identical periods as in the UV with a UV/opt amplitude ratio of 6 indicating a low order mode (Szkody *et al.* 2007).

A mid-cycle proposal for UV spectra with COS on HST was approved to determine the heated temperature of the white dwarf, the pulsation mode during this heated interval, and whether the atmosphere was rotating as slowly as the quiescent mode splitting implied. An early attempt to observe failed due to spacecraft jitter but was successfully redone on August 15, 7 months after outburst. The AAVSO data showed V386 Ser had declined to about 0.5 mag above quiescence. The resulting spectrum can be generally fit with a white dwarf near 20,000±2000K, about 6000K hotter than at quiescence. However, the blue wing of Lyα is lower than models predict. If this is not due to calibration issues with the new grating setting of the G140L, it implies a high helium abundance, exactly what is predicted by Arras, Townsley & Bildsten (2006). The fit also shows that the outer layer of the white dwarf atmosphere is rotating at a fast rate, about 150±50 km/s, similar to most accreting white dwarfs, and not at the very slow rate implied by the pulsation splitting.

Using the time-tag data acquisition, a light curve was constructed by summing and binning the spectral data for the two HST orbits. The power spectrum of this light curve reveals a very strong signal at 104.28±0.05 sec with an amplitude of 2.5%. Ground based photometry obtained at McDonald Observatory on July 25 shows a similar period of 104.01±0.08 sec with an amplitude of 0.6%. After more than a decade of trying, this is the first clear confirmation of the driving of shorter period modes after a dwarf nova outburst. The period of 100 sec is also interesting in that it is what Arras, Townsley & Bildsten (2006) show to be driven for a temperature of ∼19,000K and a helium abundance Y = 0.58.

3. Conclusions

Our recent HST and ground observations of the accreting pulsating white dwarf in V386 Ser 7 months after its large amplitude dwarf nova outburst reveal a temperature about 6000K hotter than quiescence, a pulsation period about 6 times shorter than quiescence and an atmospheric rotation rate comparable to other accreting white dwarfs and much faster than the slow rotation rate implied from asteroseismology. Future observations as it continues to cool to its quiescent temperature can track how a deepening convection zone begins to drive pulsation modes with longer periods.

This work was partially supported by HST grant GO-14912, GO-15703 and NSF grant 1514737.

References

Arras, P., Townsley, D., & Bildsten, L. 2006, *ApJ*, 643, L119
Gianninas, A., Bergeron, P., & Fontaine, G. 2007, *ASP-CS.*, 372, 577
Godon, P. *et al.* 2006, *ApJ*, 642, 1018
Mukadam, A. S. *et al.* 2010, *ApJ*, 714, 1702
Mukadam, A. S. *et al.* 2011, *ApJ*, 728, L33
Sion, E. M. 1995, *ApJ*, 438, 876
Szkody, P. *et al.* 2002, *AJ*, 123, 430
Szkody, P. *et al.* 2007, *ApJ*, 658, 1188
Szkody, P. *et al.* 2010, *ApJ*, 710, 64
Szkody, P. *et al.* 2013a, *ApJ*, 775, 66
Szkody, P. *et al.* 2013b, *AJ*, 145, 121
Toloza, O. *et al.* 2016, *MNRAS*, 459, 3929
Woudt, P. & Warner, B. 2004, *MNRAS*, 348, 599

White Dwarfs as Probes of Fundamental Physics: Tracers of Planetary, Stellar and Galactic Evolution
Proceedings IAU Symposium No. 357, 2019
M. A. Barstow, S. J. Kleinman, J. L. Provencal & L. Ferrario, eds.
doi:10.1017/S1743921320001398

QPOs from post-shock accretion column of strongly magnetized accreting white dwarfs

Prasanta Bera
Mathematical Sciences and STAG Research Centre, University of Southampton,
Southampton SO17 1BJ, U.K.
email: pbera.phy@gmail.com

Abstract. An extended magnetosphere of a strongly magnetized accreting white dwarf (known as a polar) prevents the formation of an accretion disk and the matter is channelled to the magnetic pole(s). A few such sources show quasi-periodic oscillations in their optical light curves. These high-frequency oscillations are thought to be generated from the post-shock accretion column. The kinetic energy of the accretion flow is finally emitted from this post-shock region and the involved radiation processes decide the state of the matter. Here we study the structure and the dynamical properties of such accretion columns and compare the results with the observational characteristics.

Keywords. magnetic accretion, white dwarf, magnetic field, accretion column

1. Introduction

A set of polars (e.g. AN UMa, VV Puppis, V834 Cen) show quasi-periodic oscillations in their optical light curves in various observations (Middleditch 1982; Mouchet *et al.* 2017). The presence of cyclotron lines and Zeeman splitting in their line spectra indicates the signature of a strong magnetic field at the surface of the white dwarf. A strong magnetic field at the near surroundings of a white dwarf in a binary stellar system modifies the motion of the accretion flow. The ionized accretion flow gains kinetic energy as it moves closer to the white dwarf and follows the magnetic field lines within the magnetosphere of the compact star. The freely falling accretion matter impacts at the magnetic pole of the compact star after forming a shock. Along with the motion of the flow, the kinetic energy of the pre-shocked material is transformed into the thermal energy of the post-shock phase. This thermal energy is eventually radiated via some processes e.g. bremsstrahlung, emitted photons from the accelerated charged particles in a field of ions; and cyclotron, emission from the gyrating electrons about the magnetic field Chanmugam & Wagner (1979). The stellar parameters indicate the high energy X-ray photons generated via bremsstrahlung emission escape freely from the shocked matter but the optical photons cause the thermal radiation from the outer surface of the post-shock region.

The presence of rapid quasi-periodic oscillation in these systems indicates that they should originate from a region with a small length scale affected by high speed. Different studies show the characteristics of the emitted radiation from the post-shocked region (Langer *et al.* 1981). The theoretical models indicate the possibilities of QPOs in both the X-ray and optical energy bands as they are generated from the same post-shock region. However, X-ray observations have not confirmed any QPOs to date (Imamura *et al.* 2000; Bonnet-Bidaud *et al.* 2015). Hence to develop a consistent model for the dynamics of the post-shock requires us to study the overall post-shock region considering the involved radiation processes.

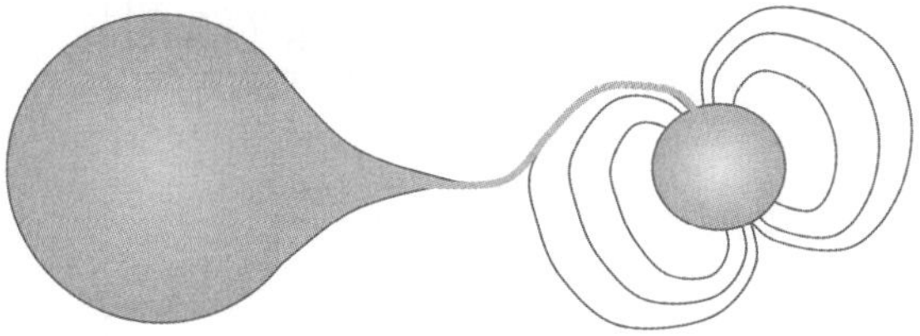

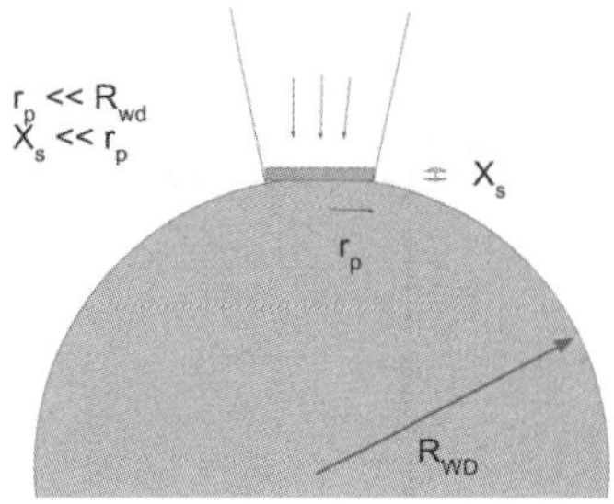

Figure 1. *Left*: A cartoon image of a binary stellar system where one of the stars is a white dwarf (smaller in size) with a strong magnetic field. The other star overflows matter which is being accreted by the white dwarf at its magnetic pole. *right*: The accretion column forms on the white dwarf surface at the magnetic pole. The distance of shock location from the surface x_s is very small compared to the length of the cross-section of the accretion column.

2. Model setup

We consider matter accreted from the companion star to one of the magnetic poles of the white dwarf without the formation of an accretion disk (Fig. 1). We further assume that at the magnetic pole, the field lines are vertical to the surface and we can exclude the local field geometry. The typical accretion parameters (e.g. inflow matter density, magnetic field strength) of the accreting polars suggest the height of the shock from the stellar surface is much less than the characteristic length scale of the cross-sectional area. In this condition, we study the vertical variations only.

The dynamics of accretion matter is described by the following hydrodynamic equations 2.1, 2.2 and 2.3 corresponding to the conservation of mass, momentum and energy respectively (Hameury *et al.* 1983).

$$\frac{\partial \rho}{\partial t} + \nabla \cdot (\rho \mathbf{v}) = 0 \tag{2.1}$$

$$\rho\left(\frac{\partial \mathbf{v}}{\partial t} + (\mathbf{v} \cdot \nabla)\mathbf{v}\right) = -\nabla p - \rho \nabla \Phi \tag{2.2}$$

$$\frac{\partial (E + \rho\Phi)}{\partial t} + \nabla \cdot (\mathbf{v}(E + p + \rho\Phi)) = -\rho \mathbf{v} \cdot \nabla \Phi - \Lambda \tag{2.3}$$

here ρ: matter density, $\mathbf{v}$: velocity, p: pressure, $E = \frac{1}{2}\rho \mathbf{v}\cdot\mathbf{v} + \rho\epsilon$: total energy density, $\epsilon = \frac{1}{\rho}\frac{p}{\gamma - 1}$: internal energy per unit mass with adiabatic index γ, Λ : cooling function, Φ : stellar gravitational potential, t: time and ∇: spatial gradient. We solve these equations for the provided boundary conditions in one-spatial dimension i.e. along the vertical direction.

3. Results

In the steady accretion condition, for the provided boundary values the structure can be obtained from the solution of the above equations neglecting the time derivative terms. For a negligible inflow velocity at the base of accretion column, Fig. 2 shows the vertical variation of the flow velocity, density and temperature in terms of the corresponding inflow values. In the strong shock assumption, the jump in density has a discontinuity of a factor of four. The flow density increases towards the stellar surface but the temperature reduces to balance the pressure and the cooling processes. The bremsstrahlung process becomes significant at the base of the column where matter density is higher

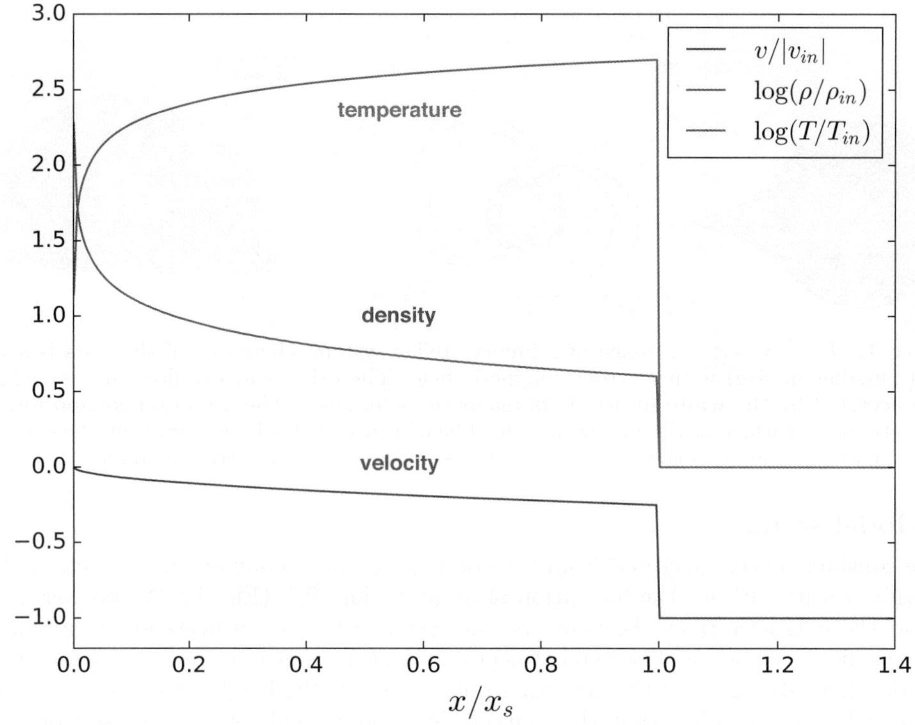

Figure 2. The steady equilibrium solution of velocity, density and temperature at the post-shock region corresponding to a fixed accretion inflow and a similar mass absorption rate at the base. Here, bremsstrahlung is the only cooling process considered.

as $\Lambda_{brem} \sim \rho^2 T^{1/2}$. The temperature-dependent cyclotron cooling ($\Lambda_{cycl} \sim \rho^{3/20} T^{5/2}$) becomes dominating just below the shock. Hence, the relative importance of the cooling processes decides the dynamics of the post-shock region (Bera & Bhattacharya 2018).

The temporal variability time scale is dependent on the cooling length and the inflow speed i.e. the characteristic cooling time scale. A perturbation of the steady solution indicates the characteristic oscillation time scale. We calculate the oscillation time scale from linear perturbation study and verify it from non-linear evolution (Mignone 2005; Bera & Bhattacharya 2018). The non-linear evolution case provides the oscillation amplitude that is used to compare with the observational characteristics. Fig. 3 shows the expected model-dependent bremsstrahlung and cyclotron rms amplitudes and QPO frequencies for different cross-sectional size of the accretion column. The stellar parameters are used similar to those for V834 Cen and the results show consistency with observations.

4. Summary

The vertical structure and the dynamical properties of the post-shock accretion column formed on the surface of a magnetic white dwarf are studied here. We found that the characteristic time of oscillation is connected to the involved radiation processes. A lower accretion rate or a larger surface coverage of the accretion column reduces QPO frequency. The rms amplitude of the oscillation is maximum when bremsstrahlung is the only dominating cooling process.

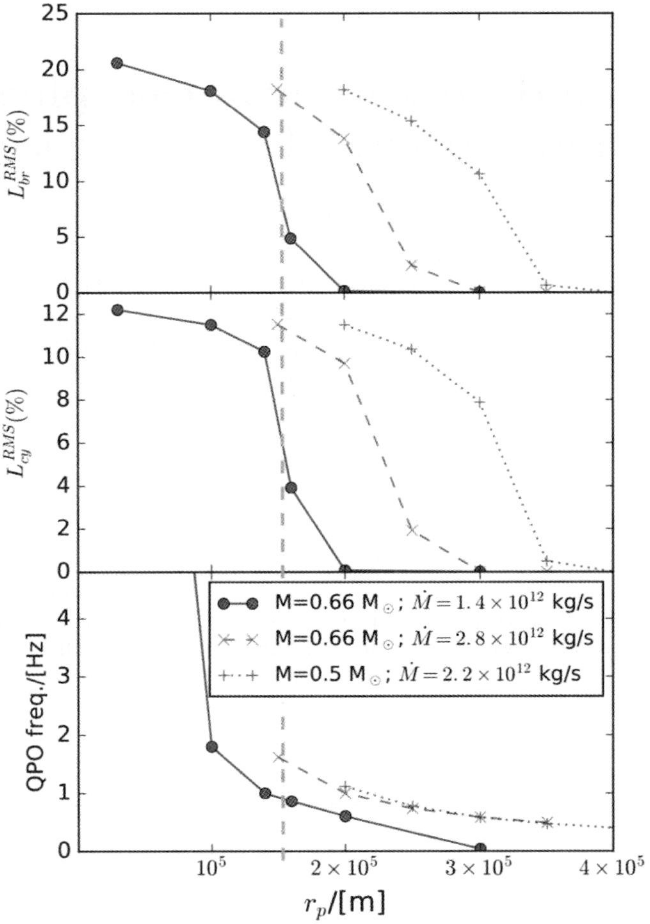

Figure 3. Expected rms variability in bremsstrahlung, cyclotron radiation and corresponding QPO frequency from a dynamical evolution study of the accretion column for different polar cap radii. The accretion parameters are similar to the source V834 Cen. The results corresponding to $r_p = 1.5 \times 10^5$ m on the magnetic pole of a 0.66 $M_\odot$ mass white dwarf with the accretion rate 1.4×10^{12} kg/s match well with the observed values.

Acknowledgement

I acknowledge the financial support from the organizers in terms of a travel grant to present this work at IAUS 357, Hilo.

References

Bera, P. & Bhattacharya, D. 2018, *MNRAS*, 474, 1629

Bonnet-Bidaud, J. M., Mouchet, M., Busschaert, C., *et al.* 2015, *A&A*, 579, A24

Chanmugam, G. & Wagner, R. L. 1979, *ApJ*, 232, 895

Hameury, J. M., Bonazzola, S., Heyvaerts, J., *et al.* 1983, *A&A*, 128, 369

Imamura, J. N., Steiman-Cameron, T. Y., & Wolff, M. T. 2000, *PASP*, 112, 18

Langer, S. H., Chanmugam, G., & Shaviv, G. 1981, *ApJ*, 245, L23

Middleditch, J. 1982, *ApJ*, 257, L71

Mignone, A. 2005, *ApJ*, 626, 373

Mouchet, M., Bonnet-Bidaud, J.-M., Van Box Som, L., *et al.* 2017, *A&A*, 600, A53

White Dwarfs as Probes of Fundamental Physics: Tracers of Planetary, Stellar and Galactic Evolution
Proceedings IAU Symposium No. 357, 2019
M. A. Barstow, S. J. Kleinman, J. L. Provencal & L. Ferrario, eds.
doi:10.1017/S1743921320000873

White dwarfs as advanced physics laboratories. The axion case

Jordi Isern[1,2]

[1]Institut de Ciéncies de l'Espai (ICE,CSIC),
c/Can Magrans, Campus UAB, 08193 Cerdanyola, Spain
email: isern@ice.cat

[2]Institut d'Estudis Espacials de Catalunya (IEEC),
Edifici Nexus 201, c/Gran Capità 2-4, 08034 Barcelona, Spain
email: isern@ieec.cat

Abstract. The shape of the luminosity function of white dwarfs (WDLF) is sensitive to the characteristic cooling time and, therefore, it can be used to test the existence of additional sources or sinks of energy such as those predicted by alternative physical theories. However, because of the degeneracy between the physical properties of white dwarfs and the properties of the Galaxy, the star formation history (SFH) and the IMF, it is almost always possible to explain any anomaly as an artifact introduced by the star formation rate. To circumvent this problem there are at least two possibilities, the analysis of the WDLF in populations with different stories, like disc and halo, and the search of effects not correlated with the SFH. These procedures are illustrated with the case of axions.

Keywords. white dwarfs, axions

1. Introduction

Often non-standard theories predict the existence of particles which existence and properties cannot be tested in the terrestrial laboratories as a consequence of the large energies involved. One way to alleviate the problem is the use of stars to constrain their properties, either looking for a decay or a change of these properties during the trip, or examining the perturbations they introduce into the *normal* evolution of stars (Raffelt 1996).

Because the evolution of white dwarfs is a relatively simple process of cooling, the basic ingredients necessary to predict their behavior are well identified, and there is a solid observational background to test the theoretical results, these stars have proved to be excellent laboratories for testing new ideas in Physics (Isern & García-Berro 2008). This procedure has allowed to put bounds on the mass of axions (Raffelt 1986; Isern *et al.* 1992, 2008), on the neutrino magnetic momentum (Blinnikov & Dunina-Barkovskaya 1994), the secular drift of the Newton gravitational constant (Vila 1976; García-Berro *et al.* 1995), the density of magnetic monopoles (Freese 1984) and WIMPS (Bertone & Fairbairn 2008), as well as constraints on properties of extra dimensions (Malec & Besiada 2001), on dark forces (Dreiner *et al.* 2013), on modified gravity (Saltas *et al.* 2019), and formation of black holes by high energy collisions (Giddings & Mangano 2008). In this talk only axions will be discussed.

According to the standard theory, white dwarfs are the last stage of the evolution of low and intermediate mass stars. Since their core is almost completely degenerate, they cannot obtain energy from nuclear reactions and their evolution is just a process of contraction and cooling†. The main sources of energy are the gravo-thermal readjustment of the structure, represented by the first two terms of the r.h.s. of Equation 1.1, the gravitational settling of heavy species like ^{22}Ne, g_s, the latent heat and sedimentation associated to crystallization, times the crystallization rate $\dot{m}_s$ and any other exotic source or sink of energy ($\dot{\varepsilon}_e$). The l.h.s. of Equation 1.1 contains the sinks of energy, photons and neutrinos. This equation has to be complemented with a relationship connecting the temperature of the core with the luminosity of the star. Typically $L \propto T_c^{\alpha}$ with $\alpha \approx 2.5 - 2.7$.

$$L + L_\nu = -\int_{M_{WD}} c_V \frac{dT_C}{dt} dm - \int_{M_{WD}} T\left(\frac{\partial P}{\partial T}\right)_{V,x} \frac{dV}{dt} dm + g_s + (l_s + e_s)\,\dot{m}_s \pm (\dot{\varepsilon}_e) \quad (1.1)$$

1.1. *Energy losses by neutrinos*

The importance of neutrino losses in the evolution of white dwarfs was early recognized by Vila (1965) and Savedoff *et al.* (1969). During the first stages of cooling, when the star is still very hot, the energy losses are dominated by the plasma and photo neutrino processes and, as soon as the temperature decreases, by neutrino bremsstrahlung (Iben & Tutukov 1984; D'Antona & Mazzitelli 1989). Besides their role in the cooling, neutrinos force the thermal structures produced by AGB stars to converge towards a unique one, guaranteeing in this way the uniformity of models dimmer than $\log(L/L_\odot) < -1.5$. However, in spite of the enormous progress experienced by the physics of neutrinos, several questions still remain. For instance, are neutrino Dirac or Majorana particles, do sterile neutrinos exist, which is their mass spectrum, do they have magnetic momentum?. This last problem, for instance, is specially important since the existence of a magnetic dipole momentum can notably enhance the neutrino losses in white dwarfs (Blinnikov & Dunina-Barkovskaya 1994).

1.2. *Influence of the DA non-DA character*

The luminosity strongly depends on the properties of the envelope (mass, chemical composition and structure) as well as on the total mass and radius of the white dwarf. The main characteristics of the envelope is its tendency to become stratified, the lightest elements tending to be placed on top of the heaviest ones as a consequence of the strong gravitational field. However, this behavior is counterbalanced by convection, molecular diffusion and other processes that tend to restore the chemical homogeneity. In any case, the $\sim 80\%$ of white dwarfs shows the presence of H–lines in their spectra while the remaining $\sim 20\%$ not. This proportion is not constant along the cooling sequence. The first ones are generically called DAs and the second ones non–DAs. The most common interpretation is that the DAs have a double layered envelope made of H ($M_{\rm H} \sim 10^{-4} M_{\rm WD}$) and He ($M_{\rm He} \sim 10^{-2} M_\odot$) while the non-DAs have just a single He layer or an extremely thin H layer. An additional complication is that the initial conditions at the moment of formation are not well known and for the moment it is not possible to disentangle which part of this behavior is inherited and which part is evolutive, although probably both are playing a role (Althaus *et al.* 2010). In principle, it is possible to adjust the parameters of the AGB progenitors to obtain 25% of white dwarfs completely devoided of the hydrogen layer. But, since the relative number of DA/non-DA stars changes during their evolution, a mechanism able to transform this character must exist (Shipman 1997).

† See Iben & Tutukov (1984); Koester & Shoenberner (1986); D'Antona & Mazzitelli (1989); Isern *et al.* (1998); Fontaine *et al.* (2001); Hansen (2004); Althaus *et al.* (2010); García-Berro & Oswalt (2016) for detailed descriptions of the cooling process.

It is commonly accepted that DAs start as the central star of a planetary nebula and asteroseismological data suggest they are born with a hydrogen layer of a mass in the range of $10^{-8} - 10^{-4}$ $M_{\odot}$. As the star cools down, the outer convection zone deepens and, depending on the mass, completely mixes the hydrogen layer with the larger helium layer in such a way that DAs turn out into non-DAs and, consequently, the ratio DA/non-DA decreases with the effective temperature.

The evolution of non-DAs is more complex. They are born as He-rich central stars of planetary nebulae and, as they cool down they look as PG 1159 stars first and DO after. The small amount of hydrogen present in the envelope floats up to the surface and when the temperature is $\sim 50,000$ K forms an outer layer thick enough to hide the helium layer to the point that the star becomes a DA. When the temperature goes below 30,000 K, the convective helium layer engulfs the hydrogen one and the white dwarf recovers the non-DA character, now as a DB, and, as it continues to cool down, it becomes a DC. Notice that a fraction of DCs has a DA origin. Besides the phenomenological differences between DA and non-DA families, the most important property is that they cool down at different rates since hydrogen is more opaque than helium.

2. The axion case

The so called *strong CP problem*, i.e. the existence in the Lagrangian of Quantum Chromodynamics of a term, not observed in Nature, that violates the charge-parity symmetry, is one of the most important problems that has to face the Standard Model. One possible solution consists on the introduction of a new symmetry that breaks at energies of the order of $f_a \sim 10^9 - 10^{11}$ GeV and gives raise to a new particle, the *axion*. This particle is a boson and has a mass $m_{\rm a} = 6\,{\rm meV}(10^9{\rm GeV}/f_{\rm a})$ that is not fixed by the theory. The larger is the mass, the larger is the interaction with matter (Raffelt 1996).

The interaction with photons and fermions is described as:

$$L_a = -\frac{1}{4} g_{a\gamma} F_{\mu\nu} \tilde{F}^{\mu\nu} - \sum_{fermions} g_{ai} a \overline{\psi_i} \gamma_5 \psi_i \tag{2.1}$$

where $g_{a\gamma\gamma}({\rm GeV}^{-1})$ and g_{ai} are the corresponding coupling constants, F is the electromagnetic field tensor and γ_5 is the corresponding Dirac field. The values that these constants take are model dependent.

There are several models of axions (Dias *et al.* 2014). Here only the DFSZ one is considered (Dine, Fishler & Srednick 1981; Zhimitskii 1980) since it predicts a profuse production of axions in the hot and dense interior of white dwarfs as a consequence of the interaction with electrons. In this model, the adimensional axion-electron coupling constant is related to the mass of the axion through

$$g_{\rm ae} = 2.8 \times 10^{-14} m_{\rm a}[{\rm meV}] \cos^2 \beta \tag{2.2}$$

where $\cos^2 \beta$ is a free parameter. For white dwarfs in the luminosity range $8.5 \leqslant M_{\rm bol} \leqslant 12.5$ the production of neutrinos and axions is dominated by bremsstrahlung.

The emission of thermal axions is similar to that of thermal neutrinos†. There are, however, subtle differences introduced by the boson character of axions. One is that when the white dwarf cools down, neutrino emission, $\dot{\epsilon}_{\rm bremss} \propto T^7$ (Itoh *et al.* 1996), is quenched before axionic emission, $\dot{\epsilon}_{\rm a} \propto T^4$ (Nakagawa *et al.* 1987, 1988), as it can be seen in Fig. 1‡. The other effect is that when white dwarfs are hot, axion emission modifies the temperature profile and reduces the neutrino losses (Miller Bertolami *et al.* 2014).

† With the exception of the Primakoff effect.

‡ Altherr *et al.* (1994) have claimed that these values were understimated by a factor ~ 4. If this were correct this would mean that g_{ae} should be divided by a factor ~ 2

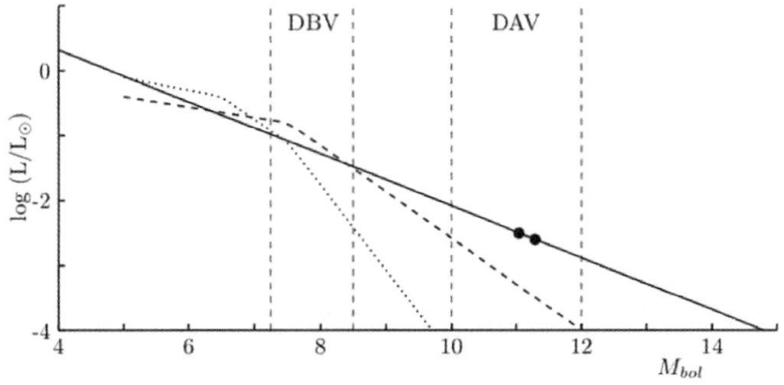

Figure 1. Evolution of the photonic (solid), nautrinic (dotted), and axionic (dashed) luminosities as the white dwarf fades. The vertical bands display the loci of the different families of variables.

3. Variable white dwarfs

During the process of cooling, white dwarfs cross some regions of the H-R diagram where they become unstable and pulsate. The multifrequency character and the size of the period of pulsation ($10^2 - 10^3$ s) indicate that they are g-mode pulsators†. As a variable white dwarf cools down, the oscillation period, P, changes as a consequence of the changes in the thermal and mechanical structure. This secular drift can be approximated by (Winget *et al.* 1983):

$$\frac{\dot{P}}{P} = -a\frac{\dot{T}}{T} + b\frac{\dot{R}}{R} \tag{3.1}$$

where a and b are positive constants of the order of unity. The first term of the r.h.s. reflects the decrease of the Brunt-Väisälä frequency with the temperature, while the second term reflects the increase of the frequency induced by the residual gravitational contraction.

There are at least three types of variable white dwarfs, the DOV, DBV, and DAV. In the first case, gravitational contraction is still significant and the second term of Eq. 3.1 is not negligible, for which reason $\dot{P}$ can be positive or negative. The DBV stars are characterized by the lack of H-layer, their effective temperatures are in the range of $23,000 - 30,000$ k and the secular drift is always positive and in the range of 10^{-13}–10^{-14} ss^{-1}. The DAV white dwarfs, also known as ZZ Ceti stars, are characterized by the presence of a thin atmospheric layer of pure hydrogen. Their effective temperature is in the range of $12,000 - 15,000$ K and their period drift, always positive, is of the order of $\sim 10^{-15}$ ss^{-1}. Therefore, these secular drifts can be used to test the predicted evolution of white dwarfs and, if the models are reliable enough, to test any physical effect able to change the pulsation period of these stars. In order of magnitude,

$$\frac{L_0 + L_x}{L_0} \approx \frac{\dot{P}_{obs}}{\dot{P}_0} \tag{3.2}$$

where $\dot{P}_{obs}$ is the observed period drift, L_0 and $\dot{P}_0$ are obtained from standard models and L_x is the extra luminosity necessary to fit the observed period (Isern *et al.* 1992).

G117-B15A is a ZZ Ceti star discovered by McGraw & Robinson (1976) that has been monitored since then. The first measurement of $\dot{P}$ gave a value that was a factor two larger than expected (Kepler *et al.* 1991). The temperature of such star is low enough to neglect the radial term of Eq. 3.1 and the influence of neutrinos. These two facts led to Isern *et al.* (1992) to postulate axions of the DSKZ type with $g_{ae} \approx 2.2 \times 10^{-13}$ as responsible

† Radial pulsations have shorter periods, $P \sim 10$ s.

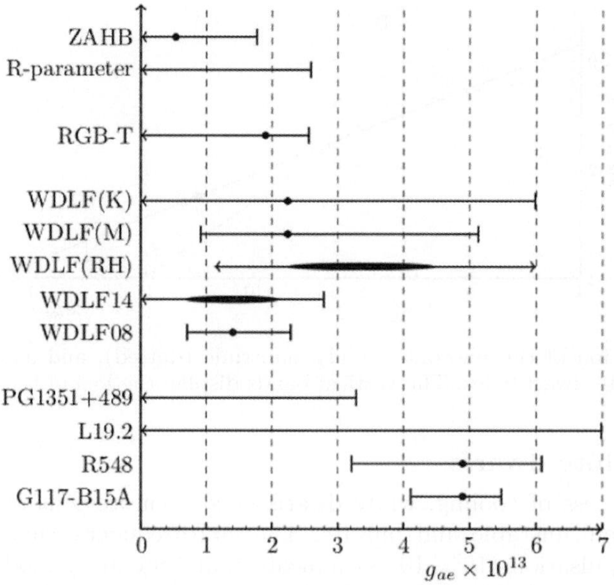

Figure 2. Axion-electron strengths obtained from different stellar constraints.

of the anomalous cooling. The analysis of Córsico *et al.* (2001) and Bischoff-Kim *et al.* (2008) showed that the presence of axions accelerates the drift but not changes the period of pulsation. The most recent estimation of the drift of the 215 s pulsation period is $\dot{P}_{\text{obs}} = (4.07 \pm 0.61) \times 10^{-15}$ s s^{-1} (Kepler *et al.* 2012), which suggests $g_{\text{ae}} = 4.9 \times 10^{-13}$ (Fig. 2) from full model fitting (Córsico *et al.* 2012a).

Similar anomalous drifts have been found in R548, the ZZ Ceti star itself (Mukadam *et al.* 2013), and in L19.2 (Sullivan & Chote 2015), which could be accounted introducing axions with $g_{\text{ae}} \times 10^{13} = 4.8$ and < 7 respectively (Córsico *et al.* 2012b, 2016).

These analysis indicate that the cooling rate is larger than expected if the pulsation modes are trapped in the outer envelope. This poses a problem since the trapping strongly depends on the detailed chemical gradients and the chemical structure of these layers, which are built during the AGB phase, is very sensitive to the methods used to treat convection and pulses during this epoch. Other problems come from the fact that these regions are partially degenerate and not all the physical inputs, specially axion emissivities, are correctly computed in this regime. Furthermore, there are still many uncertainties in the equation of state and opacities.

In the case of DB white dwarfs the drift has been measured in PG1351+489 (Redaelli *et al.* 2011), $\dot{P}_{\text{obs}} = (2.0 \pm 0.9) \times 10^{-13}$ s s^{-1}, which provides a bound of $g_{ae} \times 10^{13} < 3.3$ (Battich *et al.* 2016). See Figure 2. Notice that at these temperatures, neutrinos are still active and their emission can be affected by axions or even by the existence of a hypothetical magnetic momentum.

4. The luminosity function

The luminosity function (LF) is defined as the number density of white dwarfs of a given luminosity per unit magnitude interval:

$$n\left(L\right) = \int_{M_l}^{M_u} \Phi\left(M\right) \Psi\left(T_G - t_{cool} - t_{ps}\right) \tau_{cool} dM \tag{4.1}$$

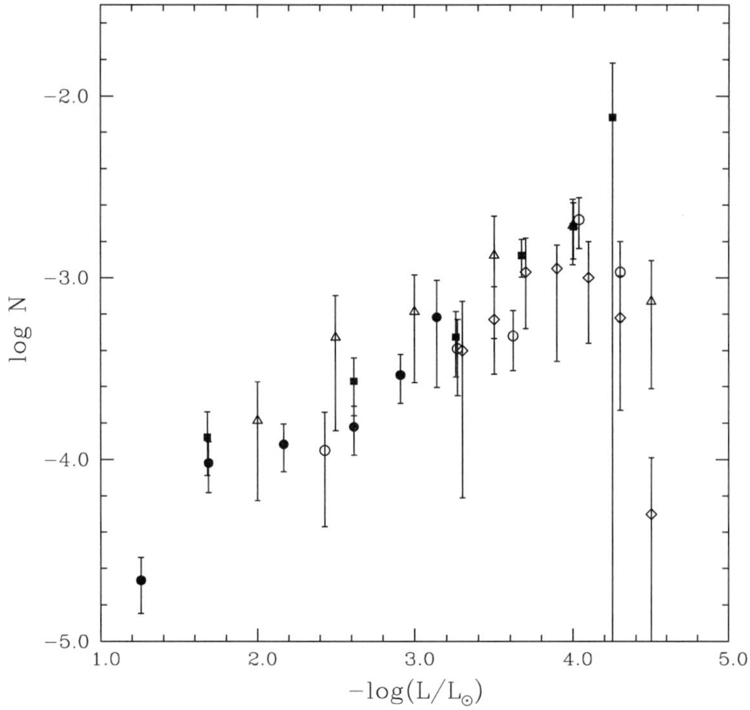

Figure 3. Luminosity functions obtained before the large cosmological surveys – Liebert, Dahn & Monet (1988), full circles; Evans (1992), full squares; Oswalt *et al.* (1996), open triangles; Legget *et al.* (1998), open diamonds; Knox, Hawkins & Hambly (1999), open circles.

where M is the mass of the parent star (for convenience all white dwarfs are labeled with the mass of the zero age main sequence progenitor), t_{cool} is the cooling time down to luminosity L, $\tau_{cool} = dt/dM_{bol}$ is the characteristic cooling time of the white dwarf, t_{ps} is the lifetime of the progenitor of the white dwarf and T_G is the age of the Galaxy or the population under study, and M_u and M_l are the maximum and the minimum mass of the main sequence stars able to produce a white dwarf, therefore M_l satisfies the condition $T_G = t_{cool}(L, M_l) + t_{ps}(M_l)$. $\Phi(M)$ is the initial mass function and $\Psi(t)$ is the star formation rate (SFR) of the population under consideration. Additionally, hidden, there is an initial-final mass function connecting the properties of the progenitor with those of the white dwarf. In order to compare theory with observations, and since the total density of white dwarfs is not yet well known, the computed luminosity function is usually normalized to the bin with the smaller error bar, usually $\log L/L_\odot \simeq 3$.

This equation contains three sets of terms, the observational ones, $n(L)$, the stellar ones, $t_{\rm cool}, \tau_{\rm cool}, t_{\rm PS}, M_{\rm U}, M_{\rm i}$, plus the initial final mass function (Catalan *et al.* 2008), and the galactic ones Φ and Ψ.

The first empirical luminosity function was obtained by Weidemann (1968) and was improved by several authors during the 1990s (Figure 3) proving in this way that the evolution of white dwarfs was just a cooling process and that there was a cut-off in the distribution caused by the finite age of the Galaxy. The position of the cut-off is sensitive to the cooling rate and, consequently, it can be used to constrain any new theory or hypothesis implying the introduction of an additional source or sink of energy. However the low number of stars in the samples, few hundreds, and the uncertainties in the position of the cut-off prevented anything else than obtaining upper bounds.

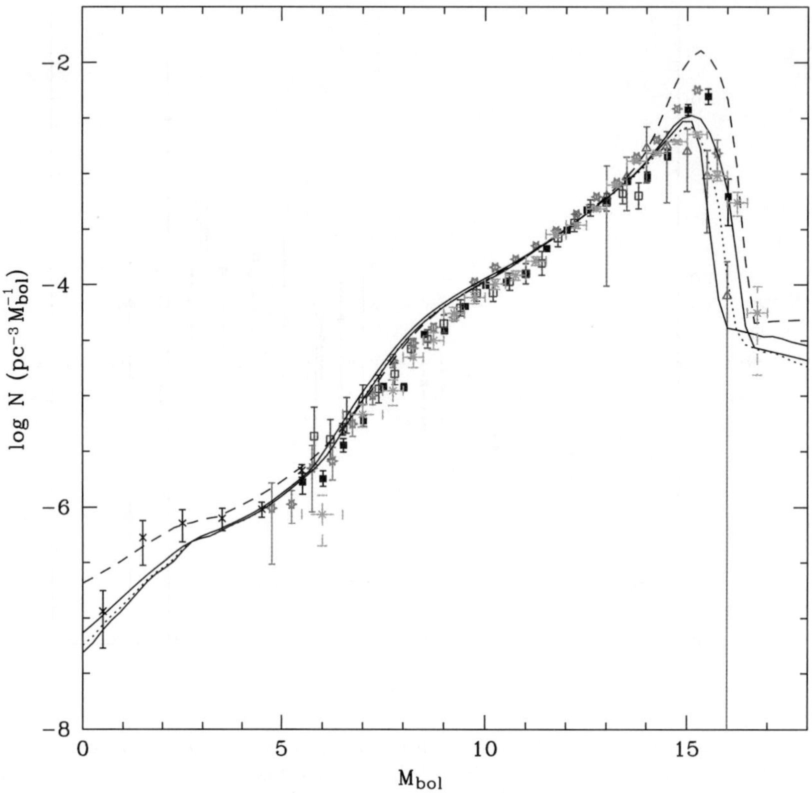

Figure 4. Luminosity functions obtained from the SDSS catalogue: Black solid squares (Harris *et al.* 2006), open blue squares, only DAs (DeGennaro *et al.* 2008), black crosses (Krzesinski *et al.* 2009), and green stars (Munn *et al.* 2017). Magenta stars were obtained from the SCSS catalogue (Rowell & Hambly (2011)) and contain DA and non–DA stars. The two solid black lines are the theoretical luminosity function, obtained assuming a constant SFR and an age of 10 and 13 Gyr, respectively, while the dashed line corresponds to a decreasing exponential SFR and the dotted line to a constant SFR with an exponential tail.

The advent of large cosmological surveys like the Sloan Digital Sky Survey (SDSS) and the Super COSMOS Sky Survey, both completely independent, introduced a noticeable improvement in the precision and accuracy of the luminosity function since they allowed to increase the sample size to several thousands of stars. Figure 4 displays both functions normalized to the $M_{\rm bol} \approx 12$ bin. As can be seen, they nearly coincide over a large part of the brightness interval. At large brightness, $M_{\rm bol} < 6$, both luminosity functions show a large dispersion, not plotted in the figure, as a consequence of the fact that the proper motion method is not appropriate there. One way to circumvent this problem relies on the UV-excess technique (Krzesinski *et al.* 2009). The results obtained in this way are represented by black crosses after matching their dim region with the corresponding bright segment of the Harris *et al.* (2006) distribution. As a complement, the luminosity function of the dimmest white dwarfs obtained by Legget *et al.* (1998) has been included (red triangles). The discrepancies at low luminosities are due to the difficulty to separate DAs from non-DAs and to the different behavior of the envelope.

The quality of these new luminosity functions allows, for the first time, a determination of their shape and the use of the slope as a tool to test new physical theories. If an additional source or sink of energy is added, the characteristic cooling time is modified and its imprint appears in the luminosity function, as can be seen in Fig. 4, where a change of slope is evident at magnitudes $M_{\rm bol} \sim 8$. This change is caused by the transition from a cooling dominated by neutrinos to one dominated by photons. As an example, this technique was used by Isern *et al.* (2008) to suggest that axions of the DFSZ type could be contributing to the cooling of white dwarfs.

The main problem when using Eq. 4.1 is that the star formation rate has to be obtained independently from white dwarfs in order to break the degeneracy between the galactic properties and stellar models. Fortunately, the luminosity function has an important property. The shape of the bright branch, stars brighter than $M_{bol} \approx 13$, is almost independent of the assumed star formation rate as it can be seen in Fig. 4 where several theoretical white dwarf luminosity functions are displayed. The two solid black lines have been obtained assuming a constant SFR but two ages of the Galaxy, 10 and 13 Gyr respectively, the dashed line assuming a decreasing exponential SFR, $\Psi \propto \exp(-t/\tau)$, $\tau = 3\,{\rm Gyr}$, where t is the age of the Galaxy, and the dotted line assuming an almost constant SFR with an exponentially decreasing tail that represents models where the star formation propagated from the center to the periphery, $\Psi \propto (1+\exp[(t-t_0)/\tau])^{-1}$, $\tau = 3\,{\rm Gyr}$, $t_0 = 10\,{\rm Gyr}$, where t is the looking back time. As can be seen, in the region $M_{\rm bol} \approx 6 - 13$, all luminosity functions overlap as far as the SFR is smooth enough. The differences due to the shape of the SFR only appear in the regions containing cool or very bright white dwarfs. Unfortunately, the observational uncertainties in these regions prevent at present to discriminate among the different possibilities.

This behavior can be understood in the following way: Eq. (4.1) can be written as:

$$n\,(l) \propto \langle \tau_{cool} \rangle \int_{M_i}^{M_{\rm max}} \Phi\,(M)\,\Psi\,(T_G - t_{cool} - t_{ps})dM \tag{4.2}$$

Restricting this equation to the bright white dwarfs – namely, those for which $t_{\rm cool}$ is small – the lower limit of the integral is satisfied by low-mass stars and, as a consequence of the strong dependence of the main sequence lifetimes with mass, it adopts a value that is almost independent of the luminosity under consideration. Therefore, if Ψ is a well-behaved function and T_G is large enough, the lower limit is almost independent of the luminosity, and the value of the integral is incorporated into the normalization constant in such a way that the shape of the luminosity function only depends on the averaged physical properties of the white dwarfs (Isern & García-Berro 2008). This average is dominated by low mass white dwarfs and, as far as the mass spectrum is not strongly perturbed by the adopted star formation or the initial mass function, it is representative of the intrinsic properties of white dwarfs (Isern *et al.* 2009). This shape, however, can be modified by a recent burst of star formation since, in this case, low-mass main sequence stars have no time to become white dwarfs and M_I in Eq. 4.2 becomes luminosity dependent. On the contrary, if the burst is old enough, the corresponding luminosity functions are barely modified.

Another important property is that in the bright region considered here, the slope of the relationship between the luminosity and the core temperature of DA and non–DA white dwarfs almost coincide, and both luminosity functions almost overlap in this luminosity interval after normalization. This is the reason why the luminosity function of DeGennaro *et al.* (2008) containing only DAs coincides after normalization with those containing DAs and non–DAs as can be seen in Fig. 4. Therefore, Eq. 4.2 offers the possibility to use the slope of the bright branch of the luminosity function to detect the presence of unexpected additional sinks or sources of energy in white dwarfs. In the case of axions, this method was used for the first time by Isern *et al.* (2008) who obtained $g_{ae} = (1.4^{+0.9}_{-0.8}) \times 10^{-13}$, see WDLF08 label in Fig. 2, using a preliminary version of the Harris *loc cit.* Miller Bertolami *et al.* (2014) reexamined this result using all the luminosity functions of

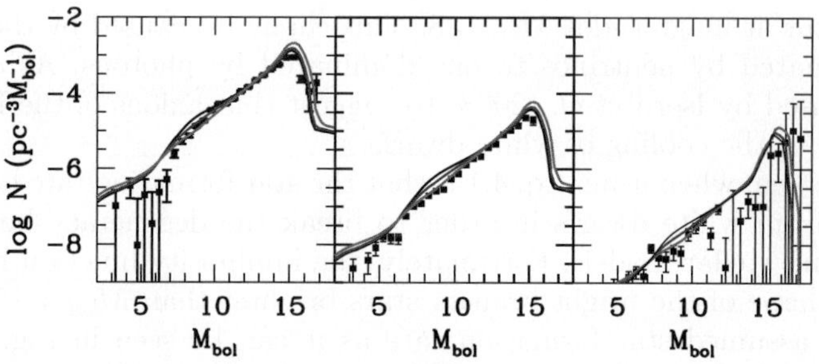

Figure 5. Luminosity functions of DA and non-DA white dwarfs of the thin and thick discs and halo (Rowell & Hambly 2011). The black line represent the case without axions. The blue and red lines correspond to the cases where DFSZ axions with coupling constants $g_{ae} \times 10^{13} = 2.24$ and 4.48 are included in the cooling model.

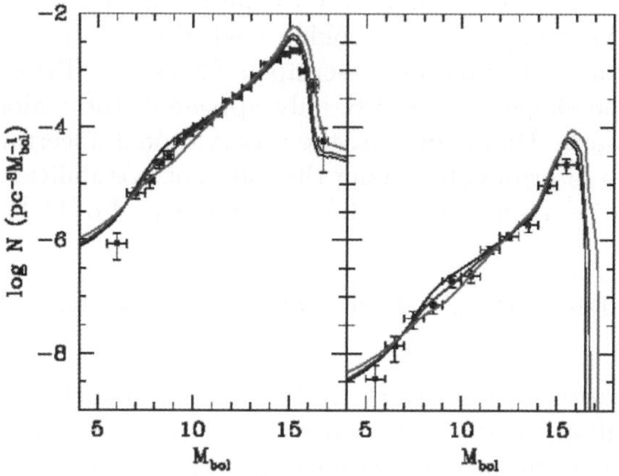

Figure 6. Luminosity functions of DA and nDA white dwarfs belonging to the disk (thin and thick) and the halo (Munn *et al.* 2017). The meaning of the solid lines is the same as in Fig. 5.

Fig. 4 (but with a consolidated Harris *loc cit.* luminosity function) and with a self consistent treatment of the neutrino cooling and concluded that DFSZ axions with a coupling constant g_{ae} in the range of $(0.7-2.1) \times 10^{-13}$ could exist (WDLF14 label in Fig. 2). All these results, however, have to be regarded as qualitative, since the uncertainties plaguing the determination of both, observed and computed, luminosity functions are still large.

Given the degeneracy between the stellar and galactic terms in Eq. 4.2, it is natural to wonder if the changes in the shape of the luminosity function attributed to axions are an artifact introduced by the star formation rate. One way to break this degeneracy and decide if it is necessary to introduce or not new physical effects is to examine the luminosity function of populations that have different formation histories. If axions exist and can modify the cooling of white dwarfs, their imprint will be present in all the luminosity functions at roughly the same luminosities. Furthermore, it is well known that the adopted white dwarf scale height above the galactic plane s has a noticeable effect on the shape of the luminosity function (García-Berro *et al.* 1988; Harris *et al.* 2006; Kilic *et al.* 2017). This argument reinforces the convenience of analyzing the luminosity function of white dwarfs with different scale heights and independent star formation histories (Isern *et al.* 2018).

Rowell & Hambly (2011) provided, for the first time, a luminosity function for the thin and thick disks and noticeably improved that of the halo (Fig. 5). Table 1 shows

Table 1. Reduced χ^2 obtained from the fitting of the observed white dwarf luminosity functions in the brightness interval $6 < M_{\rm bol} < 12.5$ with different intensities of the coupling between electrons and axions (Isern *et al.* 2008).

$g_{\rm ae} \times 10^{13}$	0.00	1.12	2.24	4.48
$m_{\rm a} \cos^2 \beta$ (meV)	0	4	8	16
Rowell & Hambly (2011)				
thin disk	18.59	15.33	7.33	21.52
thick disk	30.72	26.61	11.73	1.43
halo	3.31	2.94	2.36	1.87
Munn *et al.* (2017)				
thin+thick disk	4.89	2.87	1.46	6.13
halo	2.46	1.50	0.65	1.42
Kilic *et al.* (2017)				
$\Phi_{200-900}$	3.25	2.09	1.05	6.11
$\Phi_{200-700}$	3.53	2.32	1.14	5.80
$\Phi_{200-500}$	4.02	2.73	1.28	2.26
Φ_{300}	6.11	4.95	2.26	4.98

how the discrepancies between theoretical calculations and observations decrease in all three cases when axions are included. The fit favors axions within the interval $g_{ae} = (2.24 - 4.48) \times 10^{-13}$ with some tension between the halo and the disk results, although the $2\,\sigma$ bounds are compatible. See Fig. 2, label WDLF(RH).

Munn *et al.* (2017) improved the Harris *et al.* (2006) luminosity function resolving the peak and increasing the precision of the brighter branch assuming a constant scale height above the Galactic plane (Fig. 6). They also computed the luminosity function for the halo selecting white dwarfs with $200 \leqslant v_{\rm tan} \leqslant 500$ km s^{-1}. As before, the inclusion of axions improves the concordance between theory and observations both in the disk and the halo. The best fit is obtained for $g_{ae} \approx 2.24 \times 10^{-13}$, label WDLF(M) Fig. 2, with ($2\,\sigma$) upper bounds of $g_{ae} < (4.2 \text{ and } 14) \times 10^{-13}$ coming, respectively, from the disk and halo luminosity functions.

Kilic *et al.* (2017) reexamined the Munn *et al.* (2017) data assuming three variable scale heights above the galactic plane going linearly from 200 pc now to 900, 700 and 500 pc in the past, respectively, and one with a fixed scale height of 300 pc. They concluded that the slight discrepancies in the region $6 \leqslant M_{\rm bol} \leqslant 12.5$ were caused by the use of a fixed scale height. This argument is correct but, as it was shown by Isern *et al.* (2008), see Fig. 7, the inclusion of axions improves the agreement. In these cases the best fit was also around $g_{ae} \approx 2.24 \times 10^{-13}$, Table 1, and a $2\,\sigma$ global upper bound $g_{ae} < 6 \times 10^{-13}$ (Fig. 2, label WDLF(K)).

Recently a second possibility has appeared. Tremblay *et al.* (2019) have been able to build a reliable and precise luminosity function of massive white dwarfs that belong to the solar neighborhood ($d \leqslant 100$ pc) using the data provided by Gaia (Fig. 8, left panel). If the luminosity function, Eq. 4.1, is restricted to massive white dwarfs, i.e. those for which it is possible to neglect the lifetime of the progenitor in front of the cooling time, and Ψ is smooth enough, the age of any bin and the star formation rate corresponding to this time can be computed as

$$\langle t \rangle = \frac{\int_{\Delta M} \Phi(M) t dM}{\int_{\Delta M} \Phi(M) dM} \tag{4.3}$$

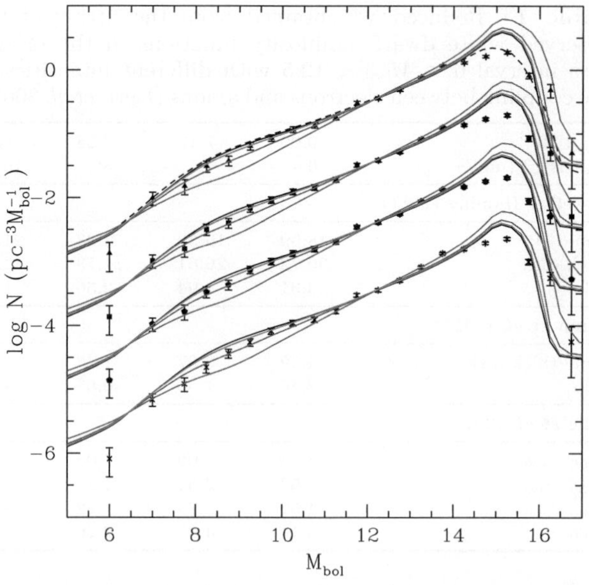

Figure 7. White dwarf luminosity functions of the disk assuming different scale heights as proposed by Kilic *et al.* (2017). From top to bottom $\Phi_{200\text{–}900}+3$, $\Phi_{200\text{–}700}+2$, $\Phi_{200\text{–}500}+1$ and Φ_{300}. Solid lines correspond to $g_{ae}\times 10^{13}=0, 1.12, 2.24, 4.48$ (black, magenta, blue, red, respectively). The dashed line corresponds to a case with no axions and an SFR constant plus an exponential tail as in Fig 4.

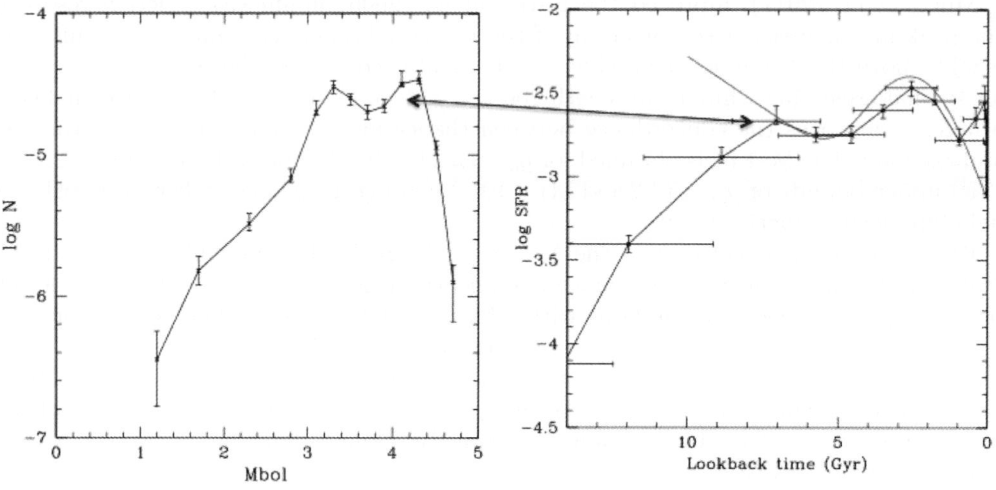

Figure 8. Left panel: empirical luminosity function of massive ($M_{\rm WD}\leqslant M/M_\odot \leqslant 1.1$) DA white dwarfs in the solar neighborhood (Tremblay *et al.* 2019). Right panel: in black, the corresponding star formation rate (Isern 2019), in red, the Galactic disc star formation (Mor *et al.* 2019).

and

$$\langle\Psi\rangle = \frac{n(l)\Delta l}{\int_{\Delta M} \Phi\,(M)\,\Delta t_{cool}\,(l,M)\,dM} \tag{4.4}$$

with $\Delta t_{cool} = t_{cool}\,(l+0.5\Delta l, M) - t_{cool}\,(l-0.5\Delta l, M)$, $l = -\log(L/L_\odot)$ and Δl the width of the luminosity function bin (Isern 2019) and this allows to reconstruct the star

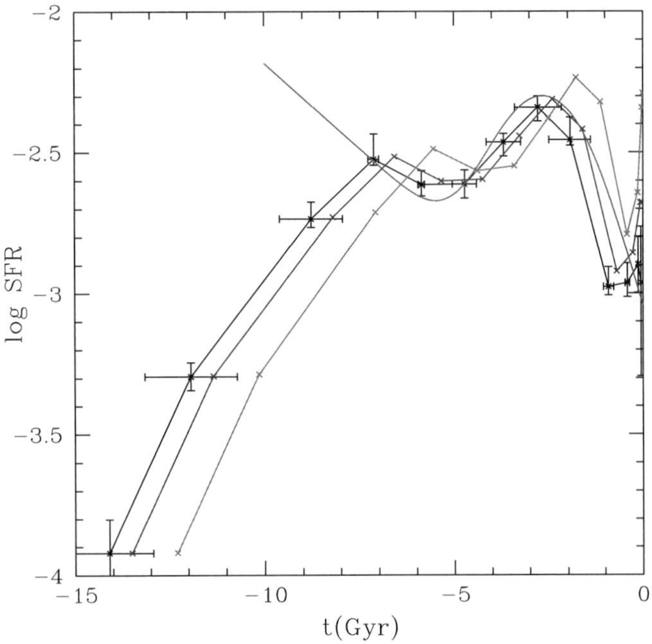

Figure 9. Star formation rate ($M_\odot Gyr^{-1}pc^{-3}$) in the solar neighborhood obtained from massive white dwarfs assuming no axions (black line). The blue and magenta lines correspond to the case when axions are present and have a coupling constant $g_{ae} = 2.24$ and 4.48×10^{-13}, respectively. The red line corresponds to the star formation rate obtained by Mor *et al.* (2019).

formation history of the solar neighborhood† (Fig. 8, right panel). The star formation rate obtained in this way is not constant or monotonically decreasing as it is often assumed. It grew quickly in the past, during the first epochs of the Galaxy, it roughly stabilized and started to decrease 7 to 6 Gyr ago. A noticeable feature is a prominent peak centered around 2.5 Gyr ago, the exact position being model depending.

The existing degeneracy between galactic properties and evolutionary models imply that different stellar models can lead to different star formation histories, for which reason it is necessary to compare these results with others obtained independently. Mor *et al.* (2019) computed the star formation history of the Galactic disk using main sequence stars from the Gaia DR2 and the Besançon Galaxy Model. Since this function is expressed in stars per unit of disk surface it has been divided by an arbitrary and constant height scale above the galactic plane (red line, Fig. 9). As it can be seen both methods, local and galactic, predict a concordant burst of star formation $\sim$2.5 Gyr ago but diverge at early times. This divergence may have several origins, a local delay in starting the star formation process, a different behavior of the outer and inner disks, a vertical dilution caused by a galaxy collision or just the conversion of DA white dwarfs into non-DAs. The peak that appears at $\sim$0.2 Gyr is in the limit of applicability of the method and deserves more attention.

Figure 9 displays the star formation rate obtained when including axions (and crystallization sedimentation effects) in the cooling models. As it can be seen, the SFRs obtained in this way have the same behavior, but they are displaced towards shorter

† It is important to notice here that the star formation rate obtained in this way does not take into account the secular evolution of the sample caused by radial migrations, scale height inflation or merging of double degenerate stars.

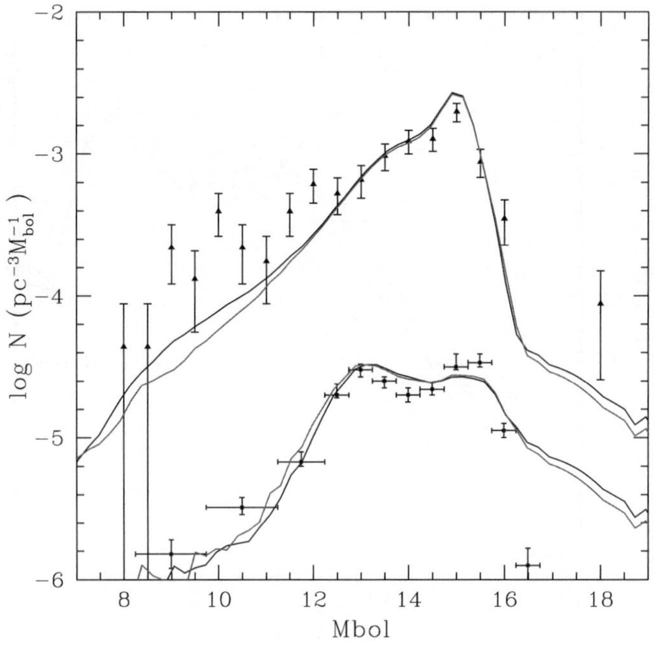

Figure 10. Theoretical versus empirical luminosity functions for massive and all white dwarfs. The empirical ones are from Tremblay *et al.* (2019) and Oswalt *et al.* (2017) respectively.The theoretical ones were obtained with the star formation rates of Fig. 9 with $g_{ae} = 0$ (black lines) and 2.24×10^{-13} (blue lines).

ages and larger rates thus providing an additional way to discriminate among theoretical models. Figure 10 displays a test of consistency where the SFR obtained from the luminosity function of the massive ones is used to compute the total one. The agreement between theory and observations is reasonable taking into account that observations were obtained independently and that the results are very sensitive to the DA/nDA properties as well as to the adopted initial mass function and initial final mass relationship. The excesses around $M_{\rm bol} \sim 9-10$ could be due to recent bursts of star formation not resolved by the present binning of the massive luminosity function and, as mentioned before, deserves a detailed analysis. Figure 9 also shows that values of $g_{ae} \gtrsim 3 \times 10^{-13}$ are not compatible with the results obtained by Mor *et al.* (2019).

5. Discussion and conclusions

The two methods presented here are complementary since they measure the cooling rate of the same object using different phenomenologies. One, the drift of the pulsation period, applies to individual stars, while the other, the luminosity function, applies to the population ensemble. As it is evident from Fig. 2, there is some tension between both sets of data but, given the uncertainties, they can be considered as qualitatively concordant.

Axions with the properties described here not only would perturb the cooling of white dwarfs but they would also modify in a subtle way the evolution of other kind of stars. An extensive review can be found in Giannotti *et al.* (2017), while Sedrakian (2019) provides a recent analysis of the cooling by DFSZ axions of the neutron star in Cas A. Here only the two more restrictive tests are presented.

The luminosity of a star in the red giant branch depends on the mass of its core and is due to the hydrogen burning in a thin shell surrounding it (Paczynski 1970). Thus, the luminosity grows with the core until He is ignited in the center. If axions were present, the core would be more massive than in the standard case and, consequently, the tip of the red giant branch would be brighter. Obviously, since H-burning occurs via CNO-cycle, this luminosity depends on the metallicity. The analysis of the red tip in M5 suggested the necessity of an extra cooling source. If this source were axions, the electron coupling constant should be $g_{ae} \sim 1.9 \times 10^{-13}$ (Viaux *et al.* 2013), but a similar analysis performed on M3 did not find the necessity of an extra cooling term and provided a more stringent upper bound of $g_{ae} < 2.57 \times 10^{-13}$ (Straniero *et al.* 2018). A similar bound has also been obtained using a set of 50 globular clusters (Arceo-Diaz *et al.* 2019). In Fig. 2 this bound and the hint found in M5 are represented by the RGB-T line. A detailed analysis of the existing uncertainties can be found in Serenelli *et al.* (2017).

The evolution of HB stars provides additional observational tests. The number of stars in a given region of the HR diagram is roughly proportional to the time spent in it. Since HB stars are the direct descendants of red giant stars, the ratio between the number of HB stars, $N_{\rm HB}$, and the red giants, $N_{\rm RGB}$, in a cluster should satisfy the relationship $R = N_{\rm HB}/N_{\rm RGB} = \Delta t_{\rm HB}/\Delta_{\rm RGB}$ and, since the densities and temperatures of both populations are very different, the inclusion of the axion emission should strongly perturb these quantities. This parameter has been measured in a large sample of globular clusters and is fairly constant at low metallicities, $R_{\rm av} = 1.39 \pm 0.03$ (Salaris *et al.* 2004).

Ayala *et al.* (2014) and Straniero *et al.* (2015) examined the influence of axions in this case and found that the best fit to the parameter R was obtained for $g_{a\gamma} = 0.29 \pm 0.18 \times 10^{-10}$ GeV^{-1}, i.e. that it was necessary an extra cooling factor, with an upper bound of $g_{a\gamma} < 0.66 \times 10^{-10}$ GeV^{-1}. In terms of the mass of the axion, these values translate to 0.12 eV and < 0.2 eV, respectively, in the case of the KSVZ axions. Since they did not include bremsstrahlung with electrons, that strongly affects the evolution of RGB stars, it is not possible to extrapolate these values to the DFSZ case. However, in an updated version, taking into account the axion-electron interaction, Straniero *et al.* (2018) obtained $g_{a\gamma} < 0.5 \times 10^{-10}$ GeV^{-1} and $g_{ae} < 2.6 \times 10^{-13}$ (these values are represented in Fig. 2). These results, however, strongly depend on the assumed He abundance in the cluster.

The *Zero Age Horizontal Branch* provides an additional test. If the axion-electron interaction is operating in RGB stars, the core is larger and the luminosity of HB stars too. Straniero *et al.* (2018) examined this case and found a best fit at $g_{ae} = 0.54 \times 10^{-13}$ and a bound at $g_{ae} < 1.78 \times 10^{-13}$. These values are plotted in Fig. 2.

Certainly, the deviations represented in Fig. 2 have a small statistical significance, but taken together they suggest that stars loose energy more efficiently than expected. Axions of the DFSZ type, with a mass of few meV could do the job, although other possibilities are open. For instance an ALP† with $g_{ae} \sim 1.5 \times 10^{-13}$ and $g_{a\gamma} \sim 1.4 \times 10^{-11}$ GeV^{-1} (Giannotti *et al.* 2016). It is evident from this discussion that a direct detection is necessary to prove the existence of axions and that white dwarfs provide an interesting hint on where to search. IAXO is a future experiment designed to detect the emission of axions by the Sun (Irastorza *et al.* 2011) that will have enough sensitivity to detect axions of the DFSZ type and mass $\gtrsim 3$ meV, according to the calculations of Redondo (2013).

Thanks to the data that is providing Gaia and will provide the LSST as well as other instruments, white dwarfs will become in a next future one of the best characterized populations of the Galaxy, and this will convert them into an excellent laboratory to explore new ideas in Physics.

† Axion-like particles (ALPs) appear in a natural way in the extensions of the Standard Model as pseudo Nambu-Goldstone bosons associated to the breaking of symmetries (Ringwald 2014). They are similar to axions but mass and photon coupling are no longer linked, thus offering a larger space of parameters.

Acknowledgements

This work has been supported by MINECO grant ESP2017-82674-R, by EU FEDER funds, and by grants 2014SGR1458 and CERCA Programe of the Generalitat de Catalunya.

References

Althaus, L. G., Córsico, A. H., Isern, J., & García-Berro, E. 2010, *AAR*, 18, 471
Altherr, T., Petitgirard, E., & Gaztelarrutia, T. 1994, *Astroparticle Phys.*, 2, 175
Arceo-Diaz, S., Schroeder, K.-P., Zuber, K., *et al.* 2019, arXiv:1910.10568
Ayala, A., Dominguez, I., Giannotti, M., Mirizzi, A., & Straniero, O. 2014, *Phys. Rev. Lett.*, 113, 191302
Battich, T., Córsico, A. H., Althaus, L. G., Miller Bertolami, M. M., *et al.* 2016, *JCAP*, 8, 62
Bertone, G. & Fairbairn, M. 2008, *Phys. Rev. D*, 77, 043515
Bischoff-Kim, A., Montgomery, M. H., & Winget, D. E. 2008, *ApJ*,675, 1512
Blinnikov, S. I. & Dunina-Barkovskaya, N. V. 1994, *MNRAS*, 266, 289
Catalan, S. *et al.* 2008, *MNRAS*, 387, 69
Córsico, A. H., Benvenuto, O. G., Althaus, L. G., Isern, J., & García-Berro, E. 1989, *New. Astron.*, 6, 197
Córsico, A. H., Altahus, L. G., Miller Bertolami, M. M., Romero, A. D., García-Berro, E., Isern, J., Kepler, S. O. 2012a *MNRAS*, 424, 2792
Córsico, A. H., Altahus, L. G., Romero, A. D., Mukadam, A. S., García-Berro, E., Isern, J., Kepler, S. O., Corti, M. A. 2012b *JCAP*, 12, 10
Córsico, A. H. *et al.* 2016, *JCAP*, 7, 36
D'Antona, F. & Mazzitelli, I. 1989, *ApJ*, 347, 934
DeGennaro, S. *et al.* 2008, *AJ*, 135,1
Dias, A. G., Machado, A. C. B., Nishi, C. C., Ringwald, A. & Vandrevange, P. 2014, *JHEP*, 2014, 37
Dine, M., Fishler, W., & Srednick, M 1981, *Phys. Lett. B*, 104, 109
Dreiner, H. K., Fortin, J-F., Isern, J., & Ubaldi, L. 2013, *Phys. Rev. D*, 88, 043517
Evans, D. W. 1992, *MNRAS*, 255, 521
Fontaine, G., Brassard, P., & Bergeron, P. 2001, *PASP*, 113, 409
Freese, K. 1984, *ApJ*, 286, 216
García-Berro, E., Hernanz, M., Mochkovitch, R. Isern, J. *et al.* 1988, *A&A*, 193, 141
García-Berro, E., Hernanz, M., Isern, J., & Mochkovitch, R. 1995, *MNRAS*, 277, 801
García-Berro, E. & Oswalt, T. D. 2016, *New Astron. Rev*, 72, 1
Giannotti, M., Irastorza, I. G., Redondo, J., & Ringwald, A. 2016, *JCAP*, 05, 057
Giannotti, M., Irastorza, I. G., Redondo, J., Ringwald, A. & Saikawa, K. 2017, *JCAP*, 10, 010
Giddings, S. B. & Mangano, M. L. 2008, *Phys.Rev.D*, 78, 035009
Hansen, B. M. 2004, *Phys. Rep.*, 399, 1
Harris, H. C. *et al.* 2006, *AJ*, 131, 571
Iben, I. & Tutukov, A. V. 1984, *ApJ*, 282, 615
Irastorza, I. G., *et al.* 2011, *JCAP*, 06, 013
Isern, J. 2019, *ApJL*, 878, L11
Isern, J., Catalan, S., García-Berro, E., & Torres, S. 2009, *J. Phys. Conf. Series*, 172, 012005
Isern, J. & García-Berro, E. 2008, *MemSAI*, 79, 545
Isern, J., García-Berro, E., Altahus, L. G., & Córsico, L. H. 2010, *A&A*, 512, 86
Isern, J., García-Berro, E., Hernanz, M., & Mochkovitch, R. 1998, *Jour. Phys. Cond. Mat*, 10, 11263
Isern, J., García-Berro, E., Torres, S., & Catalán, S. 2008, *ApJL*, 682, L100
Isern, J., García-Berro, E., Torres, S., Cojocaru, R., & Catalán, S. 2018, *MNRAS*, 478, 2569
Isern, J., Hernanz, M., & García-Berro, E. 1992, *ApJL*, 392, L23
Itoh, N., Hayashi, H., Nishikawa, A., & Kohyama, Y. 1988, *ApJS*, 102, 411
Kepler, S. O. *et al.* 1991, *ApJL*, 378, L45

Kepler, S. O. *et al.* 2012, *ApJ*, 757, 177

Kilic, M., Munn, J. A., Harris, H. C. *et al.* 2017, *ApJ*, 837, 162

Knox, R. A., Hawkins, M. R. S., & Hambly, N. C. 1999, *MNRAS*, 306, 736

Koester, D. & Shoenberner, D. 1986, *A&A*, 154, 125

Krzesinski, J. *et al.* 2009, *A&A*, 508, 339

Malec, B. & Besiada, M. 2001, *Acta Phys. Pol. B*, 32, 3683

McGraw, J. T. & Robinson, E. L. 1976, *ApJ*, 205, 155

Miller Bertolami, M. M., Mlendez, B., Althaus, L. G., & Isern, J. 2014, *JCAP*, 1410, 069

Mor, R., Robin, A. C., Figueras, F., Roca-Fàbrega, S., & Luri, X. 2019, *A&A* 624, L1

Mukadam, A. S. *et al.* 2013, *ApJ*, 771, 17

Munn, J. A. *et al.* 2017, *AJ*, 153, 10

Nakagawa, M, Kohyama, Y., & Itoh, N. 1987, *ApJ*, 322, 291

Nakagawa, M., Adachi, T., Kohyama, Y. & Itoh, N. 1988, *ApJ*, 326, 241

Legget, S. K., Ruiz, M. T., & Bergeron, P. 1998, *ApJ*, 497, 294

Liebert, J., Dahn, C. C., & Monet, D. G. 1988, *ApJ*,332, 891

Oswalt, T. D., Holberg, J., & Sion, E. 2017, *20th European White Dwarf Workshop*, Ed. P.-E. Tremblay *et al.*, ASP Conference Series 509, 59

Oswalt, T. D., Smith, J. A., Wood, M. A., & Hintzen, P. A. 1996, *Nature* 382, 692

Paczynski, B. 1970, *Acta Astron.*, 20, 287

Raffelt, G. G. 1986, *Phys. Lett. B*, 166, 402

Raffelt, G. G. 1996, in: K. K. Turekian, H. D. Holland & A. M. Davis (eds.), *Stars as Laboratories of Fundamental Physics* (University of Chicago Press).

Redaelli, M., Kepler, S. O., Costa, J. E.S. *et al.* 2011, *MNRAS*, 415, 1220

Redondo, J. 2013, *JCAP*, 12,008

Ringwald, A. 2014, arXiv:14070546

Rowell, N. & Hambly, N. C. 2011, *MNRAS*, 417, 93

Salaris, M., Riello, M., Cassisi, S., & Piotto, G. 2004, *A&A*, 420, 911

Saltas, I. C., Sawicki, I., & Lopes, I. 2019, *JCAP*, 05, 028S

Savedoff, H. P., Van Horn, H. M., & Vila, S. C.1969, *ApJ*, 155, 221

Sedrakian, A. 2019, *Phys. Rev. D*, 99, 3011

Serenelli, A., Weiss, A., Cassisi, S., Salaris, M., & Pietrinferni, A. 2017 arXiv: 1706.09910

Shipman, H. L. 1997, in: *White Dwarfs*, Proceedings of the 10th European Workshop on White Dwarfs. Ed. J. Isern, M.Hernanz & E. García-Berro (Dordrecht: Kluwer Academic Press), 214, 165

Straniero, O., Ayala, A., Giannotti, M., Mirizzi, A., & Dominguez, I. 2015, *Proceedngs 11th Patras Workshop on Axions, WIMPs and WISPs*, Zaragoza, Spain, p.77

Straniero, O., Dominguez, I., Giannotti, M., & Mirizzi, A. 2018, *PATRAS17*, arXiv:1802.10357 *& slides*

Sullivan, D. J. & Chote, P. 2015, *White Dwarfs*, Proceedings of the 10th European Workshop on White Dwarfs. Ed. P.Dufour, P. Bergeron, G. Fontaine (ASP Conf. Series) 493, 199

Tremblay, P.-E., Fontaine, G., Fusillo, N. P.G., Dunlap, B. H., *et al.* 2019, *Nature*, 565, 202

Viaux, N., Catelan, M., Stetson, P. B., *et al.* 2013, *Phys. Rev. Lett.*, 111, 231301

Vila, S. C. 1965, *AJ*, 70, 695

Vila, S. C. 1976, *ApJ*, 206, 213

Weidemann, V 1968, *ARAA*, 6, 351

Winget, D. E., Hansen, C. J., & van Horn, H. E. 1983, *Nature*, 303, 781

Zhimitskii, 1980, *Sov. J. Nuc. Phys.*, 31, 260

White Dwarfs as Probes of Fundamental Physics: Tracers of Planetary, Stellar and Galactic Evolution
Proceedings IAU Symposium No. 357, 2019
M. A. Barstow, S. J. Kleinman, J. L. Provencal & L. Ferrario, eds.
doi:10.1017/S174392132000040X

The real time evolution of post-AGB stars

Marcin Hajduk

University of Warmia and Mazury, ul.Oczapowskiego 2, 10-719 Olsztyn, Poland,
email: marcin.hajduk@uwm.edu.pl

Abstract. Evolution of post-AGB stars is extremely fast. They cross the HR diagram vertically on a timescale of hundreds to some ten thousands of years to reach maximum temperature in their lifetime. This is reflected in an increasing excitation of planetary nebulae on a timescale of years and decades. Since evolutionary timescale of post-AGB stars is very sensitive to their mass, observed changes can be used to determine model dependent central star masses. If an additional parameter is determined (e.g. luminosity or dynamic age), the observed evolution of planetary nebulae can be utilized for observational verification of theoretical models.

Keywords. stars: AGB and post-AGB, planetary nebulae: general, white dwarfs, stars: evolution, stars: fundamental parameters, Hertzsprung-Russell diagram

1. Introduction

Post-AGB stars evolve from the stars with initial masses from 1 to 8 $M_{\odot}$. The star loses most of its envelope on the tip of the asymptotic giant branch (AGB). Once the star is hot enough, the ionization front expands throughout the ejected shell, creating a planetary nebula (PN). A post-AGB star crosses the HR diagram horizontally on a timescale τ_{cross} ranging from tens to tens thousands of years. After the star reaches maximum temperature in its lifetime (Herwig 2005), it gradually descend on the white dwarf (WD) cooling track.

Only a few sets of evolutionary models had been available till recently, including Bloecker (1995) and Vassiliadis & Wood (1994). The models demonstrate that τ_{cross} depends critically on the central star mass. More massive remnants evolve three or more orders of magnitude faster than the least massive ones. The duration of the post-AGB phase is determined by the remaining mass of the hydrogen envelope after AGB phase. For low mass remnants, the envelope is consumed at the H burning rate. For more massive remnants, mass loss due to the stellar wind can significantly shorten the evolutionary timescale (Bloecker 1995).

Gesicki *et al.* (2014) needed to accelerate the Bloecker's post-AGB tracks by a factor of three, so that central stars, derived from stellar-mass - age relation, could match the local WD masses. New models by Miller-Bertolami (2016) confirmed their results and predicted 3 to 10 times faster evolution than the previous authors. Models computed by Miller-Bertolami (2016) leave AGB with significantly reduced H envelope mass compared to previous models. This difference was attributed to an updated treatment of the constitutive microphysics and updated description of the mixing processes and winds. Post-AGB timescales only show a mild dependence on metallicity in Miller-Bertolami (2016), contrary to the previous authors.

2. Observational verification of evolutionary models

The crossing time τ_{cross} calculated in post-AGB models depends critically on the central star mass. τ_{cross} is too long to be observed. Instead, we can use heating rate $\dot{T}_{eff}$.

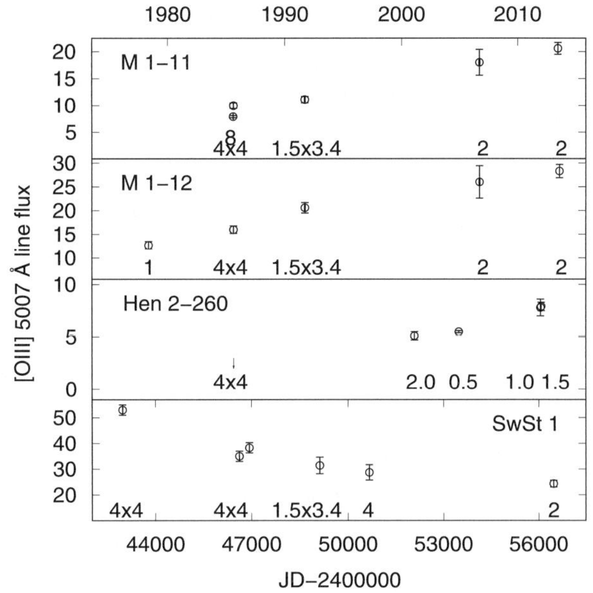

Figure 1. Temporal evolution of the [O III] 5007Å / Hβ flux ratio in selected PNe observed at more than two epochs. The numbers indicate aperture widths used in each observation.

Hajduk *et al.* (2015) showed that $\dot{T}_{eff}$ does not change significantly throughout the horizontal part of the evolution. Post-AGB stars evolve at constant luminosity till they reach T_{eff} of 10^4 K or more. Luminosity is also sensitive to the mass of post-AGB star. Both $\dot{T}_{eff}$ and $L_\star$ have to be determined for the same star to verify evolutionary models. If evolutionary model is correct, both $\dot{T}_{eff}$ and $L_\star$ should indicate the same mass of the post-AGB star.

Heating rate can be derived from observations of the central star or/and the nebula covering a time span of several years by means of atmosphere or/and photoionization modeling. Temperature evolution was measured for a few stars experiencing late or very late thermal pulse: FG Sge (Jeffery & Schönberner 2006), V605 Aql Clayton (2006), V4334 Sgr (van Hoof *et al.* 2007), and recently SAO 244567 (Reindl *et al.* 2017), which largely contributed to our understanding of this special case of post-AGB evolution.

In this paper we focus on the study of post-AGB stars which are not known to experience late thermal pulses. We included hydrogen deficient post-AGB central stars, which evolutionary status is not yet fully understood. Although their evolution is slower, $\dot{T}_{eff}$ was also determined for a few central stars which are not known to have experienced (late) thermal pulse. Hydrogen deficient stars are hydrogen burners, while hydrogen rich post-AGB stars can be either helium or hydrogen burners. Evolution of helium burning post-AGB stars proceeds slower that of hydrogen burning stars.

Zijlstra *et al.* (2008) determined $\dot{T}_{eff}$ of $156 \pm 36\,\mathrm{K\,yr^{-1}}$ for the central star of NGC 7027, and Hajduk *et al.* (2014) $45 \pm 7\,\mathrm{K\,yr^{-1}}$ for the central star of Hen 2-260. The nebular variability was reported in IC 4997 and NGC 6572 by Feibelman *et al.* (1992) and in M 1-11 and M 1-6 by Kondratyeva (2005) without giving the heating rates. Hajduk *et al.* (2015) observed increase of the excitation of many young planetary nebulae on a timescale of years and decades (Fig. 1). Unexpectedly, the central star of the PN SwSt 1 showed decreasing temperature, inconsistent with evolutionary models (Hajduk *et al.*, in prep.).

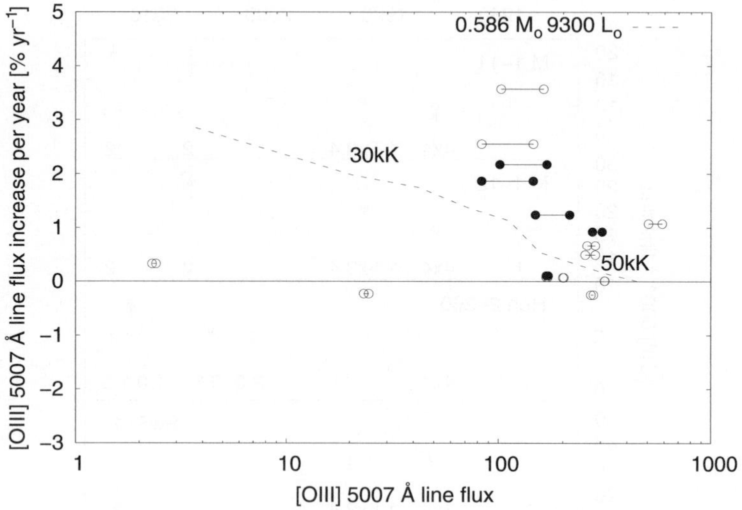

Figure 2. Temporal evolution of the [O III] 5007Å / Hβ flux ratio in Magellanic Clouds PNe. The dashed lines plot the evolutionary tracks by Miller-Bertolami (2016). Filled symbols show post-AGB stars which show emission lines in their spectra.

3. New observations and preliminary results

We observed a new sample of young southern PNe with the New Technology Telescope and South African Large Telescope. We focused on low excitation F([O III] 5007Å) / $F(H\beta) < 5$) and compact (d < 5 arc sec) PNe. Small observed diameters of the PNe allow to avoid systemic errors related with the ion stratification within the nebula, as the slit width probes most of or all the nebular light. Each PN was previously observed at least one time, most of them by Acker *et al.* (1992). Unfortunately, only six of the Galactic PNe in the sample have a precise distance determination with GAIA better than 30% (Kimeswenger & Barria 2018) (the observations were performed before the launch of GAIA mission). Accurate determination of the luminosities is necessary for verification of the evolutionary tracks. Some 20 SMC and LMC PNe were observed as well. These objects can be used for the verification of evolutionary tracks as their distances and luminosities are known.

Fig. 2 shows clear evidence for evolution of the [O III] 5007Å / Hβ flux ratio for some of the Magellanic Cloud PNe. Interestingly, the evolution is significantly faster than predicted for the $0.586 M_\odot$ evolutionary track. Two coolest post-AGB stars do not show evidence for flux evolution, which may indicate that they are very low mass stars.

We obtained a model using photoionization code CLOUDY (Ferland *et al.* 2017) for two of the observed PNe so far. The measured line fluxes and other data available in the literature served as input parameters. In this preliminary work we used reference spectra, abundances, and luminosities provided by Leisy & Dennefeld (2006) and Dopita & Meatheringham (1991).

Initially, we found the best solution for temperature for one epoch. Then we varied the central star temperature to fit optical line fluxes observed in other epoch. Luminosity remains constant for young PNe, so the changing nebular flux ratios reflect the evolution of stellar temperature and to much smaller extent expansion. As a result, we derived the difference of the stellar temperature between the two epochs. It is illustrated in Fig. 3. The determined $\dot{T}_{eff}$ for LHA 120 N199 of 30 K yr^{-1}, and for SMP SMC 1 even higher – of 130 K yr^{-1}. They exceed the expected heating rates expected from their luminosities.

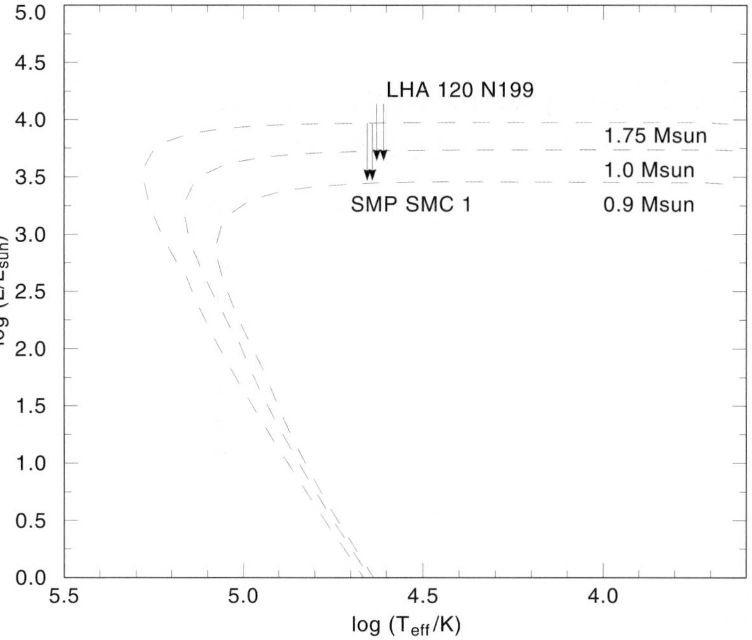

Figure 3. Temporal evolution of two central stars of PNe in Magellanic Clouds in the HR diagram. Two arrows for each object correspond to two different epochs. The dashed lines plot the evolutionary tracks by Miller-Bertolami (2016).

Acknowledgments

We gratefully acknowledge financial support from National Science Centre, Poland, grant No. 2016/23/B/ST9/01653. Some of the observations reported in this paper were obtained with the Southern African Large Telescope (SALT). Polish participation in SALT is funded by grant No. MNiSW DIR/WK/2016/07. Based on observations collected at the European Organisation for Astronomical Research in the Southern Hemisphere under ESO programmes 093.D-0182 and 095.D-0614.

References

Acker, A. *et al.* 1992, *The Strasbourg-ESO Catalogue of Galactic Planetary Nebulae*
Bloecker, T. 1999, *A&A*, 299, 755
Clayton, G. 2006, *ApJ*, 646L, 69
Dopita, M. A. & Meatheringham, S. J. 2006, *ApJ*, 377, 480
Feibelman, W. A. *et al.* 1992, *PASP*, 104, 339
Ferland, G. *et al.* 2017, *RMxAA*, 53, 385
Gesicki, K., Zijlstra, A. A., Hajduk, M., & Szyszka, C. 2014, *A&A*, 566A, 48
Hajduk, M. *et al.* 2014, *A&A*, 567A, 15
Hajduk, M., van Hoof, P. A. M., & Zijlstra, A. A. 2015, *A&A*, 573A, 65
Herwig, F. 2005, *ARA&A*, 43, 435
Jeffery, C. S. & Schönberner, D. 2006, *A&A*, 459, 885
Kimeswenger, S. & Barria, D. 2018, *A&A*, 616L, 2
Kondratyeva, L. N. 2005, *Astronomical and Astrophysical Transactions*, 24, 291
Leisy, P. & Dennefeld, M. 2006, *A&A*, 456, 451
Miller-Bertolami, M. M. 2016, *A&A*, 588A, 25
Reindl, N. *et al.* 2017, *MNRAS*, 464, 51
Vassiliadis, E. & Wood, P. R 1994, *ApJS*, 92, 125
van Hoof, P. A. M. *et al.* 2007, *A&A*, 471L, 9
Zijlstra, A. A. *et al.* 2008, *ApJ*, 681, 1296

White Dwarfs as Probes of Fundamental Physics: Tracers of Planetary, Stellar and Galactic Evolution
Proceedings IAU Symposium No. 357, 2019
M. A. Barstow, S. J. Kleinman, J. L. Provencal & L. Ferrario, eds.
doi:10.1017/S1743921320000617

(Pre)-white dwarf stars as measuring tools for yields of AGB nucleosynthesis

Lisa Löbling

Institut für Astronomie und Astrophysik, Sand 1, 72076 Tübingen, Germany
email: loebling@astro.uni-tuebingen.de

Abstract. In the helium-rich intershell region of asymptotic giant branch (AGB) stars, slow neutron-capture nucleosynthesis produces heavy elements beyond iron. If the stars experience a final-flash of the He-burning shell, a pulse-driven convection zone establishes, the stars become hydrogen-deficient and exhibit former intershell material at their surfaces. In their subsequent evolution towards the white-dwarf cooling sequence, but still at constant luminosity, a strong stellar wind prevents diffusion to wipe out the information about AGB yields. We present and interpret the analysis results of hydrogen-rich and -deficient post-AGB stars, discuss difficulties in their analysis and review the implications on the understanding of post-AGB evolution.

Keywords. atomic data, techniques: spectroscopic, stars: abundances, stars: AGB and post-AGB, stars: atmospheres, white dwarfs, planetary nebulae

1. Introduction

Asymptotic giant branch (AGB) stars are a significant contributor to the enrichment of galaxies with elements beyond iron (trans-iron elements, TIEs, Busso *et al.* 1999). Due to the neutron source ^{13}C, the helium (He) rich region between the hydrogen (H)- and He-burning shells provides perfect conditions for the formation of heavy elements via neutron captures on seed elements like iron (s-process). If these nucleosynthesis outcomes are transported to the stellar surface by convection, they become detectable. A possible mechanism is the final thermal pulse (FTP). Here, a final flash of the He-burning shell happens after the descent from the AGB. It can mix the entire H-rich envelope to the stellar interior resulting in a H-deficient or even H-free white dwarf (WD). This scenario can explain the detection of 16 TIEs in the photosphere of the DO-type WD RE 0503–289 (Rauch et al. submitted, Rauch *et al.* 2017). The strong enrichment of up to 100 000 times solar is due to efficient radiative levitation and, thus, does not carry any information about the abundance pattern produced by AGB nucleosynthesis. Another mechanism that mixes traces of nucleosynthesis products to the surface is the third dredge up (TDU), when the surface convection zone overshoots into the He-rich region after thermal pulses of the He-burning shell on the AGB. This process should happen for stars with initial masses above $1\,M_\odot$ (Karakas & Lugaro 2016) which translates to final WD masses of $\gtrsim 0.5\,M_\odot$ (Cummings *et al.* 2018). To constrain the s-process yields, stars in a phase before becoming a WD, i.e., before diffusion can act, need to be analyzed. For central stars of planetary nebulae (CSPNe), the stellar wind is strong enough to be chemically homogeneous and to preserve the s-process yields (Unglaub 2008).

2. Objects, observations, and model atmospheres

The two hybrid PG 1159-type CSPNe Abell 43 and NGC 7094 are known as spectroscopic twins with the same atmospheric parameters of $T_{\rm eff}=115$ kK and $\log g=5.6$

(Fig. 2). Both are H-deficient with surface H mass fractions of $25 \pm 3\,\%$ and $15 \pm 3\,\%$, respectively (Löbling *et al.* 2019b). Nucleosynthesis calculations predict the AGB s-process to be most efficient for stars with initial masses of about $3\,M_\odot$ which translates into post-AGB masses of $0.7 - 0.8\,M_\odot$. The extremely hot ($T_{\rm eff}$=150 kK, log g=5.7) PG 1159-type central star of NGC 246 (Rauch & Werner 1997) falls into that region. The hot DAO-type WD BD−22°3467, the CS of Abell 35, is H-rich. Ziegler *et al.* (2012) derived a very low mass of $0.53^{+0.04}_{-0.03}\,M_\odot$ from interpolation of evolutionary tracks with the atmospheric parameters $T_{\rm eff}$=80 kK and log g=7.2 (Fig. 2, Ziegler *et al.* 2012).

The results were obtained from analyses of high resolution spectra from ultraviolet to the optical or even infrared regime, i.e., from FUSE (910 – 1180 Å), STIS (1150 – 1730 Å) or GHRS (1140 – 1435 Å) as well as UVES (3280 – 4560 , 4580 – 6690 Å). For NGC 246, we present recent XSHOOTER (3000 – 25000 Å) spectra. To our knowledge, no optical observations of Abell 35 are available with a sufficient signal-to-noise (S/N) ratio to detect features of the hot DAO-type WD in this regime that is dominated by the cool (3700 K, Herald & Bianchi 2002) companion.

The analyses were performed using the model atmosphere code of the Tübingen NLTE Model atmosphere package (TMAP). The input atomic data for the light elements (atomic number $Z < 20$) was taken from the Tübingen Model Atom Database (TMAD, http://astro.uni-tuebingen.de/~TMAD/), for all elements with $Z \geqslant 20$, statistical atoms (Rauch & Deetjen 2003) were constructed. The atomic data for the iron group elements ($20 \leqslant Z \leqslant 28$) was taken from Kurucz's line list (http://kurucz.harvard.edu/atoms.html) and for the TIEs from the Tübingen Oscillator Strengths Service (TOSS, http://dc.g-vo.org/TOSS).

3. Selected results and their discussion

For the two hybrid PG 1159-type CSPNe Abell 43 and NGC 7094, upper abundance limits for the s-process elements could be determined (Fig. 1). Although that does not help constraining AGB yields, the fact that these upper limits lie below the strong enrichment determined in the DO-type WD RE 0503−289 (Hoyer *et al.* 2017), confirms the understanding of stellar wind and diffusion. While diffusion can accumulate s-process elements in the atmosphere of post-AGB stars in the regime of hot WDs, the strong, coupled stellar wind of CSPNe prevents this process. The abundance determination was hampered by the lack of atomic data of highly ionized (stages VIII–IX) s-process elements. This was also a reason why no reasonable upper abundance limits can be determined for s-process elements in NGC 246 (Fig. 1). Here, the problem becomes more severe due to the higher temperature. Another problem is the strong line broadening observed in its spectrum. This is consistent with a rotational velocity of 75 km/s (Löbling 2018) but could also be of different origin. While Abell 43 and NGC 7094 also show broadened lines, some PG 1159 stars, e.g. K 1 – 16, in the same $T_{\rm eff}$ –log g regime show no signatures of rotation (Werner *et al.* 2010). We speculate whether this might be connected to the finding, that NGC 246 is a pulsator (Ciardullo & Bond 1996) while K 1 – 16 does not pulsate (Sowicka *et al.* 2018).

The analysis of XSHOOTER spectra of NGC 246 reveals that the profiles of the He II $4 - n$ lines with $n = 6 - 11$ can be reproduced with a model atmosphere with $T_{\rm eff}$=150 kK and log g=5.7. Using these values with the V magnitude and the mass $0.74^{+0.19}_{-0.23}\,M_\odot$ (Löbling 2018), we find a spectroscopic distance of 970 ± 200 pc. The distance based on Gaia parallaxes is 506 ± 25 pc (Bailer-Jones *et al.* 2018). This value would require log g=6.0 − 6.1. A higher surface gravity is in disagreement with the He II line profiles (Fig. 3) and cannot be solved by varying the He abundance. The same phenomenon

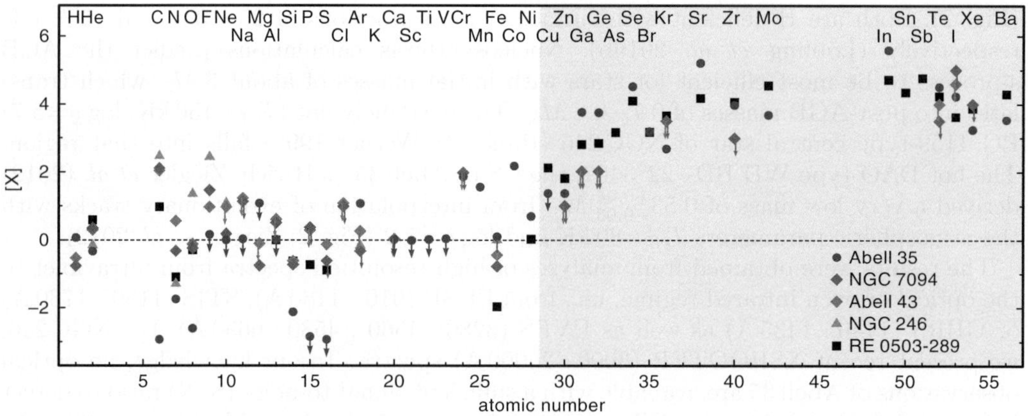

Figure 1. Abundances [X] = log (mass fraction/solar mass fraction) of the CS of Abell 35, Abell 43, NGC 7094, and NGC 246 from Löbling (2018); Löbling *et al.* (2019a,b). Upper limits are indicated with arrows. For comparison, the abundances of RE 0503–289 are shown (Hoyer *et al.* 2017).

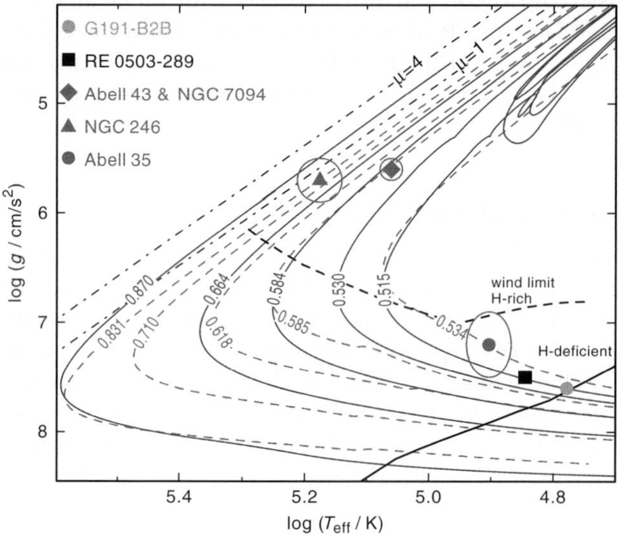

Figure 2. Location of the CS of Abell 35, Abell 43, NGC 7094, and NGC 246 (red) with error ellipses in the $T_{\rm eff}-\log g$ diagram. Included are post-AGB tracks for H-burners (green dashed, Miller Bertolami 2016) and final flashers (blue, Miller Bertolami & Althaus 2007) labeled with the stellar mass (in $M_\odot$). The Eddington limits for H- and He-rich atmospheres are indicated (black, dashed-dotted). The wind limit for H-rich and H-deficient post-AGB stars is indicated (Unglaub & Bues 2000).

of too high spectroscopic distances was found for other NLTE analyses (Schönberner *et al.* 2018; Schönberner & Steffen 2019). Their argument of missing metal line blanketing does not hold in our case, since opacities of 17 elements up to Ni are included in the model-grid calculations. The line profiles are calculated based on broadening tables of Schoening & Butler (1989a,b). It needs to be investigated, whether line broadening theory can cause the discrepancy. It is highly desirable to solve this tension, since a precise knowledge of $T_{\rm eff}$ and $\log g$ is required to perform reliable abundance measurements.

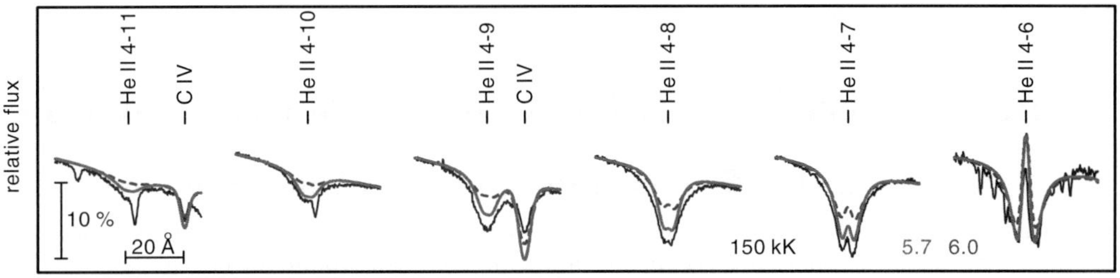

Figure 3. Synthetic spectra calculated with $T_{\text{eff}}=150\,\text{kK}$ and $\log g=5.7$ (red) and 6.0 (blue, dashed), compared with the He II lines in the XSHOOTER observations of NGC 246 (gray).

For the DAO-type CS of Abell 35, a high enrichment of the level of RE 0503−289 with s-process elements was found (Fig. 1). While RE 0503−289 evolved through a FTP and shows intershell material at the surface, Abell 35 still has its H-rich envelope. With its very low mass ($0.533^{+0.040}_{-0.025}\,M_{\odot}$, Ziegler *et al.* 2012), it is unclear, whether it ever experienced TDU that could pollute the atmosphere. This result raises the question, whether FTP mixing is necessary for the peculiar abundance pattern or whether it is the sole result of effective diffusion accumulating initial material.

Acknowledgement

The author would like to thank the IAU for awarding a travel grant to attend the symposium and Scot Kleinman for the careful review of this work. This research has been supported by the German Research Foundation (DFG) under grant WE 1312/49–1. The GAVO project at Tübingen had been supported by the Federal Ministry of Education and Research (BMBF) at Tübingen (05 AC 6 VTB, 05 AC 11 VTB). Based on observations collected at the European Southern Observatory under ESO programme 0102.D-0144.

References

Bailer-Jones, C. A. L. *et al.* 2018, *AJ*, 156, 58
Busso, M. *et al.* 1999, *ARA&A*, 37, 239
Ciardullo, R. & Bond, H. E. 1996, *AJ*, 111, 2332
Cummings, J. D. *et al.* 2018, *ApJ*, 866, 21
Herald, J. E. & Bianchi, L. 2002, *ApJ*, 580, 434
Hoyer, D. *et al.* 2017, *A&A*, 598, A135
Karakas, A. I. & Lugaro, M. 2016, *ApJ*, 825, 26
Löbling, L. 2018, *Galaxies*, 6, 65
Löbling, L. *et al.* 2019a, arXiv e-prints, arXiv:1911.09573
Löbling, L. *et al.* 2019b, *MNRAS*, 489, 1054
Miller Bertolami, M. M. 2016, *A&A*, 588, A25
Miller Bertolami, M. M. & Althaus, L. G. 2007, *A&A*, 470, 675
Rauch, T. & Deetjen, J. L. 2003, *ASPCS*, Vol. 288, 103
Rauch, T. *et al.* 2017, *A&A*, 606, A105
Rauch, T. & Werner, K. 1997, The Third Conference on Faint Blue Stars, L. Davis Press, 217
Schoening, T. & Butler, K. 1989a, *A&AS*, 78, 51
Schoening, T. & Butler, K. 1989b, *A&A*, 219, 326
Schönberner, D. *et al.* 2018, *A&A*, 609, A126
Schönberner, D. & Steffen, M. 2019, *A&A*, 625, A137
Sowicka, P. *et al.* 2018, *MNRAS*, 479, 2476
Unglaub, K. 2008, *A&A*, 486, 923
Unglaub, K. & Bues, I. 2000, *A&A*, 359, 1042
Werner, K. *et al.* 2010, *ApJ*, 719, L32
Ziegler, M. *et al.* 2012, *A&A*, 548, A109

White Dwarfs as Probes of Fundamental Physics: Tracers of Planetary, Stellar and Galactic Evolution
Proceedings IAU Symposium No. 357, 2019
M. A. Barstow, S. J. Kleinman, J. L. Provencal & L. Ferrario, eds.
doi:10.1017/S1743921320000289

The spectral evolution of hot white dwarfs

Antoine Bédard, Pierre Bergeron and Gilles Fontaine

Département de Physique, Université de Montréal, Montréal, QC H3C 3J7, Canada
emails: bedard@astro.umontreal.ca, bergeron@astro.umontreal.ca

Abstract. As they evolve, white dwarfs undergo major changes in their atmospheric composition, a phenomenon known as spectral evolution. In particular, most hot He-rich (DO) stars transform into H-rich (DA) stars as they cool off, most likely as a result of the float-up of residual H. We investigate this DO-to-DA transition by taking advantage of the extensive spectroscopic dataset provided by the Sloan Digital Sky Survey (SDSS). Using our new state-of-the-art non-LTE model atmospheres, we perform a spectroscopic analysis of 1882 hot ($T_{\rm eff} > 30,000$ K) white dwarfs identified in the SDSS. We find that at least 15% of all white dwarfs are born with a He-dominated atmosphere. Among these, $\sim$2/3 turn into H-rich stars before they reach $T_{\rm eff} \sim 40,000$ K, while the remaining $\sim$1/3 maintain their He-rich surface throughout their entire evolution. We speculate on the origin of these two groups of objects.

Keywords. diffusion, stars: abundances, stars: atmospheres, stars: evolution, white dwarfs

1. Introduction

White dwarf stars can be divided into two main groups based on their atmospheric composition, either H-rich or He-rich. In both cases, the atmosphere usually displays a high degree of chemical purity, resulting from the high efficiency of gravitational settling in the strong gravitational field of white dwarfs. The corresponding optical spectra are thus dominated either by H lines (spectral class DA) or by He lines (spectral classes DO for $T_{\rm eff} \gtrsim 45,000$ K and DB for $T_{\rm eff} \lesssim 45,000$ K).

Interestingly, the distribution of spectral types along the white dwarf cooling sequence is far from uniform. In particular, the hot ($T_{\rm eff} > 30,000$ K) white dwarf population exhibit two striking features: a deficiency of DA stars at the very hot end of the cooling sequence ($T_{\rm eff} > 90,000$ K), and a deficiency of DB stars within the so-called DB gap ($45,000 \text{ K} > T_{\rm eff} > 30,000$ K). Between the two regimes, the fraction of He-rich objects is observed to decrease monotonously (Fontaine & Wesemael 1987). This suggests that the main atmospheric constituent of a white dwarf can change radically as it cools, a phenomenon referred to as spectral evolution. More specifically, the above-mentioned trend is best explained by a scenario in which most white dwarfs are born with a He-dominated atmosphere, but eventually develop a H-dominated atmosphere.

The physical mechanism most likely responsible for such a transformation is described by the so-called float-up model. It is generally believed that hot He-rich white dwarfs descend from stars that experience a late He-shell flash, following which the outer H layer is mixed deeply into the He envelope and hence burned (Werner & Herwig 2006). According to the float-up model, a small amount of H survives this evolutionary episode, but initially remains invisible (since it is thoroughly diluted within the He envelope). However, under the influence of gravitational settling, this residual H gradually diffuses upward and accumulates at the surface, ultimately causing a DO star to turn into a DA star (Fontaine & Wesemael 1987).

The float-up model, while qualitatively accounting for the spectral evolution of hot white dwarfs, has seldom been quantitatively investigated. Most notably, it is still unclear what fraction of white dwarfs undergo the DO-to-DA transition, and what fraction retain a H-rich or He-rich atmosphere throughout their life. As the spectral evolution of a degenerate star depends sensitively on its total H content, such measurements can provide global constraints on the amount of H inherited from previous evolutionary phases. We propose to tackle this issue by studying the statistical variations of atmospheric composition with effective temperature for a large sample of hot stars.

2. Sample

The largest spectroscopic sample of white dwarfs is provided by the Sloan Digital Sky Survey (SDSS). Starting from the available SDSS white dwarf catalogs (Kleinman *et al.* 2013; Kepler *et al.* 2015, 2016), we first defined a color-selected ($u - g < 0$) sample of 6270 objects, whose optical spectra were then retrieved from the SDSS database. For classification purposes, we performed preliminary fits to all spectra with both our pure-H and pure-He model atmospheres (see below). All fits were visually inspected, and stars with $T_{\rm eff} < 30,000$ K or $\log g < 6.5$ were rejected. Our final sample contains 1882 white dwarfs, including 1713 DA, 95 DO, and 74 DB stars. Also, note that 128 objects show a hybrid spectrum with both H and He lines (spectral type DAO, DAB, DOA, or DBA).

3. Model Atmospheres

We employed the codes TLUSTY and SYNSPEC (Hubeny & Lanz 1995) to compute non-LTE H+He model atmospheres and synthetic spectra. We constructed two sets of models differing by the assumed chemical configuration, which we refer to as homogeneous and stratified. In our chemically homogeneous models, H and He are distributed uniformly throughout the atmosphere, in which case the He-to-H number ratio (He/H) is a free parameter. Our homogeneous grid covers $30,000\ {\rm K} < T_{\rm eff} < 150,000$ K, $6.5 < \log g < 9.5$, and $-6.0 < \log {\rm He/H} < 6.0$. In our chemically stratified models, a thin H layer floats on top of a He envelope in diffusive equilibrium, in which case the mass of the H layer ($M_{\rm H}/M_*$) is a free parameter. The equilibrium chemical profile is obtained from the formalism of Vennes *et al.* (1988). Our motivation for calculating these models arises from the float-up model: as the float-up process builds a thin H layer at the surface, a chemically stratified atmosphere is indeed expected. Our stratified grid covers $30,000\ {\rm K} < T_{\rm eff} < 60,000$ K, $6.5 < \log g < 9.5$, and $-13.0 < Q < -5.0$ where $Q \equiv 3\log g + 2\log M_{\rm H}/M_*$. Note that models at the composition boundaries of our grids are practically identical to pure-H and pure-He atmospheres. Finally, it is worth mentioning that we implemented in TLUSTY and SYNSPEC the state-of-the-art treatment of Stark broadening of He I lines from Beauchamp *et al.* (1997), which is essential to properly reproduce the observed spectrum of He-rich white dwarfs.

4. Spectroscopic Analysis

For each star in our sample, we fitted the optical spectrum with our model atmosphere grids in order to derive the effective temperature, the surface gravity, and the atmospheric composition. We relied on our usual two-step fitting procedure: we first normalized the observed and synthetic spectra to a continuum set to unity, and then minimized the difference between these normalized spectra using the non-linear least-square Levenberg-Marquardt algorithm. Objects showing only H lines or only He lines were simply analyzed with pure-H or pure-He models, respectively. Hybrid white dwarfs were analyzed with both our homogeneous and stratified grids, and we adopted the best-fitting solution.

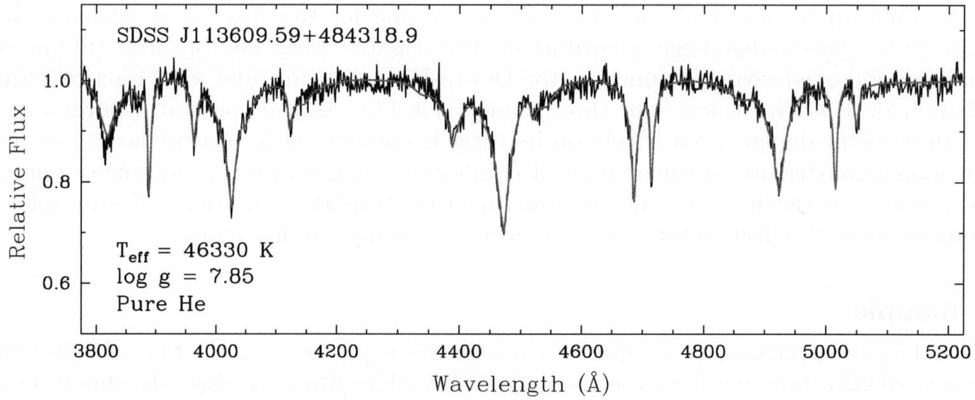

Figure 1. Example of a spectroscopic fit for a DO star in our sample. The observed and synthetic spectra are shown as the black and red lines, respectively, and the best-fitting atmospheric parameters are given in the figure.

Indeed, the H and He line profiles can be quite sensitive to the outer chemical configuration, which could thus be determined unambiguously in most cases (Manseau *et al.* 2016). Figure 1 displays an example of a fit for a DO star in our sample.

5. Results and Discussion

The left panel of Figure 2 presents the fraction of He-dominated white dwarfs as a function of decreasing effective temperature for our sample. The fraction of He-rich stars falls between 55 and 70% at the beginning of the cooling sequence, but drops sharply to ∼15% at ∼90,000 K. Taken at face value, this result seems to imply that between 40 and 55% of all white dwarfs are born with a He-rich atmosphere and then develop a H-rich atmosphere very early in their evolution, perhaps because they contain a relatively large amount of residual H (say, $M_{\rm H}/M_* \sim 10^{-6}$) that rapidly floats to the surface. Nevertheless, the large proportion of He-rich objects observed at very high temperatures might also be an artifact of the difference in evolutionary timescales between H-atmosphere and He-atmosphere white dwarfs. As discussed in recent studies, DA stars evolve faster than their DO counterparts at the beginning of the cooling sequence, hence the apparent lack of very hot DA white dwarfs (Werner *et al.* 2019).

Going down the temperature scale, we find that the fraction of He-rich objects is roughly constant between 90,000 and 60,000 K, and then decreases gradually from ∼15% at ∼60,000 K to ∼5% at ∼40,000 K. This suggests that, among the ∼15% of stars whose atmosphere is still He-dominated at ∼60,000 K, ∼2/3 contain a small amount of residual H (say, $M_{\rm H}/M_* \sim 10^{-10}$) that slowly floats to the surface, while ∼1/3 contain no H whatsoever. Therefore, below ∼40,000 K, the former become DA white dwarfs, while the latter remain DB white dwarfs. This interpretation is fairly robust, contrary to the case of hotter objects discussed in the previous paragraph, since the cooling ages of H-rich and He-rich stars become similar at lower temperatures. Thus, we can confidently assert the following: *at least* 15% of all white dwarfs are born with a He-dominated atmosphere, and those comprise two distinct groups, namely, H-deficient objects and H-free objects. Interestingly enough, this is somewhat reminiscent of the two known evolutionary channels feeding the He-rich white dwarf population, each associated with a different progenitor: the He-, C- and O-rich PG 1159 stars produced by a late He-shell flash, and the He-rich O(He) stars formed through a merger process (Reindl *et al.* 2014). We speculate that these two channels produce the two groups of hot He-rich white dwarfs that we have identified. More specifically, we propose that white dwarfs descending from

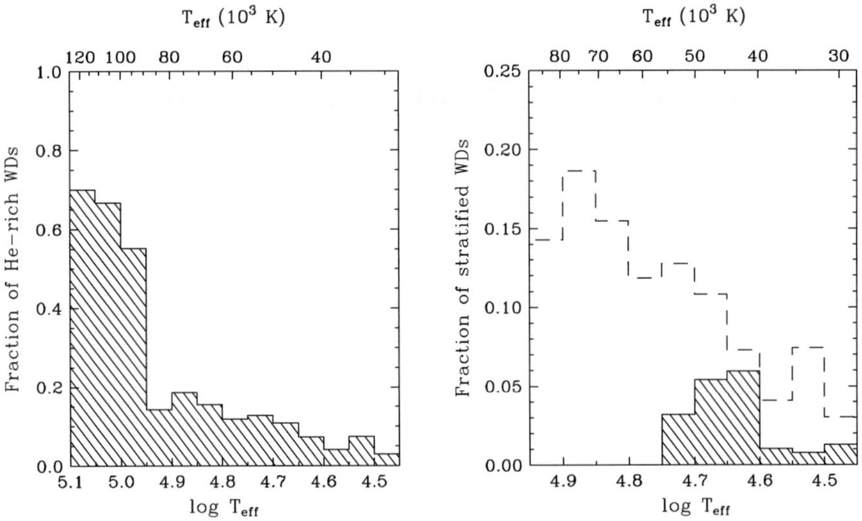

Figure 2. *Left panel:* Fraction of He-rich white dwarfs as a function of effective temperature. *Right panel:* Fraction of chemically stratified white dwarfs as a function of effective temperature. The histogram from the left panel is shown as the dashed line for comparison.

PG 1159 stars initially are He-rich but eventually become H-rich through the float-up of residual H, while white dwarfs descending from O(He) stars retain a He-dominated atmosphere throughout their entire evolution (perhaps because the whole H content was lost in the merger event). According to our results, these two evolutionary channels account for $\sim$10% and $\sim$5% of the total white dwarf population, respectively.

Finally, our spectroscopic analysis allowed us to identify 27 white dwarfs with a chemically stratified atmosphere. The right panel of Figure 2 displays the fraction of stratified objects as a function of decreasing effective temperature. The fraction of He-rich stars (the histogram from the left panel) is also reproduced for comparison. It is interesting to note that the stratified white dwarfs are found mainly in the temperature range where the fraction of He-rich white dwarfs decreases, just above the DB gap ($55,000$ K $> T_{\rm eff} > 40,000$ K). This confirms the idea that they are transitional objects currently undergoing the DO-to-DA conversion, and consequently lends strong support to the float-up model.

References

Beauchamp, A., Wesemael, F., & Bergeron, P. 1997, *ApJS*, 108, 559
Fontaine, G. & Wesemael, F. 1987, *IAU Colloq. 95: 2nd Conference on Faint Blue Stars*, 319
Hubeny, I. & Lanz, T. 1995, *ApJ*, 439, 875
Kepler, S. O., Pelisoli, I., Koester, D., *et al.* 2015, *MNRAS*, 446, 4078
Kepler, S. O., Pelisoli, I., Koester, D., *et al.* 2016, *MNRAS*, 455, 3413
Kleinman, S. J., Kepler, S. O., Koester, D., *et al.* 2013, *ApJS*, 204, 5
Manseau, P. M., Bergeron, P., & Green, E. M. 2016, *ApJ*, 833, 127
Reindl, N., Rauch, T., Werner, K., *et al.* 2014, *A&A*, 572, A117
Vennes, S., Pelletier, C., Fontaine, G., *et al.* 1988, *ApJ*, 331, 876
Werner, K. & Herwig, F. 2006, *PASP*, 118, 183
Werner, K., Rauch, T., & Reindl, N. 2019, *MNRAS*, 483, 5291

White Dwarfs as Probes of Fundamental Physics: Tracers of Planetary, Stellar and Galactic Evolution
Proceedings IAU Symposium No. 357, 2019
M. A. Barstow, S. J. Kleinman, J. L. Provencal & L. Ferrario, eds.
doi:10.1017/S1743921320000277

The spectral evolution of cool white dwarfs

Simon Blouin[1] and Patrick Dufour[2]

[1]Los Alamos National Laboratory, P.O. Box 1663, Mail Stop B265,
Los Alamos, NM 87545, USA
email: sblouin@lanl.gov

[2]Département de physique, Université de Montréal, Montréal, QC H3C 3J7, Canada
email: dufourpa@astro.umontreal.ca

Abstract. Empirically characterizing the spectral evolution of cool white dwarfs is a prerequisite to understanding the physical processes that shape the evolution of these old objects. However, the high photospheric densities of cool helium-rich atmospheres seriously complicate the study of those stars. We have recently developed an updated atmosphere code that is appropriate for high densities and that can model any cool white dwarf (including DZs and DQpecs). Here, we present recent advances in our understanding of the spectral evolution of cool white dwarfs that were made possible thanks to these improved models. We discuss in particular the evolution of the hydrogen-rich to helium-rich ratio at low effective temperatures as well as the DQ→DQpec transition.

Keywords. stars: abundances, stars: atmospheres, stars: evolution, white dwarfs

1. Introduction

Many different physical processes can affect the surface composition of a white dwarf throughout its evolution. These include diffusion, convective mixing, convective dilution, convective dredge-up from the core, radiative levitation, accretion from the interstellar medium and accretion of rocky debris. The competition between these mechanisms leads to the spectral evolution of white dwarfs, a topic that has been studied for decades. As larger and more accurate surveys become available, our understanding of spectral evolution and the underlying physical processes become gradually more complete (see in particular A. Bédard et al. and T. Cunningham et al. in these proceedings).

However, by lack of appropriate atmosphere models, the spectral evolution of the coolest white dwarfs has remained nebulous. Here, we describe the recent advances that were made possible thanks to our new cool white dwarf atmosphere models (see Blouin *et al.* 2018 and references therein for a complete description of the improved physics of our updated code).

2. The hydrogen-rich to helium-rich ratio

Previous attempts to constrain the ratio of H-rich to He-rich atmospheres for the coolest white dwarfs can be roughly divided into two categories. The first set of studies was based on the Montreal atmosphere code (Bergeron *et al.* 1997, 2001, Kilic *et al.* 2010) and the second set of studies relied on P. Kowalski's code (Kowalski 2006, Kilic *et al.* 2009). The major advantage of the second code is that it includes many high-density nonideal effects that need to be taken into account when modeling the fluidlike atmospheres of cool, He-rich white dwarfs. However, contrarily to the studies based on the Montreal code, studies that relied on P. Kowalski's code were limited to DA and DC stars, thus ignoring the He-rich DQs and DZs from their analyses.

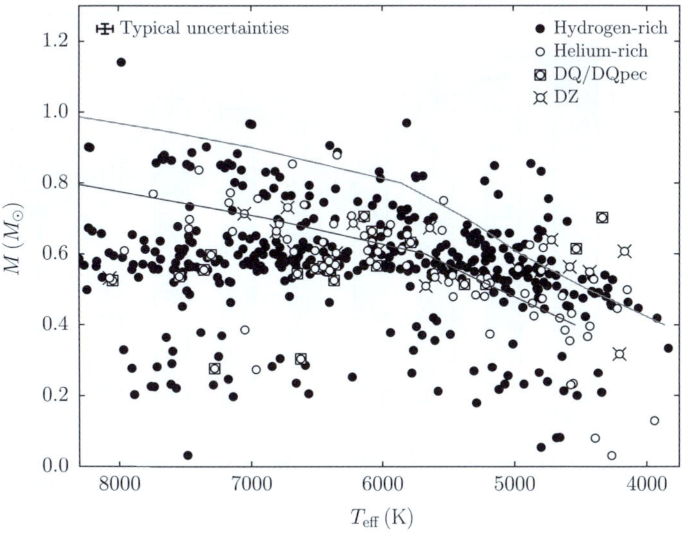

Figure 1. Mass of white dwarfs in our sample as a function of their effective temperatures. The blue and red curves indicate, respectively, the locations where 10% and 80% of the total mass has crystallized.

As we recently developed an updated atmosphere code that includes all relevant non-ideal effects and is suitable for any cool white dwarf, we were able to solve the issues that affected previous studies. We used this code to revisit the question of the H-rich to He-rich ratio of cool white dwarfs and we did so using an unbiased sample, without any discrimination on the spectral type. This section is a summary of the main results of this study, the details can be found in Blouin *et al.* (2019).

An $M - T_{\rm eff}$ diagram with all 501 objects contained in our sample is shown in Figure 1. As expected, most objects are clustered either around $0.6\,M_\odot$ or within the crystallization sequence where a pile-up is expected due to latent heat release and extra thermal energy resulting from the first phase of convective coupling (see also Bergeron *et al.* 2019 and Tremblay *et al.* 2019).

Figure 2 shows how the fraction of white dwarfs with H-rich atmospheres evolves with decreasing $T_{\rm eff}$. The first important trend that can be noticed is the decrease of the H-rich fraction between ≈ 8000 and 6000 K. This can be interpreted as the conversion of H-dominated white dwarfs into He-dominated objects due to convective mixing. Then, between ≈ 6000 and 5000 K, there seems to be an increase of the H-rich fraction. This feature is reminiscent of the non-DA gap, already described in various studies (e.g., Bergeron *et al.* 1997, 2001, Kilic *et al.* 2010). The origin of this increase remains unknown, although we note that our results allow us to eliminate the hypothesis according to which it is due to accretion of hydrogren from the interstellar medium (Kowalski 2006). If accretion of hydrogen were important, we would expect the increase of the H-rich fraction to continue down to 4000 K. On the contrary, we find that it stops at ≈ 5000 K (note that the apparent decrease of the H-rich fraction below 5000 K is probably mainly due to the faster cooling rate of He-atmosphere white dwarfs, as we find that the H-rich fraction stays approximately constant if plotted as a function of the cooling time).

While we took great care to avoid or correct any selection effect in our sample, we recognize that magnitude-limited spectroscopically identified samples can be prone to intractable biases (mainly due to complex SDSS selection effects). On the one hand, except for objects cooler than 5000 K (where differences are expected due to differences in model atmospheres), it is comforting that our results are in agreement with those of

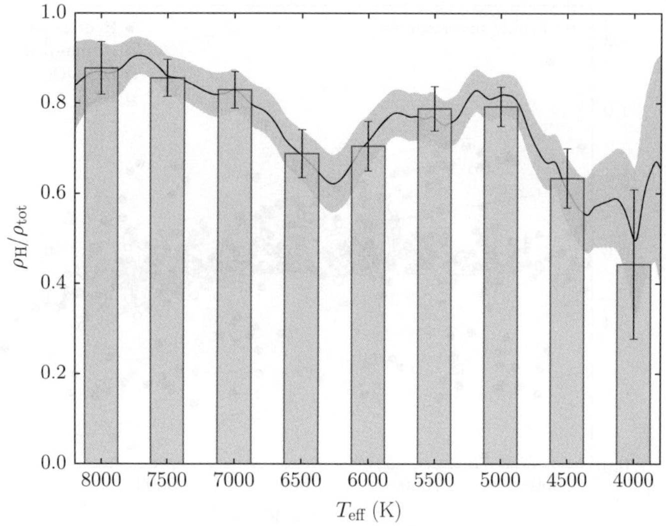

Figure 2. Fraction of H-rich white dwarfs as a function of effective temperature for fixed 500 K bins (in blue) and for a 500 K moving bin (in gray). The error bars indicate the 1σ uncertainty associated with the finite number of objects in each bin.

Limoges *et al.* (2015). Their sample was selected from reduced proper motion diagrams and is therefore not subject to biases in favor of H-rich or He-rich objects. On the other hand, the disagreement with the results of Cunningham *et al.* (2019) at the hotter hand of our sample is troubling, since their study is based on a volume-limited photometric sample for which selection effects are well understood. Hopefully, a large enough volume-complete sample of spectroscopically identified white dwarfs will soon allow us to investigate these differences.

3. Carbon-polluted white dwarfs

At low $T_{\rm eff}$, the C_2 Swan bands visible in carbon-polluted white dwarfs are distorted and shifted due to high-pressure effects (Kowalski 2010), which explains the DQpec phenomenon. Our code contains all the necessary physics to obtain satisfactory fits to those objects and, in Blouin & Dufour (2019), we present the first comprehensive star-by-star analysis of DQpecs. Our results allow us to extend our understanding of the evolution of carbon-polluted white dwarfs down to very low $T_{\rm eff}$.

At higher $T_{\rm eff}$, DQs are clustered along two distinct evolutionary sequences (Coutu *et al.* 2009; Koester & Kepler 2019). Most of them are normal-mass white dwarfs located along a track that is well explained by the dredge-up scenario (Figure 3). The second sequence, made of more polluted and generally more massive objects, is thought to be at least partially made of the descendants of Hot DQs.

In Figure 3, we show where DQpec white dwarfs (roughly, objects cooler than 6500 K) are located with respect to the two DQ sequences in a log C/He – $T_{\rm eff}$ diagram. The main result of this analysis is that DQpecs are simply the natural continuation of the DQ phenomenon, no matter what was the origin of carbon in the atmosphere. Most of them are the likely descendants of DQs that have dredged-up carbon from their core and a smaller fraction, located close to the second sequence, could be the descendants of Hot DQs (especially the more massive ones).

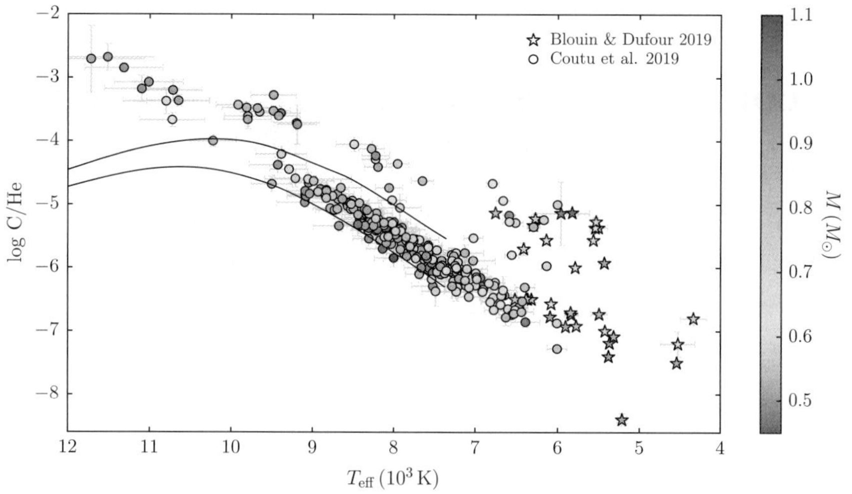

Figure 3. Carbon abundance as a function of effective temperature for white dwarfs in the Blouin & Dufour (2019, stars) and (Coutu *et al.* 2019, circles) samples. The solid lines correspond to the evolutionary models of Fontaine & Brassard (2005) assuming $M_{\rm WD} = 0.6\,M_\odot$ and $\log q({\rm He}) = -3$ (top) or $\log q({\rm He}) = -2$ (bottom).

Acknowledgments

Research presented in this conference proceeding was partially supported by the Laboratory Directed Research and Development program of Los Alamos National Laboratory under project number 20190624PRD2.

References

Bergeron, P., Ruiz, M. T., & Leggett, S. K. 1997, *ApJ*, 108, 339

Bergeron, P., Leggett, S. K., & Ruiz, M. T. 2001, *ApJ*, 133, 413

Bergeron, P., Dufour, P., Fontaine, G., Coutu, S., Blouin, S., Genest-Beaulieu, C., Bédard, A., & Rolland, B. 2019, *ApJ*, 876, 67

Blouin, S., Dufour, P., & Allard, N. F. 2018, *ApJ*, 863, 184

Blouin S., Dufour P., Thibeault C., & Allard N. F. 2019, *ApJ*, 878, 63

Blouin S. & Dufour P. 2019, *MNRAS* (in press)

Coutu, S., Dufour, P., Bergeron, P., Blouin, S., Loranger, E., Allard, N. F., & Dunlap, B. H. 2019, *ApJ*, 885, 74

Cunningham, T., Tremblay, P.-E., Gentile Fusillo, N. P., Hollands, M. A., & Cukanovaite, E. 2019, eprint arXiv:1911.00014

Fontaine, G. & Brassard, P. 2005, in Koester D., Moehler S., eds, Astronomical Society of the Pacific Conference Series Vol. 334, 14th European Workshop on White Dwarfs. p. 49

Kilic, M., Kowalski, P. M., & von Hippel, T. 2009, *ApJ*, 138, 102

Kilic, M., Leggett, S. K., Tremblay, P.-E., von Hippel, T., Bergeron, P., Harris, H. C., Munn, J. A., Williams, K. A., Gates, E., & Farihi, J. 2010, *ApJS*, 190, 77

Koester, D. & Kepler, S. O. 2019, *A&A*, 628, 102

Kowalski, P. M. 2006, PhD thesis, Vanderbilt University

Kowalski, P. M. 2010, *A&A*, 519, 8

Limoges, M.-M., Bergeron, P., & Lépine, S. 2015, *ApJS*, 219, 19

Tremblay, P.-E., Fontaine, G., Gentile Fusillo, N. P., Dunlap, B. H., Gänsicke, B. T., Hollands, M. A., Hermes, J. J., Marsh, T. R., Cukanovaite, E., & Cunningham, T. 2019, *Nature*, 565, 202

White Dwarfs as Probes of Fundamental Physics: Tracers of Planetary, Stellar and Galactic Evolution
Proceedings IAU Symposium No. 357, 2019
M. A. Barstow, S. J. Kleinman, J. L. Provencal & L. Ferrario, eds.
doi:10.1017/S1743921320000319

The completeness of Gaia-selected samples of white dwarfs

Terry D. Oswalt[1], Jay B. Holberg[2] and Edward M. Sion[3]

[1]Department of Physical Sciences, Embry-Riddle Aeronautical University,
1 Aerospace Blvd., Daytona Beach, FL 32114 USA
email: terry.oswalt@erau.edu

[2]Lunar & Planetary Laboratory, University of Arizona, Tucson, AZ 75201 USA
email: holberg@vega.lpl.arizona.edu

[3]Department of Astronomy & Astrophysics, Villanova University
Villanova, PA 19085 USA
email: edward.m.sion@villanova.edu

Abstract. The Gaia DR2 has dramatically increased the ability to detect faint nearby white dwarfs. The census of the local white dwarf population has recently been extended from 25 pc to 50 pc, effectively increasing the sample by roughly an order of magnitude. Here we examine the completeness of this new sample as a function of variables such as apparent magnitude, distance, proper motion, photometric color index, unresolved components, etc.

Keywords. stars, white dwarfs, Gaia

1. The Nearby White Dwarf Sample Revisited

Surveys to detect white dwarfs (WDs) have been conducted for almost a century, usually via astrometric and photometric criteria. Such samples tend to be magnitude-limited and are rarely complete. Drawing candidates from all available sources and confirming WDs via spectroscopy, Holberg *et al.* (2016) presented a 25 parsec WD sample, expanding the local WD census by about a factor of two over prior work (Holberg, Oswalt & Sion 2002). Based on number counts vs. trigonometric and spectroscopic parallaxes, this sample is about 68 percent complete and is among the most thoroughly vetted and well-defined samples of nearby WDs.

The Sloan Digital Sky Survey (SDSS) spectroscopically identified over 40,000 WDs (Kleinman *et al.* 2013; Kepler *et al.* 2019), increasing the census of WDs by nearly an order of magnitude. However, the SDSS primarily samples the youngest and most luminous WDs. It therefore provides little information on the large fraction of cool WDs and binary systems that contain WDs. Also, due to its bright magnitude limit the SDSS overlaps very little with searches for WDs within about 100 pc. For example, the 25 pc Holberg *et al.* (2016) sample contains only 5 WDs in common with the SDSS DR7.

With the advent of the Gaia mission DR2 (Gaia Collaboration 2016), the potential to construct extremely large samples of WDs is being realized. Recently, Gentile Fusillo *et al.* 2019) compiled a catalog of almost 500,000 WD candidates from the DR2. Using the SDSS sample of spectroscopically identified WDs to map the parameter space, multidimensional cuts in the Gaia astrometric and photometric data were used to identify 260,000 high confidence WD candidates.

Independently, our group extracted a sample of 2310 WD candidates within 50 pc of the Sun using Gaia DR2 trigonometric parallaxes and the properties of the Holberg *et al.* (2016) spectroscopically identified 25 pc sample. The result is shown in Figure 1, which

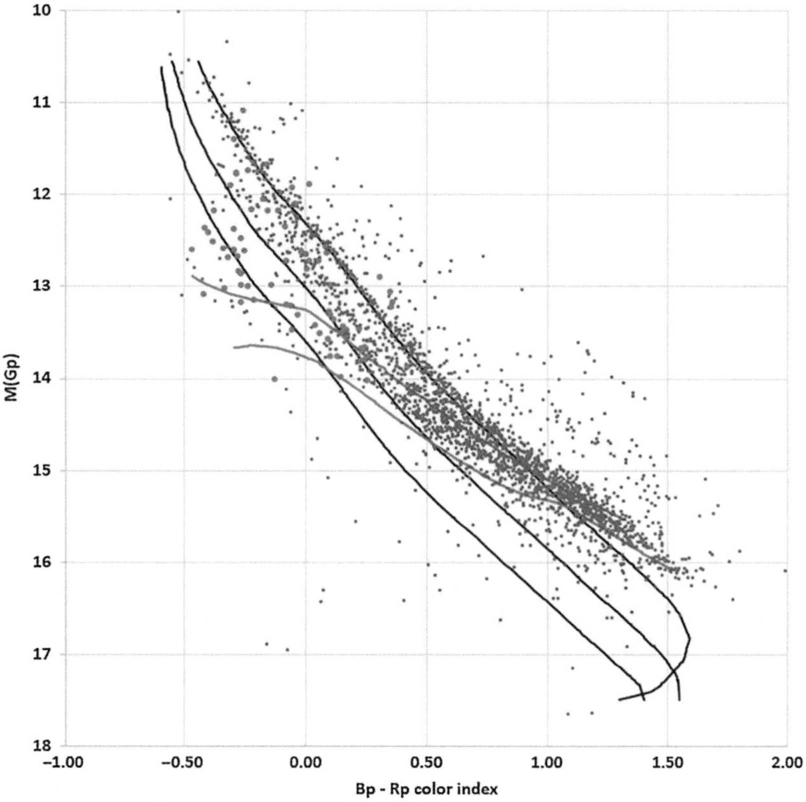

Figure 1. H-R diagram for 2310 WDs (blue dots) extracted from the Gaia DR2 using Gp magnitudes, trigonometric parallaxes and Bp - Rp color indices and spectroscopic information from Holberg *et al.* (2016). Red dots indicate known magnetic WDs identified by Kepler *et al.* (2013). Black curves indicate cooling sequences for WDs of 0.6, 0.9 and 1.1 solar mass (top to bottom, respectively) from Fontaine, Brassard & Bergeron (2001). Red curves indicate where evolutionary models by Tremblay *et al.* (2019) predict 20 percent (top) and 80 percent (bottom) of the total WD mass has crystallized.

is an H-R diagram constructed from Gaia trigonometric parallaxes, Gp magnitudes and Bp-Rp color indices. The 2310 WD candidates within 50 pc are represented by blue dots. The bifurcation of their distribution in this diagram results from the well-known effects of H-rich and He-rich atmospheres. Red dots identify magnetic WDs spectroscopically identified by Kepler *et al.* (2013) in the SDSS DR7. From top to bottom, black curves indicate cooling sequences for 0.6, 0.9 and 1.1 solar masses by Fontaine, Brassard & Bergeron (2001). This sample is large enough that a hint of the pile-up arising from the release of latent heat as the cores of WDs crystallize can be seen; red curves indicate where the evolutionary models of Tremblay *et al.* (2019) predict 20 percent (top) and 80 percent (bottom) of the total WD mass has crystallized.

Figure 1 suggests that our 50 pc sample and other Gaia DR2 WD samples are strongly affected by a number of factors. First, the scatter above the most populated 0.6 solar mass cooling curve suggests the sample includes many unresolved binaries. Based upon outliers in a Gaia color-color plot, we estimated that about 8 percent of our 50 pc sample are unresolved WD+MS binaries, comparable to the 6 percent contamination in the Holberg *et al.* (2016) 25 pc sample. Second, magnetic WDs are an important source of scatter. Unfortunately, only a small fraction of the Gaia WD candidates have spectra of sufficient

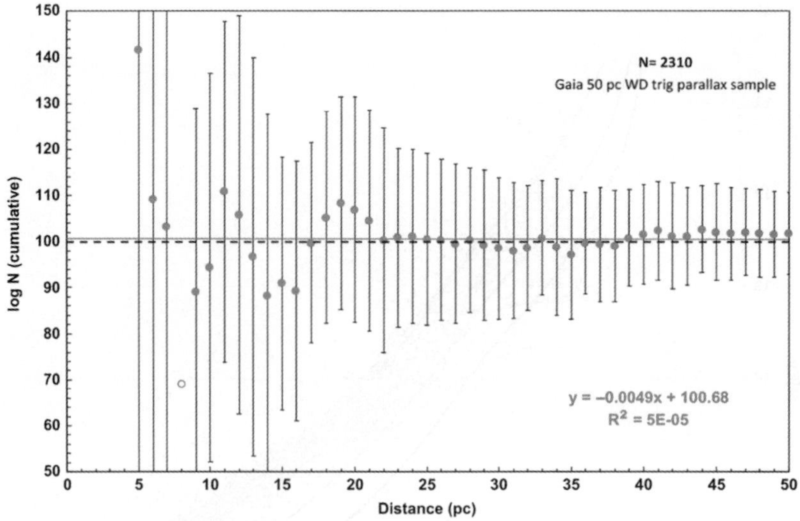

Figure 2. Completeness of the Gaia DR2 50 pc WD sample derived from cumulative counts vs. distance. Distances were directly obtained from Gaia trigonometric parallaxes. Error bars computed from $N^{-1/2}$ scatter in each bin.

resolution to detect line profile splitting and most of the WDs in the lower portion of the cooling sequence have featureless spectra. Third, stars with high tangential velocity and/or proper motion are missing from DR2; DR3 will remedy this.

Our Gaia 50 pc sample is confined well within the Galactic plane so simple star counts can be used to derive its completeness and the WD space density. As shown below, it is also large enough and complete enough to eliminate the need for the 1/Vmax method or more sophisticated statistical methods to construct the white dwarf luminosity function (WDLF), which is susceptible to small number fluctuations, observational biases, and/or incompleteness (see Garcia-Berro & Oswalt (2016) for a discussion of these problems).

2. Completeness as a Function of Distance

Assuming a uniform space density of stars, cumulative counts scale with the cube of the distance. Gaia DR2 provides an unprecedented quantity of precise trigonometric parallaxes, enabling this straightforward assessment of completeness. Oswalt, Holberg & Sion *et al.* (2017) outline several other ways, which we have also applied to our 50 pc sample. Figure 2 displays the bin-by-bin completeness of our 50 pc sample as a function of distance. Subject to the caveats outlined in the previous section, the 50 pc sample is formally 100 percent complete. The implied space density is 0.0041 WDs pc^{-3}, in very good agreement with Holberg *et al.* (2016) and the WD space densities compiled from various sources by Garcia-Berro & Oswalt (2016).

3. The White Dwarf Luminosity Function

The Gaia 50 pc sample enables the most straightforward approach to determining the WDLF: simple star counts within a volume-limited sample that is demonstrably complete. Figure 3 displays the WDLF for our Gaia 50 pc sample. The total space density implied is about 0.0044 pc^{-3}, slightly higher than that derived from number counts but well within uncertainties imposed by assuming H-rich atmospheres and bolometric corrections. The slight bump at the bright end near $M_{bol} = 9$ noted by Holberg *et al.* (2016) in the 25 pc sample is still visible. More noteworthy is a significant dip near

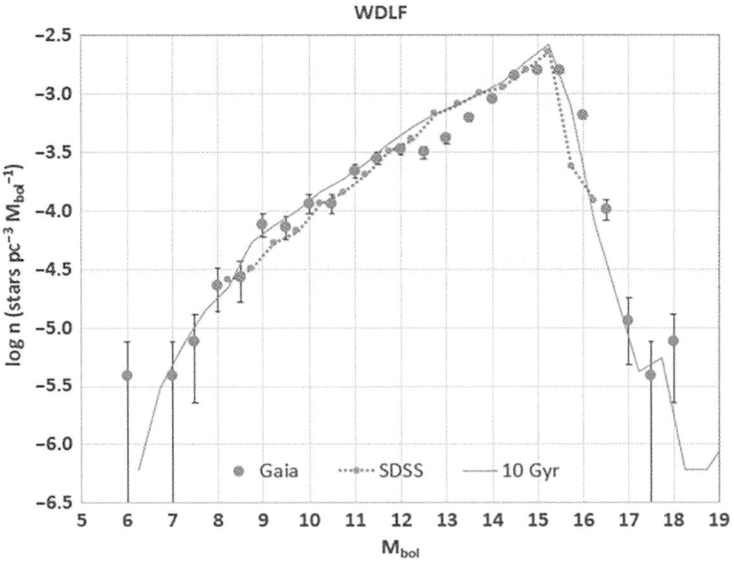

Figure 3. WDLF derived from the 50 pc Gaia sample (blue dots). The observed WDLF from the SDSS by Harris *et al.* (2016, orange dashed line) and a model WDLF for a Galactic disk age of 10 Gyr computed by Garcia-Berro & Oswalt (2016, priv. comm.; blue line) are shown for comparison. Error bars are as in Figure 1.

$M_{bol} = 13$, which may be related to crystallization among older WDs shown by Tremblay *et al.* (2019). It should be noted that unresolved pairs, Sirius-like binaries, magnetic WDs, etc., move individual objects into bins that do not reflect their actual luminosity. However, using a population synthesis approach, Rebassa-Mansergas *et al.* (2019) recently showed that unresolved double WDs and WD+MS pairs are unlikely to significantly affect the WDLF.

4. Conclusions

Our analysis of other properties, such as reduced proper motions, space motions, color indices, spectroscopic and population subtypes, etc. of the 50 pc sample is ongoing. In parallel with other groups (e.g., Jimenez-Esteban *et al.* 2018; Gentile Fusillo *et al.* 2019; Tremblay *et al.* 2019; Torres *et al.* 2019) we are also working to construct a complete volume-limited Gaia WD sample to 100 pc WD...and beyond.

We gratefully acknowledge funding for this project from NSF grants AST-1358787 and AST-1910396 to Embry-Riddle Aeronautical University, AST-1413537 to the University of Arizona and AST-1008845 to Villanova University. This work has made use of data from the European Space Agency (ESA) mission Gaia (https://www. cosmos. esa.int/gaia), processed by the Gaia Data Processing & Analysis Consortium (DPAC, https://www.cosmos.esa.int/web/gaia/dpac/consortium). Funding for DPAC has been provided by national institutions, in particular the institutions participating in the Gaia Multilateral Agreement.

References

Fontaine, G., Brassard, P., & Bergeron, P. 2001, *PASP*, 113, 409
Gaia Collaboration 2016, *A&A* 616, A1
Garcia-Berro, E. & Oswalt, T.D. 2016, *New Astron. Rev.* 72, 1

Gentile Fusillo, N., Tremblay, P.-E., Gansicke, B., Manser, C., Cunningham, T., Cukanovaite, E., Hollands, M., Marsh, T., Raddi, R., Jordan, S., Toonen, S., Geier, S., Barstow, M., Cummings, J., *et al.* 2019, *MNRAS*, 482, 4570

Harris, H. C., Munn, J. A., Kilic, M., Liebert, J., Williams, K. A., von Hippel, T., Levine, S. E., Monet, D. G., Eisenstein, D. J., Kleinman, S. J., Metcalfe, T. S., Nitta, A., Winget, D. E., Brinkmann, J., Fukugita, M., Knapp, G. R., Lupton, R. H., Smith, J. A., Schneider, D. P., *et al.* 2006, *AJ*, 131, 571

Holberg, J. B., Oswalt, T., & Sion, E. M. 2002, *ApJ*, 572, 512

Holberg, J. B., Oswalt, T. D., Sion, E. M., McCook, G. P., *et al.* 2016, *MNRAS* 462, 2295

Jimenez-Esteban, F., Torres, S., Rebassa-Mansergas, A., Skorobogatov, G., Solano, E., Cantero, C., Rodrigo, C., *et al.* 2018, *MNRAS* 480, 4505

Kepler, S., Pelisoli, I., Jordan, S., Kleinman, S., Koester, D., Kulebi, B., Pecanha, V., Castanheira, B., Nitta, A., Costa, J., Winget, D., Kanaan, A., Fraga, L., *et al.* 2013, *MNRAS* 429 2934

Kepler, S. O. *et al.* 2019, *MNRAS*, 469, 2169

Kleinman, S. J. *et al.* 2013, *ApJS*, 204, 5

Oswalt, T., Holberg, J., & Sion, E. 2017, *ASPC* 509, 59

Rebassa-Mansergas, A., Toonen, S., Torres, S., Canals, P., *et al.* 2019, *MNRAS*, (in press), astro-ph 1911.13128v1

Torres, S., Cantero, C., Rebassa-Mansergas, A., Skorobogatov, G., Jimenez-Esteban, F., Solano, S., *et al.* 2019, *MNRAS* 485, 5573

Tremblay, P.-E., Fontaine G., Gentile Fusillo, N., Dunlap, B., Gansicke, B., Hollands, M., Hermes, J., Marsh, T., Cukanovaite, E., Cunningham, T., *et al.* 2019, *Nature*, L565, 202

White Dwarfs as Probes of Fundamental Physics: Tracers of Planetary, Stellar and Galactic Evolution
Proceedings IAU Symposium No. 357, 2019
M. A. Barstow, S. J. Kleinman, J. L. Provencal & L. Ferrario, eds.
doi:10.1017/S1743921320000460

Two delays in white dwarf evolution revealed by *Gaia*

Sihao Cheng

Department of Physics and Astronomy, The Johns Hopkins University,
Baltimore, MD 21218, USA
email: s.cheng@jhu.edu

Abstract. By comparing two age indicators of high-mass white dwarfs (WDs) derived from *Gaia* data, two discoveries have been made recently: one is the existence of a cooling anomaly that produces the Q branch structure on the Hertzsprung–Russell diagram, and the other is the existence of high-mass WDs as double-WD merger products. The former poses a challenge for WD cooling models, and the latter has implications on binary evolution and type-Ia supernovae.

Keywords. white dwarfs, stars: evolution, stars: kinematics, Hertzsprung-Russell diagram, methods: statistical

1. Introduction

The unprecedented astrometric and photometric power of *Gaia* provides a unique opportunity to investigate white dwarf evolution. On the one hand, the photometric isochrone age of a large sample of white dwarfs can be accurately derived from the Hertzsprung-Russell (H–R) diagram. On the other hand, their kinematic information from *Gaia* makes it possible to infer the true age of these white dwarfs, according to the age–velocity-dispersion relation of the Milky Way disc. Because the photometric isochrone age is derived with a single-star evolution model and a standard white dwarf cooling model, any discrepancy between these two age indicators, or equivalently, the discrepancy between the observed velocity distribution and the velocity distribution predicted from the photoemtric ages and the age–velocity-dispersion relation, indicates deviations from the single-star evolution or the standard cooling model. Also, by carefully designing the comparison, one can identify more than one source of the deviation based on their different properties. In this article, I present two evolutionary delays of white dwarfs revealed by this method, which include an unexpected cooling delay and a merger delay from binary evolution, and discuss related astrophysical questions.

2. Do we understand the cooling of white dwarfs?

Most white dwarfs shine at the cost of losing their thermal energy and cooling down over cosmic time. Since the seminal work of Mestel (1952), a series of detailed cooling models have been developed, making white dwarfs accurate cosmic clocks (see e.g. Fontaine *et al.* 2001 for a review).

Unexpectedly, the *Gaia* data (Gaia Collaboration *et al.* 2018) revealed an over-density of high-mass white dwarfs on the H–R diagram, called the Q branch (Figure 1). Standard white dwarf models do not anticipate this feature. An explanation was proposed by Tremblay *et al.* (2019) that the energy release from crystallisation produces this pile-up on the H–R diagram. However, Cheng *et al.* (2019a) found that on the Q branch, an anomalously high fraction of white dwarfs have a high transverse velocity (Figure 1).

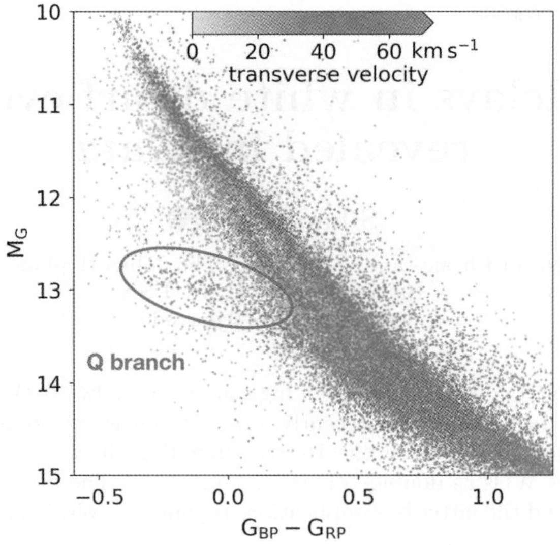

Figure 1. The H–R diagram of white dwarfs within 150 pc, colour-coded by transverse velocity. The data come from *Gaia* Data Release 2. Fast-moving high-mass white dwarfs are concentrated on the Q branch, which can only be explained by a significant slowing-down of cooling rate in a small fraction of white dwarfs.

According to the age–velocity-dispersion relation of the Milky Way disc (e.g., Holmberg *et al.* 2009), these white dwarfs must be old. As argued by Cheng *et al.* (2019a), different cooling behaviours are required to explain both the pile-up and the velocity distribution of high-mass white dwarfs (1.08–1.23 $M_{\odot}$): while most of them cool normally as the standard cooling model (which already includes crystallisation effect) predicts, about 6% almost stop their cooling on the Q branch for as long as 8 billion years. Neither crystallisation delay or merger delay alone cannot explain all observations. So, we called this delay an 'extra cooling delay' and the population the 'extra delayed population'. This anomaly is both a challenge and an opportunity for understanding white dwarf cooling.

To stop white dwarf cooling, an extra energy source is needed to provide the energy loss from shining. With a semi-analytical calculation, Cheng *et al.* (2019a) showed that the sink of ^{22}Ne in C/O-core white dwarfs is a promising candidate for the extra energy source. In this explanation, the extra delayed white dwarfs are bizarre 'sedimentars', as predicted and dubbed by Bildsten & Hall (2001); these white dwarfs shine out of the gravitational sedimentation of material instead of cooling.

There are several aspects of this cooling anomaly and the Q branch worthy of further exploration.

More evidence for the extra cooling delay: Collecting more evidence for this cooling anomaly is still necessary. LP 93-21, a high-mass white dwarf just below the Q branch but with halo kinematics (Kawka, Vennes & Ferrario 2020) may count as additional evidence. Other evidence may come from using asteroseismology to directly measure the cooling rate (as proposed by Bart Dunlap at this symposium) in the future.

Cooling models with ^{22}Ne settling: The ^{22}Ne settling effect depends on the white dwarf mass, ^{22}Ne abundance, core composition, crystallisation temperature, and the treatment of the physics of settling. Despite existing cooling models in the literature, there has not been a model incorporating both the ^{22}Ne settling effect and the updated phase diagram of crystallisation for high-mass ($> 1.1\ M_{\odot}$) C/O-core white dwarfs. Such models would test the ^{22}Ne settling hypothesis and may provide new insight into white dwarf physics.

DQ white dwarfs on the Q branch: All DQs on the Q branch belong to the extra delayed population, whereas among the extra delayed population, about half are DQs and the other half are DAs. Both the number density and carbon abundance suggest that the Q-branch DQs are evolutionary descendants of the hot-DQs discovered by Dufour *et al.* (2007). Further questions include the origin of the distinction between DQs and DAs and whether the Q-branch DQs also have magnetic field as the hot-DQs.

Merger origin: Three arguments support a double-white-dwarf (double-WD) merger origin of the extra delayed white dwarfs: the ^{22}Ne settling hypothesis for the delay (because it requires C/O-core), the possible relation to hot-DQs, and the lack of wide binary companions. Nevertheless, further definitive evidence is still needed.

Core composition: We expect that the extra delayed white dwarfs have C/O cores instead of O/Ne cores that is expected for normal high-mass white dwarfs. Future observations that can determine the core composition will be crucial to testing both the ^{22}Ne settling hypothesis and their merger origin.

3. What are the origins of high-mass white dwarfs?

High-mass white dwarfs can form from single-star evolution or from merger of stars. In particular, a considerable fraction of high-mass white dwarfs are expected to be double-WD merger products (e.g., Temmink *et al.* 2019). However, despite discussions in the literature, there is still a lack of clear evidence for the white dwarfs that are merger products. The difficulty is to reliably distinguish the merger population from singly evolved white dwarfs.

The astrometric information from *Gaia* has improved the situation significantly. Because double-WD merger products experience additional binary evolution, their true ages are younger than their photometric isochrone ages. To get rid of the influence of the cooling anomaly described in the previous section, Cheng *et al.* (2019b) selected only high-mass white dwarfs above the Q branch and used their transverse velocity distribution to infer the fraction of merger products. Thanks to the much larger sample size than what Wegg & Phinney (2012) used in a previous study with similar method, Cheng *et al.* (2019b) found clear velocity excess that corresponds to about $20 \pm 6\%$ of high-mass white dwarfs (0.8–1.3 $M_{\odot}$) being double-WD merger products. This is strong evidence for the merger origin of high-mass white dwarfs.

4. How frequently do two white dwarfs merge?

The fraction of double-WD merger products can also be translated into a double-WD merger rate in some mass range. Then, it can be compared with binary population synthesis results and other observations to improve our understanding of binary evolution, and it can also be compared with the type-Ia supernovae rate as double-WD merger is one of the promising scenarios of type-Ia supernovae.

Here, I only discuss the double-WD mergers occurring in a close binary system. In general, there are three ways to measure the double-WD merger rate: to extrapolate from pre-merger binary systems, to count merger events, and to search for merger products. Previous works were mainly focused on extrapolating the orbital distribution of pre-merger systems (e.g., Brown *et al.* 2016; Maoz *et al.* 2018), which currently provides no or low mass resolution and large uncertainties. The merger rate translated from the fraction of merger products in Cheng *et al.* (2019b) adds significant mass resolution and precision to previous measurements, and it is shown to be close to binary simulation results (Figure 2).

If it is allowed to extrapolate the observed merger rate to a higher mass range using the simulated mass distribution, one will conclude that the rate of mergers with total mass

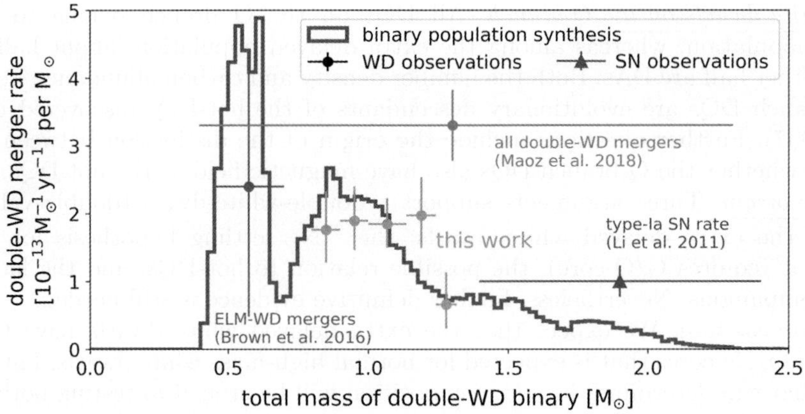

Figure 2. Estimates of the double-WD merger rate as a function of total mass, from both observational and theoretical sides. This figure is reproduced from the Figure 4 in Cheng *et al.* (2019b).

$> 1.4\,M_\odot$ is a third to a half of the type-Ia supernova rate measured for Milky-Way-like galaxies (Li *et al.* 2011). The fact that the two rates are close to each other supports the idea that massive double-WD mergers may contribute to a significant fraction of type-Ia supernovae, whereas the fact that the merger rate is lower than the supernova rate indicates other channels of type-Ia supernovae, including double-WD mergers with lower total mass and/or single degenerate progenitors.

References

Bildsten, L. & Hall, D. M. 2001, *Ap. Lett.*, 549, L219

Brown, W. R., Kilic, M., Kenyon, S. J., & Gianninas, A. 2016, *ApJ*, 824, 46

Cheng, S., Cummings, J., & Ménard, B. 2019a, *ApJ*, 886, 100

Cheng, S., Cummings, J., Ménard, B., & Toonen, S. 2019b, arXiv:1910.09558

Dufour, P., Liebert, J., Fontaine, G., & Behara, N. 2007, *Nature*, 450, 522

Fontaine, G., Brassard, P., & Bergeron, P. 2001, *PASP*, 113, 409

Gaia Collaboration, Babusiaux, C., van Leeuwen, F., *et al.* 2018, *A&A*, 616, A10

Holmberg, J., Nordström, B., & Andersen, J. 2009, *A&A*, 501, 941

Kawka, A., Vennes, S., & Ferrario, L. 2020, *MNRAS* Letters, 491, L40

Li, W., Chornock, R., Leaman, J., *et al.* 2011, *MNRAS*, 412, 1473

Maoz, D., Hallakoun, N., & Badenes, C. 2018, *MNRAS*, 476, 2584

Mestel, L. 1952, *MNRAS*, 112, 583

Temmink, K. D., Toonen, S., Zapartas, E., Justham, S., & Gänsicke, B. T. 2019, arXiv:1910.05335

Tremblay, P.-E., Fontaine, G., Fusillo, N. P. G., *et al.* 2019, *Nature*, 565, 202

Wegg, C. & Phinney, E. S. 2012, *MNRAS*, 426, 427

White Dwarfs as Probes of Fundamental Physics: Tracers of Planetary, Stellar and Galactic Evolution
Proceedings IAU Symposium No. 357, 2019
M. A. Barstow, S. J. Kleinman, J. L. Provencal & L. Ferrario, eds.
doi:10.1017/S1743921320000848

Ensemble evolutionary studies of white dwarfs in open star clusters

Kurtis A. Williams

Department of Physics & Astronomy, Texas A&M University-Commerce
P.O. Box 3011, Commerce, TX, 75402, USA
email: Kurtis.Williams@tamuc.edu

Abstract. White dwarfs (WDs) in open star clusters are a highly useful ensemble of stars. While numerous researchers use open cluster WDs to study the initial-final mass relation, numerous other evolutionary studies are also enabled by this sample of stars, including searches for stochastic mass loss, studies of binary star evolution, and measurements of metallicity impacts on WD formation and evolution. However, it is crucial to use astrometric data such as proper motions to remove contaminating field WDs from open cluster samples; multi-epoch ground based imaging is needed for most open cluster WDs. Also, the strongly correlated errors in the initial mass - final mass plane must be considered; we illustrate the importance of this consideration using a large open cluster WD sample and Monte Carlo techniques.

Keywords. white dwarfs, open clusters and associations: general, stars: evolution

1. The Utility of Open Cluster White Dwarfs

Since the start of the millennium, my collaborators and I have investigated the white dwarf (WD) populations of open star clusters. The primary advantage of open cluster WDs is that crucial stellar parameters such as distance, progenitor metallicity, and the total age (WD cooling age plus the nuclear lifetime of the progenitor) are well constrained from observations of non-WD cluster members.

This additional knowledge opens up avenues of research that are difficult, though not impossible, to pursue with field WDs. Up to now the primary use of open cluster WD samples has been determination of the semi-empirical initial-final mass relation (IFMR). Early observational studies were headed by Volker Weidemann (e.g., Weidemann 1977) and were furthered significantly by Detlev Koester and Dieter Reimers in the 1990s. Subsequently, the advent of blue-sensitive spectrographs on 8-meter and larger telescopes has allowed several groups to increase the number of known open cluster WDs by a factor of 7 from that in Weidemann's initial study; a recent compilation can be found in Cummings *et al.* (2018). This number increases if WDs in field binary systems, which are essentially two-star clusters, are included (e.g., Catalán *et al.* 2008).

Over the course of our work (e.g., Williams *et al.* 2009 and Williams *et al.* 2018), my collaborators and I recognized the ever-growing cluster WD sample can shed light on additional stellar evolutionary questions. For example, the Vogt-Russell theorem is the oft-taught conjecture that the evolutionary path of a single star from a single-metallicity population is determined almost exclusively by its mass, and it is often implicitly invoked in interpretation of WD research. Yet within NGC 6791 there is significant intrinsic scatter in mass loss (Kalirai *et al.* 2007), while in M67 no significant scatter in individual WD masses is observed (Williams *et al.* 2018).

Multiple star evolution should also affect the WD populations in open star clusters, where binary fractions are high. For example, in M67, we have identified likely WD remnants of blue stragglers, helium-core WD remnants of binary evolution, and a cataclysmic variable (Williams *et al.* 2018 and Williams *et al.* 2013). Follow-up studies of these WDs should be able to constrain model details such as interaction timescales and remnant system structure. Such work requires large sample sizes with precision measurements.

Among field WDs, atmospheric layer masses appear to vary from moderately thin (hydrogen layer masses of $\approx 10^{-7} M_{\odot}$) to the thickest layers possible without igniting a final thermal pulse ($\approx 10^{-4} M_{\odot}$). Are these variations a result of stochasticity during the final phases of post main sequence mass loss, or are they due to significant differences in the progenitor stars themselves, such as metallicity? These questions can be addressed via asteroseismic study of open cluster WDs, an idea I first heard from Antonio Kanaan and only recently feasible on large telescopes.

However, there are many important assumptions and systematic errors plaguing the above studies. Many of these were quantified by Salaris *et al.* (2009). Below I outline two other critical considerations that must be addressed in future open cluster white dwarf studies: cluster membership and correlated errors in the IFMR.

2. Cluster Membership Determination

Open cluster WD samples require that bona fide cluster member WDs be separated reliably from the significant field WD contamination. To date the primary means of this us through the use of distance modulus determinations. The observationally derived $T_{\rm eff}$and $\log g$ are used to search published WD evolutionary models to calculate the absolute magnitude for each star. We then calculate the apparent distance modulus for each WD and compare it with the open cluster's apparent distance modulus. If the two are consistent, and if the cooling age of a WD is younger than the star cluster age, we accept the WD as a likely cluster member.

This method works reasonably well, but it is not a guarantee of cluster membership. Besancon models (Robin *et al.* 2003 of typical open cluster fields show ~ 3 field WDs along the line of sight would meet these membership criteria for cluster ages $\leqslant 1$ Gyr, a significant contamination when compared to the dozen or fewer cluster member WDs typically identified. Proper motion memberships of WDs in Messier 67 confirms these contamination rates (Williams *et al.* 2018). Further, distance modulus selection eliminates unresolved double degenerates from cluster WD samples, as they generally appear significantly overluminous. While this exclusion is desirable for IFMR studies, which assume single-star evolution, it inhibits other areas of open cluster WD research.

The ideal means to determine cluster membership would be to have parallax and proper motion measurements for each WD. However, even in the final Gaia catalogs relatively few open cluster WDs will have had such kinematic information due to their intrinsic faintness ($V \geqslant 21$ beyond 1 kpc). We are just now entering the an era where sufficient time has passed since the first epoch of deep, wide-field CCD imaging of open clusters that proper motion memberships for these fainter WDs can be measured by multi-epoch ground-based imaging.

Yet proper motions alone cannot cleanly separate field and cluster WDs. The young- and intermediate-age clusters targeted in most IFMR studies are dynamically young and co-rotate with the disk field stars, so field and cluster star proper motion distributions are overlapping. However, the *dispersions* of these distributions are markedly different; field star proper motion dispersions toward kinematically young clusters are $\approx 10-15$ mas/yr (Dias *et al.* 2001).

In order to determine the minimal proper motion precision required to create a clean cluster WD sample, I have simulated ground-based WD proper motion measurements in the field of Messier 35, including the 12 known cluster WDs from Williams *et al.* (2009) and the 29 field WDs identified in that same work. I randomly drew cluster and field

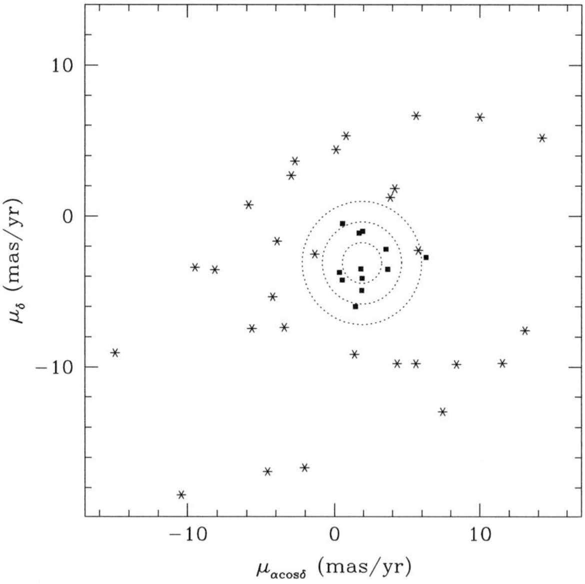

Figure 1. A simulated proper motion vector diagram for WD candidates in M35, assuming an 11 year baseline and including observational errors. Filled squares are cluster member WDs, and asterisks indicate the field stars. Dashed circles indicate membership selection criteria of 1.4 mas/yr, 2.8 mas/yr, and 4.2 mas/yr. Even though the cluster WD (filled squares) and field WD (asterisks) proper motions have similar centroids, we can obtain a low-contamination WD sample with our data.

WD proper motions from the values and dispersions in Dias *et al.* (2001) and included astrometric positional uncertainties of 10 mas per epoch.

Figure 1 shows a representative proper motion plot from this simulation. Based on 10,000 realizations, we find that a 3 mas/yr selection criteria recovers an average of ≈ 10 of the 12 cluster member WDs, with an average contamination of ≈ 2 field WDs. But this is from proper motion selection *alone*. Once the distance modulus and age criteria discussed above are applied, the field contamination is less than 1%.

3. Errors in the Semi-Empirical Initial-Final Mass Relation

The canonical way of presenting IFMR data and uncertainties is to plot each point with orthogonal error bars in the initial-final mass plane, such as we do in Williams *et al.* (2018). However, this practice fails to indicate the underlying highly correlated errors present in the initial-final mass plane, as illustrated in Figure 2. The correlated errors are not difficult to understand qualitatively but require numerical methods to quantify. The observables leading to the IFMR are the $T_{\rm eff}$and $\log g$ measured from WD spectra. These measurements are only weakly correlated for the WDs in the current open cluster IFMR and can be ignored, albeit grudgingly.

The final mass (WD mass) is primarily determined by an assumed mass-radius relationship and the measured $\log g$, with a weak dependence on $T_{\rm eff}$, especially among hotter WDs. The initial mass (progenitor star mass) is derived by subtracting the WD cooling age from the cluster age; this difference is the nuclear lifetime of the progenitor star. One can then use a favorite stellar evolutionary models to derive the star's zero-age main sequence mass. Note the WD cooling age depends strongly on both $T_{\rm eff}$and $\log g$.

The quantitative correlation errors in the initial-final mass plane depend on the relative size of the uncertainty in the WD cooling age to the star cluster age and the slope of

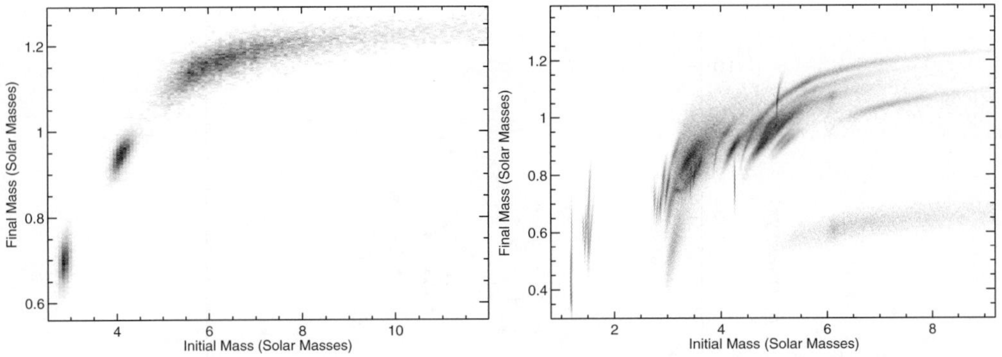

Figure 2. *Left:* Monte Carlo simulations of the IFMR for three WDs with very high S/N observations: WD 0421+162 (bottom left), NGC 3532 WD 1 (center), and M35 WD 2 (top right). *Right:* Monte Carlo simulations of the current open cluster semi-empirical IFMR from Canton (2018). Swooshes indicate the uncertainty regions for individual white dwarfs. All cluster ages are fixed at the values used by Cummings *et al.* (2018). Note that orthogonal error bars oft plotted in IFMR determinations are often not representative of the actual uncertainties.

the progenitor's nuclear lifetime as a function of initial mass. The left panel of Figure 2 shows a Monte Carlo simulation of the initial-final mass relation for three individual WDs with similar $T_{\rm eff}$ and similar observational uncertainties. The shape of the uncertainty distribution is clearly not described by orthogonal error bars in two of the three cases.

In the right panel of Figure 2, I show the uncertainty distributions in the IFMR plane for all of the open cluster WD spectra reanalyzed by Paul Canton in his Ph.D. thesis (Canton 2018). This sample has significant overlap with that of Cummings *et al.* (2018) but was obtained independently; the observational parameters measured from the WD spectra are fully consistent within stated errors. This plot strongly suggests that IFMR studies must not ignore the correlated errors of individual points.

An additional important source of uncertainty in the semi-empirical IFMR is not included in my simulations above but also cannot be ignored, and this is the potential for significant changes in the functional form of the IFMR due to uncertainties in individual star cluster ages. An error in the assumed cluster age will systematically move all of the WDs in that cluster parallel to the initial mass axis as well as potentially change the shape of each WD's individual uncertainty distribution.

Most semi-empirical IFMR studies publish the uncertainty in initial masses due to cluster age uncertainties. However, it is crucial to remember that the changes in initial mass due to these will be systematic, not random errors, within a given cluster. Advanced techniques such as the Bayesian hierarchical modelling of Si *et al.* (2018) should therefore be used in IFMR analysis.

References

Canton, P. 2018, Ph.D. Thesis

Catalán, S., Isern, J., García-Berro, E., & Ribas, I. 2008, *MNRAS*, 387, 1693

Cummings, J. D., Kalirai, J. S., Tremblay, P.-E., Ramirez-Ruiz, E., & Choi, J., 2018, *ApJ*, 866, 21

Dias, W. S., Lépine, J. R. D., & Alessi, B. S. 2001, *A&A*, 376, 441

Kalirai, J. S., Bergeron, P., Hansen, B. M. S., Kelson, D. D., Reitzel, D. B., Rich, R. M., & Richer, H. B. 2007, *ApJ*, 671, 748

Robin, A. C., Reylé, C., Derrière, S., & Picaud, S. 2003, *A&A*, 409, 523

Salaris, M., Serenelli, A., Weiss, A., & Miller Bertolami, M. 2009, *ApJ*, 692, 1013

Si, S., van Dyk, D. A., von Hippel, T., *et al.* 2018, *mnras*, 480, 1300

Weidemann, V. 1977, *A&A*, 59, 411
Williams, K. A., Bolte, M., & Koester, D. 2009, *ApJ*, 693, 355
Williams, K. A., Canton, P. A., Bellini, A., Bolte, M., Rubin, K. H. R., Gianninas, A., & Kilic, M. 2013, *ApJ*, 867, 62
Williams, K. A., Canton, P. A., Bellini, A., Bolte, M., Rubin, K. H. R., Gianninas, A., & Kilic, M. 2018, *ApJ*, 867, 62

White Dwarfs as Probes of Fundamental Physics: Tracers of Planetary, Stellar and Galactic Evolution
Proceedings IAU Symposium No. 357, 2019
M. A. Barstow, S. J. Kleinman, J. L. Provencal & L. Ferrario, eds.
doi:10.1017/S1743921320001003

New population synthesis approach: The golden path to constrain stellar and galactic physics

Nadège Lagarde and Céline Reylé

Institut UTINAM, CNRS UMR 6213, Univ. Bourgogne Franche-Comté,
OSU THETA Franche-Comté-Bourgogne, Observatoire de Besançon, BP 1615, 25010,
Besançon Cedex, France
email: nadege.lagarde@utinam.cnrs.fr

Abstract. The cornerstone mission of the European Space Agency, Gaia, has revealed properties of 260 000 white dwarfs in the Galaxy, allowing us for the first time to constrain the evolution of white dwarfs with a large sample. Complementary surveys (CoRoT, *Kepler*, K2, APOGEE and Gaia-ESO), will revolutionize our understanding of the formation and history of our Galaxy, providing accurate stellar masses, radii, ages, distances, and chemical properties for very large samples of stars across different Galactic stellar populations. To exploit the potential of the combination of spectroscopic, seismic and astrometric observations, the population synthesis approach is a very crucial and efficient tool. We develop the Besançon Galaxy model (BGM, Lagarde *et al.* 2017) for which stellar evolution predictions are included, providing the global asteroseismic properties and the surface chemical abundances along the evolution of low- and intermediate-mass stars. For the first time, the BGM can explore the effects of an extra-mixing occurring in red-giant stars (Lagarde *et al.* 2019), changing their stellar properties. The next step is to model a consistent treatment of giant stars and their remnants (e.g., white dwarfs). This kind of improvement would help us to constrain stellar and Galactic evolutions.

Keywords. Galaxy: stellar content; stars: evolution; stars: white dwarfs

---

1. Observational context

The Gaia space mission provided astrometry and photometry for one billion stars, with typical uncertainties of about 0.3 mas for the positions and parallaxes, and about 1 mas/yr for the proper motions (Data release 2, Gaia Collaboration 2018a). The last data release of Gaia has already provided precise astrometric and photometric data for 260,000 high-confidence white dwarf candidates (Gentile Fusillo *et al.* 2019), and the future data releases should highlight $\sim$ 400,000 white dwarfs in different Galactic regions (Jordan 2007). As shown by the Gaia Collaboration (2018b), the incredible accuracy of Gaia data allows us to distinguish the hydrogen white dwarfs from the helium white dwarfs in the colour-magnitude diagram.

Indeed, a broad effort is ongoing with **large spectroscopic surveys** such as the Sloan Digital Sky Survey (SDSS, Kepler *et al.* 2016), discovering more than 30 000 white dwarfs in the data release 12 (Kepler *et al.* 2016). The SDSS has allowed numerous detailed spectroscopic studies of white dwarfs, specifically for DA and DB white dwarfs (e.g., Eisenstein *et al.* 2006; Kepler *et al.* 2007; Tremblay *et al.* 2011; Koester & Kepler 2015; Genest-Beaulieu & Bergeron 2019).

Finally, **asteroseismology data** of stars observed by the space missions CoRoT (Baglin & Fridlund 2006), *Kepler* (Gilliland *et al.* 2010), and K2 and TESS (Ricker *et al.* 2015) provide crucial constraints on stellar properties such as masses, radii, evolutionary

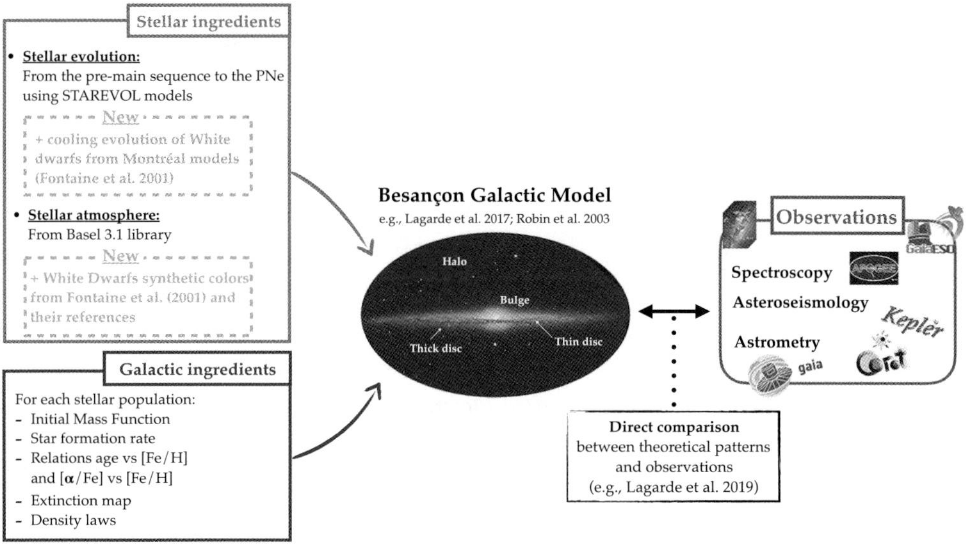

Figure 1. The schematic view of the Besançon Galactic stellar population synthesis model. The new improvements concerning the treatments of white dwarfs are shown in orange.

states, and internal rotation profile especially for red-giant stars (e.g. Stello *et al.* 2008; Mosser *et al.* 2012b; Bedding *et al.* 2011; Vrard *et al.* 2016; Mosser *et al.* 2012a; Beck *et al.* 2012). Most white dwarfs go through a pulsating stage allowing the study of their internal properties using asteroseismology (e.g., Fontaine & Brassard 2008; Winget & Kepler 2008; Althaus *et al.* 2010; Vauclair 2013; Córsico *et al.* 2019, and more detail in Corsico's review in this volume). All of these data have been and will be collected in different regions of our Galaxy, providing a unique opportunity to constrain stellar and Galactic models.

2. Galactic stellar populations synthesis

To exploit their full potential, it is crucial to perform a combined analysis of these different kinds of observations. The population synthesis approach is a powerful tool for such analysis, allowing the computation of mock catalogues under various model hypothesis, and to statistically compare these simulated catalogues with any type of large survey data. The stellar population synthesis approach is a useful tool to compare directly theoretical stellar and Galactic models with observations (see Fig. 1). The Besançon Galaxy model (BGM) is a stellar population synthesis model (e.g., Robin *et al.* 2003; Czekaj *et al.* 2014; Lagarde *et al.* 2017) intended to meld the formation and evolution scenarios of the Galaxy, stellar formation and evolution theory, and models of stellar atmospheres, as well as dynamical constraints, in order to make a consistent picture of the Galaxy in comparison with available observations (photometry, asteroseismology, astrometry, and spectroscopy) at different wavelengths.

To benefit from the combination of recent asteroseismic and spectroscopic surveys, Lagarde *et al.* (2017) have updated the stellar evolution models used in the BGM, to be able to simulate the seismic and chemical properties for all evolutionary stage. This new version of the BGM can also take into account the effect of extramixing occurring in giant stellar interiors. On the one hand, a first comparison with the large spectroscopic survey

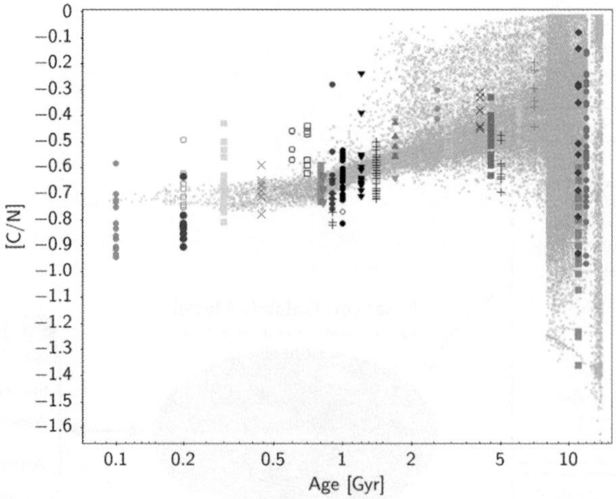

Figure 2. *Figure from Lagarde et al. (2019)* : [C/N] as a function of stellar ages for synthetic populations computed with the BGM including the effects of thermohaline instability (gray dots).The observed [C/N] abundances observed by UVES at the surface of giant stars members of open and globular clusters are also shown (each symbol represents a cluster, see table 1 of Lagarde *et al.* (2019)).

Gaia-ESO survey has been investigated in Lagarde *et al.* (2019), showing the importance of extra mixing to explain the carbon and nitrogen abundances observed at the surface of giant stars for different metallicity, masses, and ages (see Fig. 2). On the other hand, Cabral *et al.* (2019) coupled the BGM with a stoichiometric model to provide and study the chemical properties of planets building blocks in different stellar populations.

What about white dwarfs?

To treat white dwarfs and their progenitors, we improve the BGM treatment of white dwarfs, including the cooling evolutionary models from the Montréal group and their corresponding synthetic colors (e.g., Fontaine *et al.* 2001, and references therein, see Fig. 1). This improvement allows us to take into account pure C-core or C/O core white dwarfs models, as well as colors for DA and DB spectral types. The corresponding cooling models are computed for white dwarfs mass between 0.2 and 1.3 $M_\odot$, corresponding to an initial mass between $\sim$ 0.7 and 6.0 $M_\odot$ (using the initial-final mass relation from Cummings *et al.* 2018). In this new way to treat the white dwarf in the BGM, we also propose to generate white dwarfs in the Milky Way using the same initial-mass function and star formation rates as taken for all others stars. In a forthcoming paper, we plan to include cooling evolutionary models with more physics (e.g., mixing), different metallicities, and to be more consistent cooling models computed according to the evolutionary sequence of their progenitors. Those simulations will be compared with combined data from astrometric, asteroseismic and spectroscopic surveys. All these new improvements will be crucial to exploit the future large number of data concerning white dwarfs and to combine all of these different kinds of observations to constrain Galactic and stellar physics.

References

Althaus, L. G., Córsico, A, H., Isern, J., García-Berro, E. 2010, *A&A Rev.* 18, 471

Baglin, A. & Fridlund, M. 2006, *in The CoRoT Mission Pre-Launch Status, Stellar Seismology and Planet Finding, eds. M. Fridlund, A. Baglin, J. Lochard, & L. Conroy, ESA SP,* 1306, 11

Beck, P. G., Montalban, J., Kallinger, T., *et al.* 2012, *Nature*, 481, 55

Bedding, T. R., Mosser, B., Huber, D., *et al.* 2011, *Nature*, 471, 608
Cabral, N., Lagarde, N., Reylé, C., Guilbert-Lepoutre, A., Robin, A., *et al.* 2019, *A&A*, 622, A49
Cummings, J. D., Kalirai, J. S., Tremblay, P.-E., Ramirez-Ruiz, E., Choi, J., *et al.* 2018, *ApJ*, 866, 21
Córsico, A. H., Althaus, L. G., Miller Bertolami, M. M., & Kepler, S. O. 2019, *A&A Rev.*, 27, 7
Czekaj, M. A., Robin, A. C., Figueras, F., Luri, X., & Haywood, M. 2014, *A&A*, 564, A102
Eisenstein, D. J., Liebert, J., Koester, D., *et al.* 2006b, *AJ*, 132, 676
Fontaine, G., Brassard, P., & Bergeron, P. 2001, *PASP*, 113, 409
Fontaine, G. & Brassard, P. 2008, *PASP*, 120, 1043
Gaia Collaboration, *et al.* 2018a, *A&A*, 616, A1
Gaia Collaboration, Babusiaux, C., van Leeuwen, F., *et al.* 2018b, *A&A*, 616, A10
Genest-Beaulieu, C. & Bergeron, P. 2019, *ApJ*, 796, 128
Gentile Fusillo, N. P., Tremblay, P.-E., Gänsicke, B. T., *et al.* 2019, *MNRAS*, 482, 4570
Gilliland, R. L., Brown, T. M., Christensen-Dalsgaard, J., *et al.* 2010, *PASP*, 122, 131
Jordan, S. 2007, *Astronomical Society of the Pacific Conference Series*, Vol. 372, 15th European Workshop on White Dwarfs, ed. R. Napiwotzki & M. R. Burleigh, 139
Kepler, S. O., Kleinman, S. J., Nitta, A., *et al.* 2007, *MNRAS*, 375, 1315
Kepler, S. O., Pelisoli, I., Koester, D., *et al.* 2016, *MNRAS*, 455, 3413
Koester, D. & Kepler, S. O. 2015, *A&A*, 583, A86
Lagarde, N., Robin, A., Reylé, C., Nasello, G., *et al.* 2017, *A&A*, 601, A27
Lagarde, N., Reylé, C., Robin, A. C., *et al.* 2019, *A&A*, 621, A24
Mosser, B., Goupil, M. J., Belkacem, K., *et al.* 2012a, *A&A*, 548, A10
Mosser, B., Goupil, M. J., Belkacem, K., *et al.* 2012b, *A&A*, 540, A143
Ricker, G. R. *et al.* 2015, *Journal of Astronomical Telescopes, Instruments, and Systems*, 1, 014003
Robin, A. C., Reylé, C., Derrière, S., & Picaud, S. *et al.* 2003, *A&A*, 409, 523
Stello, D., Bruntt, H., Preston, H., & Buzasi, D. 2008, *ApJ*, 674, L53
Tremblay, P.-E., Bergeron, P., & Gianninas, A. 2011, *ApJ*, 730, 128
Vauclair G 2013, *Constraints on white dwarfs structure and evolution from asteroseismology. In: Alecian G, Lebreton Y, Richard O, Vauclair G (eds) EAS Publications Series, EAS Publications Series*, vol 63, pp 175?183
Vrard, M., Mosser, B., & Samadi, R. 2016, *A&A*, 588, A87
Winget, D. E. & Kepler, S. O. 2008, *ARA&A*, 46, 157

White Dwarfs as Probes of Fundamental Physics: Tracers of Planetary, Stellar and Galactic Evolution
Proceedings IAU Symposium No. 357, 2019
M. A. Barstow, S. J. Kleinman, J. L. Provencal & L. Ferrario, eds.
doi:10.1017/S1743921320000472

A catalog of 159,238 white dwarf ages

Ted von Hippel[1], Adam Moss[2], Isabelle Kloc[1], Natalie Moticska[1], Jimmy Sargent[1], Elliot Robinson[1], David Stenning[3], David van Dyk[3], Elizabeth Jeffery[4], Morgan Fouesneau[5] and Coryn Bailer-Jones[5]

[1]Embry-Riddle Aeronautical University, Daytona Beach, FL, USA
[2]University of Oklahoma, Norman, OK, USA
[3]Imperial College, London, UK
[4]California Polytechnic State University, San Luis Obispo, CA, USA
[5]Max Planck Institute for Astronomy, Heidelberg, Germany

Abstract. We employ Pan-STARRS photometry, Gaia trigonometric parallaxes, modern stellar evolution and atmosphere models, and our Bayesian fitting approach to determine cooling and total ages for 159,238 white dwarfs. In many cases we are able to derive precise ages (better than 5%) for individual white dwarfs. These results are meant for broad use within the white dwarf and stellar astrophysics communities and we plan to make available on-line the posterior distributions for cooling age, total age, initial stellar mass, and other parameters.

Keywords. Galaxy: stellar content, stars: evolution, stars: white dwarfs

1. Introduction

Many areas of astrophysics benefit from quality stellar ages. Gas giant exoplanets and brown dwarfs cool as they age, so if the age of a companion star can be determined, then such an object's temperature or luminosity can constrain its mass. Chemical-tagging–the determination of multiple quality elemental abundances to identify stars with a common origin–may allow a new approach to substructure analysis of our Galaxy. Adding age information for just a few stars in any such group would connect that stellar population to the assembly history of the Milky Way. Ultimately, with a large enough sample of precision stellar ages, one may assemble the star forming history of the Milky Way itself.

In this contribution, we aim to present initial results for a sample of ~150,000 white dwarfs (WDs) for which we have derived ages. The age derivation is based on Pan-STARRS (Hodapp *et al.* 2004) photometry, Gaia (Cacciari, Pancino & Bellazzini 2016) trigonometric parallaxes, up-to-date models (Bergeron, Wesemael & Beauchamp 1995; Bressan *et al.* 2012; Bischoff-Kim & Montgomery 2018) and BASE-9, a statistical technique to fit the data with these models (von Hippel *et al.* 2006; van Dyk *et al.* 2009; Stein *et al.* 2013; Stenning *et al.* 2016). Conceptually, photometry constrains the spectral energy distribution (SED) of each star, which in turn constrains its effective temperature. Parallax constrains stellar distance and thereby absolute luminosity, and for a WD with its electron degenerate mass-radius relation, this also constrains its mass. For a WD of a given effective temperature and mass, assuming we know whether its surface is hydrogen or helium-dominated and assuming single star evolution, we can precisely determine its cooling age. If we further assume an initial final mass relation (IFMR; Williams, Bolte & Koester 2009), we can derive its precursor age and thus the total age of the star.

2. Data

We started with objects identified by Gentile Fusillo *et al.* (2019) using Gaia Data Release 2 (DR2) photometry and astrometry and chose those they identified with probabilities $\geqslant 75\%$ of being WDs. We merged these data with Pan-STARRS Data Release 1 (PS1) *grizy* photometry. We further require that PS1 g and r photometry have listed errors $\leqslant 0.1$ mag. While the ground-based PS1 photometry is not as accurate as Gaia photometry, the broad Gaia filters are less constraining of the stellar SEDs. PS1 is internally consistent, well-calibrated, and covers ~75% of the sky. The result of the above merging and pruning is a catalog of 160,705 WD candidates.

3. Methodology

We have developed a Bayesian statistical method for fitting stellar evolution models to photometry, which is further aided by priors on distance, metallicity, or other parameters. Our method has been successfully applied to single stellar populations in open clusters (DeGennaro *et al.* 2009; Jeffery *et al.* 2011; Hills *et al.* 2015; Jeffery *et al.* 2016; Bossini *et al.* 2019), multiple populations in globular clusters (Wagner-Kaiser *et al.* 2016a,b), binary field stars (Fouesneau *et al.* 2019), and single WD field stars (O'Malley, von Hippel & van Dyk 2013). We refer to our code suite as BASE-9, for Bayesian Analysis of Stellar Evolution with nine parameters.

BASE-9 relies on a combination of numerical integration and an adaptive Markov-chain Monte Carlo (Stenning *et al.* 2016) to derive the posterior distribution for a range of parameters. Of particular relevance for this study, BASE-9 reports stellar $T_{\rm eff}$, $log(g)$, zero age main sequence (ZAMS) mass, cooling age, precursor age, total age, parallax, and metallicity. Each star is run individually through BASE-9 based solely on its photometry and parallax information. Absorption along the line-of-sight to these nearby stars is assumed to be zero. A metallicity prior of $[\mathrm{Fe/H}] = 0.0 \pm 0.5$ is used, which is consistent with assuming that all of these WDs are Galactic disk objects. This will be true for the majority of these objects. WDs that show thick disk or halo kinematics will be rerun with more appropriate metallicity priors in the future. For this subset of metal-poor stars our metallicity prior will not affect their cooling ages but could affect their total ages, particularly if these stars evolved from low mass precursors.

We do not yet possess spectroscopy of these 160,705 WDs nor in many cases is the photometry sufficiently constraining to determine whether the atmospheres are H- or He-dominated. We thus assumed that all of these WDs have H-dominated atmospheres, which should be true for the majority (likely $\gtrsim 70\%$) based on population statistics.

Total run-time on Embry Riddle's Cray supercomputer was equivalent to ~3.5 cpu years on a single core and created ~300 GB of output for the posterior parameters. Over 99% of these WD candidates ran successfully through BASE-9, which indicates that these stars can plausibly be modeled as single WDs with the above-listed priors. This yielded posterior distributions for the above-listed parameters for 159,238 WDs. In most cases the parallax information was minimally informative (low precision). From among this sample, 23,244 WD candidates have relative parallax errors $\sigma_{\varpi}/\varpi \leqslant 10\%$ and 2,640 have relative parallax errors $\leqslant 2\%$.

4. Results

Figure 1 presents the distribution of ZAMS masses versus cooling ages for the full sample and the sample subsets limited by 10% and 2% parallax precision. Uncertainties are not plotted in this diagram to avoid burying many of the points, yet the age precision is typically high. The three samples differ somewhat in their age distribution, yet are broadly similar and can be characterized by the mean cooling age of each sample, which

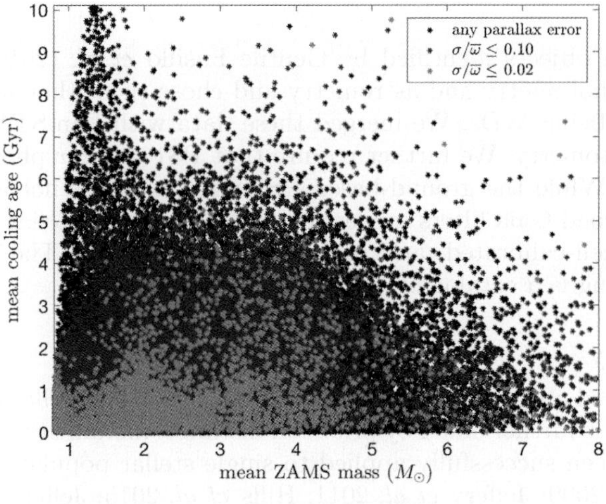

Figure 1. ZAMS mass versus cooling ages for 159,238 white dwarfs. The color version of this figure shows the full sample of stars in black, the 23,244 WDs with parallax precisions of 10% or better in blue, and the 2,640 WDs with parallax precisions of 2% or better in magenta.

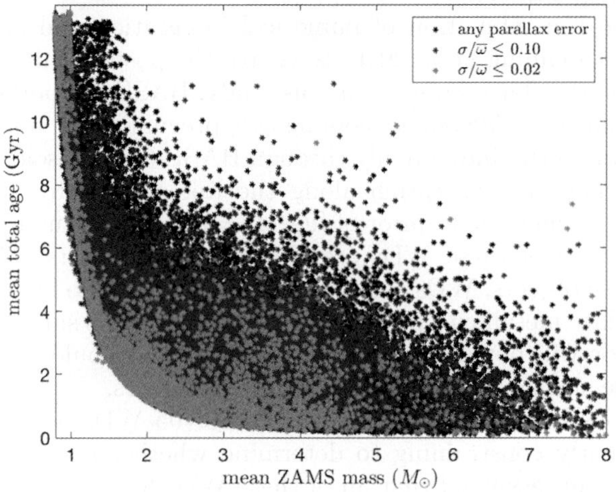

Figure 2. Same as Fig. 1, but for total stellar ages rather than cooling ages.

is 0.94, 1.04, and 0.76 Gyr, respectively. The average cooling age uncertainties for these three samples are 0.22, 0.15, and 0.03 Gyr, respectively. WDs from a very wide range of cooling ages, from significantly less than 1 Gyr up to 10 Gyr, are present in the sample over the entire precursor mass range. Even for the more limited subset of stars with reliable parallaxes (e.g., precisions at 10% or better), there are stars in the database that have been cooling for up to 10 Gyr throughout much of the precursor mass range. The deficit of higher mass and older stars is a selection effect based on their fainter apparent magnitude, which causes them to drop out of the Gaia sample (limited to G = 21).

Figure 2 presents the distribution of ZAMS masses versus total ages for these same stars. The mean total lifetimes of the stars of these three samples are 3.04, 3.63, and 3.24 Gyr. The total age uncertainties for a typical WD are now substantially larger, at 1.27, 1.06, and 0.53 Gyr, respectively. The total age depends, of course, on the precursor masses, and for low mass stars a very small uncertainty in mass can propagate to a relatively large uncertainty in precursor age. For this reason, if the sample is limited to only those WDs with precursor masses $\geqslant 2M_{\odot}$, then the uncertainty in total stellar age for the average star improves substantially to 0.71, 0.34, and 0.14 Gyr, respectively.

5. Future Work

Our goal is to make the posterior parameter distributions available via the world-wide web. It is challenging to make circa one million plots and summary statistics available, and we would be happy for ideas on how to accomplish this. We also plan to work out the selection effects that went into the input catalogs and the quality cuts and invert these to work out the star formation history for those Galactic stellar populations seen in the Solar Neighborhood.

References

Bergeron, P., Wesemael, F., & Beauchamp, A. 1995, *PASP*, 107, 1047

Bischoff-Kim, A. & Montgomery, M. H. 2018, *AJ*, 155, 187

Bossini, D. *et al.* 2019, *AA*, 623, A108

Bressan, A., Marigo, P., Girardi, L., Salasnich, B., Dal Cero, C., Rubele, S., & Nanni, A. 2012, *MNRAS*, 427, 127

Cacciari, C., Pancino, E., & Bellazzini, M. 2016, *AN*, 337, 899

DeGennaro, S., von Hippel, T., Jefferys, W. H., Stein, N., van Dyk, D. A., & Jeffery, E. 2009, *ApJ*, 696, 12

Fouesneau, M., Rix, H-W. von Hippel, T., Hogg, D. W. & Tian, H., 2019, *ApJ*, 870, 1

Gentile Fusillo, N. P. *et al.* 2019, *MNRAS*, 482, 4570

Hills, S., von Hippel, T., Courteau, S., & Geller, A. M. 2015, *AJ*, 149, 94

Hodapp, K. W. *et al.* 2004, *AN*, 325, 636

Jeffery, E. J., von Hippel, T., DeGennaro, S., Stein, N., Jefferys, W. H., & van Dyk, D. 2011, *ApJ*, 730, 35

Jeffery, E. J., von Hippel, T., van Dyk, D. A., Stenning, D. C., Robinson, E., Stein, N., & Jefferys, W. H. 2016, *ApJ*, 828, 79

O'Malley, E. M., von Hippel, T., & van Dyk, D. A. 2013, *ApJ*, 775, 1

Stein, N. M., van Dyk, D. A., von Hippel, T., DeGennaro, S., Jeffery, E. J., Jefferys, W. H. *et al.* 2013, *Statistical Analysis and Data Mining*, 6, 34

Stenning, D. C., Wagner-Kaiser, R., Robinson, E., van Dyk, D. A., von Hippel, T., Sarajedini, A., & Stein, N. 2016, *ApJ*, 826, 41

van Dyk, D. A., DeGennaro, S., Stein, N., Jefferys, W. H., & von Hippel, T. 2009, *Annals of Applied Statistics*, 3, 117

von Hippel, T., Jefferys, W. H., Scott, J., Stein, N., Winget, D. E., DeGennaro, S., Dam, A., & Jeffery, E. 2006, *ApJ*, 645, 1436

Wagner-Kaiser, R., Stenning, D., Robinson, E., von Hippel, T., Sarajedini, A., van Dyk, D. A., Stein, N., & Jefferys, W. H. 2016a, ApJ, 826, 42

Wagner-Kaiser, R., Stenning, D. C., Sarajedini, A., von Hippel, T., van Dyk, D. A., Robinson, E., Stein, N., & Jefferys, W. H. 2016b, *MNRAS*, 463, 3768

Williams, K. A., Bolte, M., & Koester, D. 2009, *ApJ*, 693, 355

White Dwarfs as Probes of Fundamental Physics: Tracers of Planetary, Stellar and Galactic Evolution
Proceedings IAU Symposium No. 357, 2019
M. A. Barstow, S. J. Kleinman, J. L. Provencal & L. Ferrario, eds.
doi:10.1017/S174392132000143X

A Bayesian analysis of white dwarfs in open clusters observed with Gaia

Elizabeth J. Jeffery[1], Ted von Hippel[2], Elliot Robinson[2], David van Dyk[3] and David Stenning[3]

[1]California Polytechnic State University, San Luis Obispo, CA, USA

[2]Embry-Riddle Aeronautical University, Daytona Beach, FL, USA

[3]Imperial College, London, UK

Abstract. We analyze individual white dwarfs in open clusters observed by Gaia. In particular, we determine ages when different model ingredients are used. We also explore fundamental properties of the white dwarfs, including temperature and mass, when using different filter combinations. Such tests are important to understanding any systematic effects when applying similar techniques to field stars.

Keywords. Galaxy: stellar content, stars: evolution, stars: white dwarfs

1. Introduction

Open clusters provide an excellent environment for many tests of stellar evolution. The coeval formation of stars in a cluster from the same parent cloud gives its members approximately the same age, distance, and metallicity. By studying the white dwarfs (WDs) in open clusters, we are able to gain additional insight into WD formation and cooling, as well as ages measured by these objects.

The *Gaia* satellite has created a wealth of data (including parallaxes and photometry) for a huge number of stars, including a vast number of WDs. These data are a treasure trove for the community in many ways. *Gaia* has observed a number of open clusters (Gaia Collaboration 2018), including WDs in some. Here we will focus our attention on the WDs observed in the Hyades and Praesepe open clusters. Specifically, we will explore and compare results of temperature and age obtained for these individual cluster WDs, using a variety of filter combinations and model ingredients.

2. Data and Methodology

Data for this study includes parallaxes from *Gaia* and photometry from *Gaia*, Sloan Digital Sky Survey (SDSS), and Pan-STARRS surveys. In total, we analyzed six WDs from the Hyades and nine WDs from Praesepe.

For analysis, we used a Bayesian statistical method for fitting stellar evolution models to photometry (see von Hippel *et al.* 2016; van Dyk *et al.* 2009; Stenning *et al.* 2016). The software package (Bayesian Analysis of Stellar Evolution in Nine Variables, or BASE-9) is freely available for download (https://github.com/argiopetech/base) and its use is described in von Hippel *et al.* (2014). BASE-9 returns parameters such as total age, metallicity, and distance (or parallax), and additional features can be employed to determine an individual WD cooling age, zero age main sequence (ZAMS) mass, $log(g)$, and T_{eff} values.

Table 1. Cluster white dwarfs included

Hyades WDs	**Praesepe WDs**	
WD0352+096 (HZ 4)	WD0833+198	WD0840+200
WD0421+162 (VR 7)	WD0836+197	WD0840+209
WD0425+168 (VR 16)	WD0836+201	WD0843+186
WD0431+125 (HZ 7)	WD0837+189	
WD0438+108 (HZ 14)	WD0837+218	
WD0406+169 (LB 277)	WD0840+190	

The required prior distributions on metallicity, parallax, and reddening are Gaussian (with reddening truncated at zero). Prior mean and sigma values on parallaxes were the *Gaia* values and errors. For the Hyades, a prior distribution on metallicity of [Fe/H] = +0.103 ± 0.05 (Taylor & Joner 2005) and reddening (A_V) of 0.0031 ± 0.0031 (Taylor 2016). For Praesepe, we used a metallicity prior of 0.01 ± 0.04 and reddening prior of 0.0837 ± 0.0124 (Taylor 2016).

Briefly, the total age of a WD is found by determining its cooling age and adding its precursor age. The cooling age can be found by combining temperature (constrained by multiband photometry) and parallax (which leads to mass when the WD's absolute luminosity is combined with temperature and the WD mass-radius relation). Once the mass of the WD has been determined, an initial-final mass relation (IFMR) is assumed to obtain the star's ZAMS mass; a main sequence (MS) model is then employed to determine the pre-WD lifetime of a star with that ZAMS mass. For more information on using BASE-9 to determine ages of individual WDs, see O'Malley, von Hippel & van Dyk (2013). and the contribution in these proceedings by von Hippel *et al.*

In our analysis we use WD cooling models of Bischoff-Kim & Montgomery (2018), and updated WD atmosphere models from Bergeron *et al.* (1995). Stellar evolution models available in BASE-9 include the Dartmouth Stellar Evolution Database (DSED, Dotter *et al.* 2008); PAdova & TRieste Stellar Evolution Code (PARSEC, Marigo *et al.* 2017); the Yale-Potsdam Stellar Isochrones (YaPSI, Spada *et al.* 2017); and the MESA Isochrones and Stellar Tracks (MIST, Dotter *et al.* 2016; Choi *et al.* 2016), both with rotation ($v/v_{crit} = 0.4$) and without rotation. Different IFMRs used included Williams *et al.* (2009), Weidemann (2000), and Salaris *et al.* (2009) linear and piecewise.

3. Results

In a recent paper, Bergeron *et al.* (2019) posit that effective temperatures and masses determined from Pan-STARRS photometry alone suffer from systematic effects, especially for hotter WDs. They compare temperatures determined from SDSS (*ugriz*) filters to those from the full Pan-STARRS filter set (*grizy*), a subset of Pan-STARRS filters (*griz*), a subset of the SDSS filters (*griz*) and a combination of SDSS *u* + Pan-STARRS *grizy*. They find that the addition of the SDSS *u* filter brings temperatures back into good agreement. Similar effects are found when they compare photomtric temperature and masses to those determined spectroscopically.

We performed a similar analysis on these cluster WDs included here and find a similar result. We show these tests in Figure 1. This figure is a modified reproduction of Bergeron's Figure 5, with corresponding points for the Hyades (diamonds) and Praesepe (stars). We note that only two of the six Hyades WDs were used because not all of the Hyades WDs fall within the footprint of SDSS observations.

Figure 2 is similar to Figure 1, but for only the cluster WDs. Here we compare the temperatures determined using Gaia photometry, Pan-STARRS, and SDSS *u* + Pan-STARRS to those determined with SDSS photometry. We note that there is more scatter when using *Gaia* photometry alone compared to the other filter combinations. Because

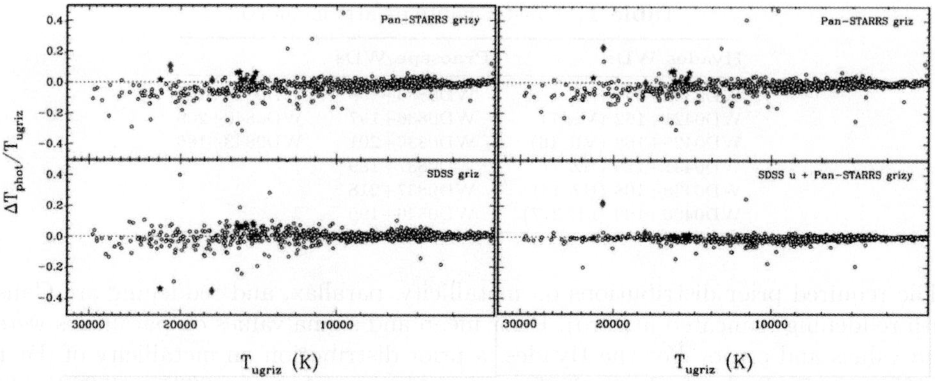

Figure 1. Modified Figure 5 from Bergeron *et al.* (2019) showing the effects of filter combination on determining temperature, compared to the temperature determined using the full SDSS filter set (*ugriz*). Overplotted are values determined for the Hyades (diamonds) and Praesepe (stars), showing similar results to those determined by Bergeron.

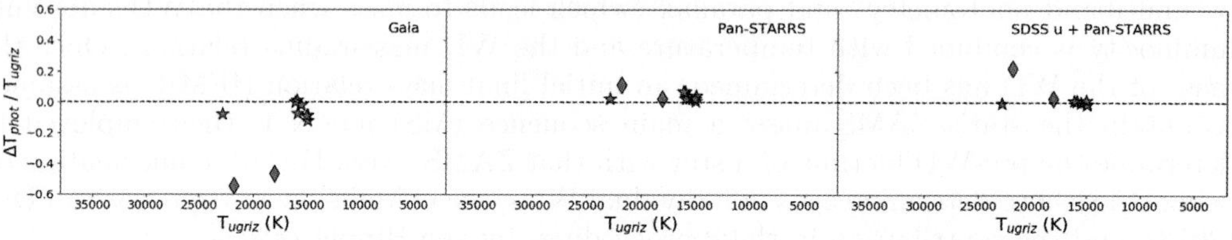

Figure 2. Temperatures determined for cluster WDs using Gaia, Pan-STARRS, and SDSS + Pan-STARRS compared to temperature determined by SDSS photometry. Temperatures from Gaia photometry alone are not as precise as those determined using other filter combinations.

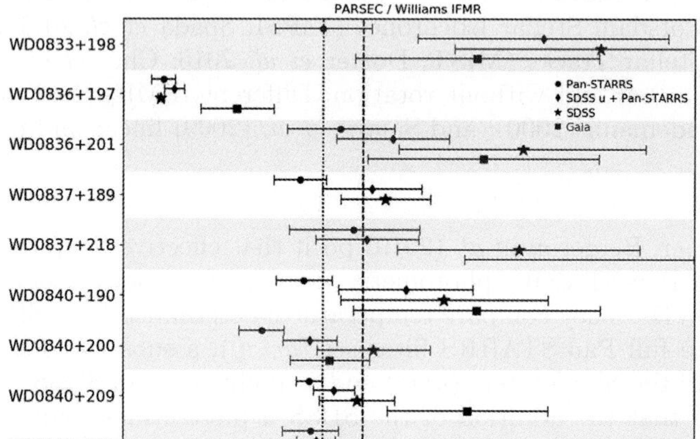

Figure 3. Total age for the individual WDs in Praesepe using different filter combinations.

of this, we would encourage other researchers to not rely on *Gaia* photometry alone in determining WD temperatures; if additional photometry is available, make use of it.

We also compare the effect of filter combination on the total age determined for the WDs. We use the IFMR from Williams *et al.* (2009) and the PARSEC MS models. We display these results in Figure 3. In this figure, the points represent the mean of the posterior distribution and the error bars are one standard deviation. The vertical dotted

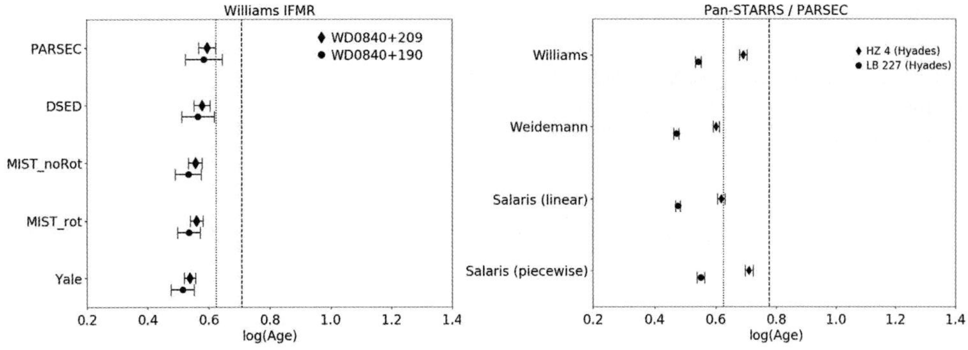

Figure 4. Left: Total age for two WDs in Praesepe, using different MS model sets. Right: Total age of two WDs in the Hyades using different IFMRs and the same MS models.

line represents a cluster age of 625 Myr (Dobbie *et al.* 2006) and the dashed line is an age of 708 Myr (log(Age) = 8.85), determined using PARSEC models (Marigo *et al.* 2017) in the Gaia H-R Diagram study (Gaia Collaboration 2018). We again note that ages determined from Gaia photometry alone tend to return older ages than other photometry sets, and have the highest standard deviation, likely due to high uncertainties in T_{eff}.

Because determining the total age of a WD is model dependent (in the use of the IFMR, along with MS timescale models), we explore effects in age when using different model ingredients. In the left panel of Figure 4, we show ages for the two coolest WDs in Praesepe using the Williams IFMR and a variety of MS models. In the right panel of Figure 4, we show results for the two coolest Hyades WDs using PARSEC MS models and use different IFMRs.

4. Summary

In this contribution we have shown the effects of different filter combinations on WD temperatures, mass, and individual age. As seen in Bergeron *et al.* (2019), we see that use of the SDSS u filter is important for photometric temperatures being consistent with spectroscopic temperatures. We find it is important to avoid using Gaia filters alone for determining WD temperatures or total age; other filters should be used, if available. Additionally, we explore the effects of using different model ingredients on total WD age.

References

Bergeron, P. *et al.* 2019, *ApJ*, 876, 67
Bergeron, P. *et al.* 1995, *PASP*, 107, 1047
Bischoff-Kim, A. & Montgomery, M. H. 2018, *AJ*, 155, 187
Choi, J. *et al.* 2016, *ApJ*, 823, 102
Dobbie, P. D. *et al.* 2006, *MNRAS*, 369, 383
Dotter, A. *et al.* 2008, *ApJS*, 178, 89
Dotter, A. *et al.* 2016, *ApJS*, 222, 8
Gaia Collaboration, 2018, *A&A*, 616, 1
Marigo, P. *et al.* 2017, *ApJ*, 835, 77
O'Malley, E. M., von Hippel, T., & van Dyk, D. A. 2013, *ApJ*, 775, 1
Salaris, M. *et al.* 2009, *ApJ*, 692, 1013
Spada, F. *et al.* 2017, *ApJ*, 838, 161
Stenning, D. C. *et al.* 2016, *ApJ*, 826, 41
Taylor, B. J. 2006, *ApJ*, 132, 2453
Taylor, B. J. & Joner, M. D. 2005, *ApJS*, 159, 100

van Dyk, D. A. *et al.* 2009, *AnApS*, 3, 117
von Hippel, T. *et al.* 2006, *ApJ*, 645, 1436
von Hippel, T. *et al.* 2014, arXiv:1411.3786
Williams, K. A. *et al.* 2009, *ApJ*, 693, 355
Weidemann, V. 2000, *A&A*, 363, 647

White Dwarfs as Probes of Fundamental Physics: Tracers of Planetary, Stellar and Galactic Evolution
Proceedings IAU Symposium No. 357, 2019
M. A. Barstow, S. J. Kleinman, J. L. Provencal & L. Ferrario, eds.
doi:10.1017/S1743921320000782

Testing white dwarf cosmochronology using wide double white dwarfs

Tyler Heintz and JJ Hermes

Institute for Astrophyiscal Research, Boston University, 725 Commonwealth Ave, Boston MA 02215, USA

Abstract. We present a sample of nearly 650 widely separated double white dwarf binaries found using Gaia DR2 astrometry. We derive preliminary total ages for each white dwarf in our sample using Gaia photometry and compare the total ages of both components of each binary in our sample. We find agreement within 3 sigma between the two ages ~85% of the time with median age uncertainties of ~ 3.5 Gyr depending on which initial-final mass relation is used. When a subsample with the most precise ages is used, the agreement within 3 sigma drops to ~70% with median age uncertainties of 300-600 Myr.

Keywords. white dwarfs, binary stars, ages

1. Introduction

In the age of Gaia, searches for widely separated (>100 AU) binary systems are uncovering large numbers of systems (e.g. >50,000 from El-Badry & Rix 2018). These systems provide great test beds for multiple areas of galactic and stellar astrophysics. For example, they have been used to study the galactic gravitational potential (i.e. Correa-Otto *et al.* 2017). In addition, wide binaries provide two coeval objects that can be used to test and calibrate current ageing techniques.

White dwarfs have been used as cosmochronometers for more than 30 years (See review by Fontaine *et al.* 2001). White dwarfs have been used to age a variety of astronomical objects including binary companions (i.e. Fouesneau *et al.* 2019) and the halo and disk of our own galaxy (i.e.Winget *et al.* 1987; Kilic *et al.* 2019). Due to its relatively simple and characteristic cooling, the mass and effective temperature of a white dwarf can be uniquely mapped to a cooling age. To determine the total age of a white dwarf from the main sequence to its current state, a relation between its current mass as a white dwarf and its zero-age main sequence mass is needed. This can be done using an empirical, cluster calibrated initial-final mass relation (i.e. Catalán *et al.* 2008; Salaris *et al.* 2009; Cummings *et al.* 2018). This method gives a model-dependent total age of the white dwarf.

In this paper, we present a sample of 642 wide white dwarf white dwarf binaries and a comparison of total ages of the two components of each binary. In Section 2, the method of finding these binaries is discussed along with the ageing technique used. In Section 3, we present a preliminary attempt to compare total ages in these wide binaries. We end with a discussion of our results and future work in Section 4.

2. Methods

2.1. *Generation of Binary Catalog*

To generate the wide double white dwarf sample, we take a similar approach to El-Badry & Rix (2018). For a full discussion of the binary search along with the relevant quality cuts and binary pair criteria, see El-Badry & Rix (2018).

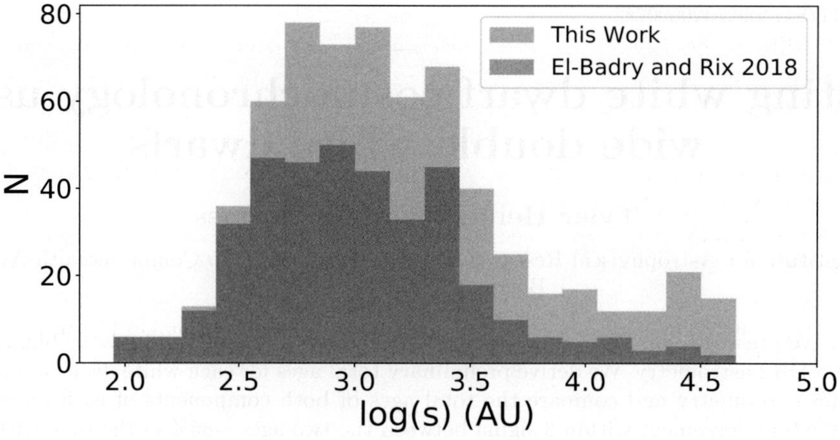

Figure 1. The separation distribution of our 642 wide WD/WD binaries in red with the wide WD/WD binaries from El-Badry & Rix (2018) in blue.

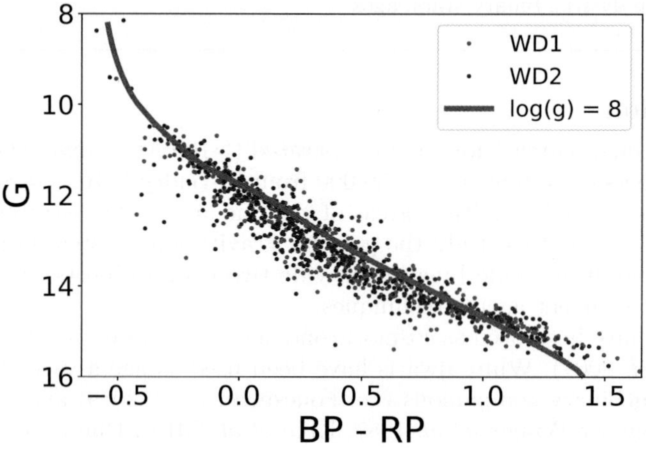

Figure 2. An H-R diagram of 580 of our 642 binaries where both white dwarfs have a probability of being a white dwarf higher than 0.8 from Gentile Fusillo *et al.* (2019). A curve for $log\ (g) = 8$ white dwarfs is provided in green as well taken from Tremblay *et al.* (2009).

We make a few changes to their search that favor white dwarfs and are aimed to generate the largest sample possible. They use the full Gaia catalog, a strict 200 parsec cutoff, and an upper bound orbit velocity with a total mass of 5 solar masses. In contrast, we use the Gaia DR2 white dwarf catalog from Gentile Fusillo *et al.* (2019), no distance cutoff, and an upper bound orbit velocity with a total mass of 2.8 solar masses.

With these changes, we increase the sample of 375 wide WD/WD binaries from El-Badry & Rix (2018) to 642. The separation distribution and an H-R diagram of our sample can be seen in Figures 1 and 2 respectively. We find more binaries at all separations and recover all but two of the 375 binaries from El-Badry & Rix (2018).

2.2. *White Dwarf Ageing*

To determine the total age of a white dwarf, a cooling age and main sequence lifetime is needed. To calculate the cooling ages of our white dwarfs, effective temperatures and masses are needed. As a preliminary age determination, we only use the photometrically determined effective temperatures and masses from the Gaia DR2 white dwarf catalog Gentile Fusillo *et al.* (2019). To get the cooling ages from these masses and effective temperatures, we use the Montréal white dwarf cooling models Tremblay *et al.* (2009); Bergeron *et al.* (2011). These cooling models need to know the spectral type of the white dwarf. Different atmospheric compositions have varying effects on the cooling rates of these stars. When a spectral type is available, we use the corresponding model. If no spectral type is known, then a DA white dwarf is assumed.

A progenitor main sequence mass must be found to determine the main sequence lifetimes of each of our white dwarfs. The main sequence masses are determined through the use of an empirical initial-final mass relation. We use three separate initial final mass relations to get a better understanding of the empirical uncertainty in the main sequence lifetimes. The three that are used are from Cummings *et al.* (2018), Salaris *et al.* (2009), and Catalán *et al.* (2008). Once the progenitor main sequence masses are determined, MESA stellar evolutionary tracks (Dotter 2016; Choi *et al.* 2016; Paxton *et al.* 2011, 2013, 2015) are used to derive the main sequence lifetimes.

To determine the uncertainties on the total ages, we take a simple approach. We use the one sigma error bars on the mass and effective temperatures to derive total ages with the upper and lower masses and effective temperatures. This gives us a range on the total age. We leave more rigorous error analysis to future work.

3. Age Comparison

The two stars in a binary system are formed at nearly the same time from the same collapsing cloud of gas. Thus, the two white dwarfs in each of the wide binaries from our sample should have the same total age.

Using the ageing methods discussed in Section 2.2, we find that the total ages of the two component white dwarfs of each binary agree within 3 sigma 90%, 80%, and 89% of the time for the initial-final mass relations from Cummings *et al.* (2018), Salaris *et al.* (2009), and Catalán *et al.* (2008), respectively. The median uncertainties from each were 3.8, 0.75, and 3.7 Gyr, respectively. This only includes 200-300 systems depending on the lower and upper cutoffs for each initial-final mass relation used. If a system has one of its components that is below or above these bounds, its ages cannot be compared. A histogram of the number of standard deviations away from zero age difference for these systems can be seen in Figure 3.

Since these median errors are so large, we construct a subsample where one of the two components of the binary has an age uncertainty less than 1 Gyr. In this subsample, we find agreement within 3 sigma 77%, 63%, and 66% of the time. The median uncertainties for this subsample are 0.625, 0.335, and 0.425 Gyr. These subsamples contain around 65-75 systems that can have the ages compared due to the bounds on the initial-final mass relations. A histogram of the number of standard deviations away from zero age difference for these systems can be seen in Figure 3.

4. Summary and Future Work

We have developed the largest wide double white dwarf catalog to date using Gaia DR2. Using similar methods to El-Badry & Rix (2018), we find 642 wide double white dwarf binaries that span most of the typical region where white dwarfs are found in the Gaia color-magnitude diagram. We make changes to their search to be tailored for finding specifically double white dwarfs. We use a total mass of 2.8 solar masses for the maximum bound orbital velocity instead of 5 solar masses, and we make no strict distance cutoff. In the future, less strict photometric cuts can be made to add additional to white dwarfs.

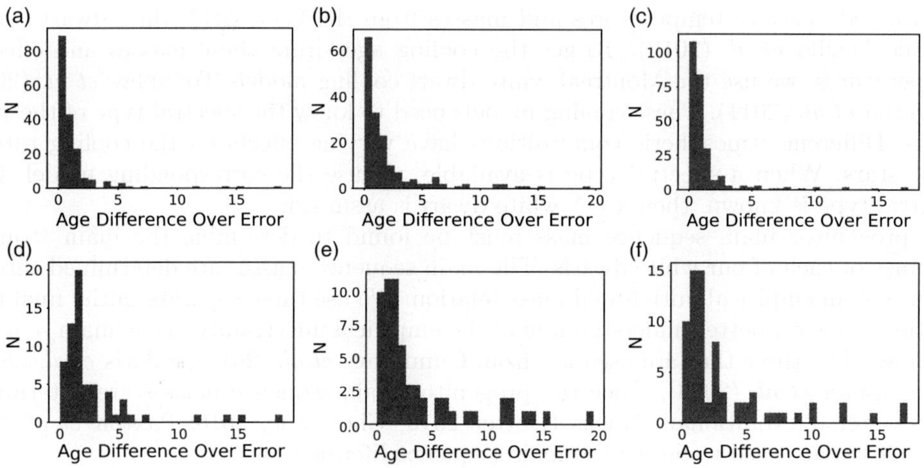

Figure 3. The top three panels show the age difference over error for the two components in each wide double white dwarf binary in our sample using the (a) Cummings *et al.* (2018), (b) Salaris *et al.* (2009), and (c) Catalán *et al.* (2008) initial-final mass relations. The histograms show 180, 168, 263 systems respectively due to the different bounds on the initial-final mass relation. The bottom three panels show the age difference over error for the two components in each wide double white dwarf binary in our subsample with more precise ages using the (d) Cummings *et al.* (2018), (e) Salaris *et al.* (2009), and (f) Catalán *et al.* (2008) initial-final mass relations. The histograms show 65, 61, 71 systems respectively due to the different bounds on the initial-final mass relation.

Since binary star systems form at approximately the same time, both white dwarfs in these systems should have the same age. We performed a preliminary check to see if our systems exhibit this age agreement by using the white dwarf ageing techniques discussed in Section 2.2. We find that the ages agree within 3 sigma 90%, 80%, and 89% of the time for the initial-final mass relations from Cummings *et al.* (2018), Salaris *et al.* (2009), and Catalán *et al.* (2008), respectively. When limiting to the systems with one white dwarf with less than a Gyr uncertainty, the 3 sigma agreement drops to 75%, 65%, and 70%, respectively. These ages are very preliminary and only provide a brief check for "ill-behaved" systems. These ill-behaved systems could provide insights into previous merger history or may provide insights into the nature of the true initial-final mass relation. Future work will be focused on studying these ill-behaved systems with follow-up observations. Additionally, future work will aim to add more photometric bandpasses to develop a broadband SED to get improved photometric surface gravities and effective temperatures which will provide more reliable and precise ages.

Acknowledgements

We thank Kareem El-Badry for helpful discussions. This work was partially supported by the Massachusetts Space Grant Consortium.

References

Bergeron, P., Wesemael, F., Dufour, P., *et al.* 2011, *ApJ*, 737, 28
Catalán, S., Isern, J., García-Berro, E., & Ribas, I. 2008, *MNRAS*, 387, 1693
Choi, J., Dotter, A., Conroy, C., *et al.* 2016, *ApJ*, 823, 102
Correa-Otto, J. A., Calandra, M. F., & Gil-Hutton, R. A. 2017, *A&A*, 600, A59
Cummings, J. D., Kalirai, J. S., Tremblay, P. E., Ramirez-Ruiz, E., & Choi, J. 2018, *ApJ*, 866, 21
Dotter, A. 2016, *ApJS*, 222, 8
El-Badry, K. & Rix, H.-W. 2018, *MNRAS*, 480, 4884

Fontaine, G., Brassard, P., & Bergeron, P. 2001, *PASP*, 113, 409
Fouesneau, M., Rix, H.-W., von Hippel, T., Hogg, D. W., & Tian, H. 2019, *ApJ*, 866, 21
Gentile Fusillo, N. P., Tremblay, P.-E., Gänsicke, B. T., *et al.* 2019, *MNRAS*, 482, 4570
Kilic, M., Bergeron, P., Dame, K., *et al.* 2019, *MNRAS*, 482, 965
Paxton, B., Bildsten, L., Dotter, A., *et al.* 2011, *ApJS*, 192, 3
Paxton, B., Cantiello, M., Arras, P. *et al.* 2013, *ApJS*, 208, 4
Paxton, B., Marchant, P., Schwab, J. *et al.* 2015, *ApJS*, 220, 15
Salaris, M., Serenelli, A., Weiss, A., & Miller Bertolami, M. 2009, ApJ, 692, 1013
Tremblay, P. E., Bergeron, P., & Gianninas, A. 2011, *ApJ*, 730, 128
Winget, D. E., Hansen, C. J., Liebert, J., *et al.* 1987, *ApJ*, 315, L77

White Dwarfs as Probes of Fundamental Physics: Tracers of Planetary, Stellar and Galactic Evolution
Proceedings IAU Symposium No. 357, 2019
M. A. Barstow, S. J. Kleinman, J. L. Provencal & L. Ferrario, eds.
doi:10.1017/S1743921320000927

Statistics of white dwarf properties in intermediate polars

Valery F. Suleimanov[1,2,3], Victor A. Doroshenko[1,3] and Klaus Werner[1]

[1]Institut für Astronomie und Astrophysik, Universität Tübingen,
Sand 1, D-72076, Tübingen, Germany
email: suleimanov@astro.uni-tuebingen.de

[2]Kazan Federal University, Kremlevskaya 18, 420008 Kazan, Russia

[3]Space Research Institute of the Russian Academy of Science,
Profsoyuznaya 84/32, 117997 Moscow, Russia

Abstract. Many intermediate polars are hard X-ray sources. The theory of their hard X-ray radiation is well developed and allows us to determine white dwarf masse in this kind of cataclysmic variables. Here we present the results of determination of the masses of 35 white dwarfs in the intermediate polars observed by observatories NuSTAR (10 sources) and *Swift/BAT* (25 sources). The corresponding mass accrerion rates and the luminosity function were also derived due to accurate distances to the sources well known now after *Gaia* DR2.

Keywords. X-rays: binaries, novae, cataclysmic variables, stars: luminosity function, mass function

1. Introduction

A typical Intermediate Polar (IP) is a semi-detached close binary system with a magnetized white dwarf (WD) as a primary (see overview by Warner 2003). The IPs are a subclass of Cataclysmic Variable Stars (CVs). The WD accretes matter from a secondary normal star which overflows its Roche lobe. The accreting matter forms an accretion disc around the WD, which is destroyed by the WD magnetic field at some magnetospheric radius $R_{\rm m}$. The matter infalls further along magnetic field lines on the WD magnetic poles and forms a strong shock above the WD surface (Aizu 1973; Fabian *et al.* 1976). Hot, optically thin, post-shock matter heated up to almost the virial temperature settles down on the WD surface. This post-shock region (PSR) is a hard X-ray source, and as a result IPs are numerous Galactic hard X-ray sources on the sky (see e.g. Barlow *et al.* 2006 and the recent review by Mukai 2017). The maximum PSR temperature and, therefore, the X-ray spectrum hardness depends on the free-fall velocity at the WD surface. The value depends both on the WD mass M and the radius R. As the WD radius is related to its mass, it is possible to determine the masses of WDs in IPs using their hard X-ray spectra (Rothschild *et al.* 1981). This method was used by many authors for WD mass determination using various PSR models and X-ray observations (see, e.g. Cropper *et al.* 1999, Suleimanov *et al.* 2005, Brunschweiger *et al.* 2009, Yuasa *et al.* 2010, and Shaw *et al.* 2018). Here we present some new results describing WD properties obtained using a new grid of PSR spectral models, new IP observations performed by observatories NuSTAR (Harrison *et al.* 2013) and *Swift*/BAT (Gehrels *et al.* 2004) and new information about distances to IPs from *Gaia* Data Release 2 (Gaia collaboration 2018).

2. Model and observational data

We used recently developed PSR models described by Suleimanov *et al.* (2016). The main features of these models are a quasi-dipole geometry and the finite initial height of the infalling matter. The latter parameter mimics a finite magnetospheric radius $R_{\rm m}$ and it is potentially important for IPs with small magnetospheres, such as EX Hya (see, e.g. Revnivtsev *et al.* 2009). The computed grid of PSR spectra is a two-parameter grid with a WD mass M and the relative magnetospheric radius $R_{\rm m}/R$ as the parameters. This grid is available in XSPEC†. **The height of the shock could be rather high in low luminous IPs. Therefore, we additionally computed a similar grid with the fixed relative PSR height $H/R = 0.25$ and used it for the low luminous IPs studies.** The magnetospheric radius of the every IP was found either from the breaking frequency in the power density spectrum (see detailed description of the approach in Suleimanov *et al.* 2016) or set equal to the corotation radius. This approach allows us to definitely determine the WD mass in the IP using the two-parameter described grid by fitting its hard X-ray spectrum.

The above method was applied to ten IPs observed with NuSTAR observatory, and 25 IPs observed with *Swift*/BAT observatory, whose spectra are available after publishing of the 105-Month Hard X-ray Survey (Oh *et al.* 2018). We used only observed spectra at energies above 15-20 keV to avoid the problems with contribution of the reflection and the complex absorption in the softer IP X-ray spectra.

3. Results

WD masses and mass accretion rates. Using observed hard X-ray spectra of 35 IPs we determined their WD masses. The distances to 34 of these IPs are now available from Gaia DR2 (Bailer-Jones *et al.* 2018). The reliable distances allow us to determine the accurate mass accretion rates, $\dot{M}$, in the IPs with the finite $R_{\rm m}$ values and to evaluate the WD magnetic field strength. All the results are described and presented in detail by Suleimanov *et al.* (2019). The resulting WD mass distribution and the relation between the IP orbital periods and $\dot{M}$ are shown in Fig. 1. The averaged WD mass obtained $\overline{M} = 0.79 \pm 0.16 M_\odot$ is close to that obtained by Yuasa *et al.* (2010) ($0.88 \pm 0.25 M_\odot$) and Bernardini *et al.* (2012) ($0.86 \pm 0.07 M_\odot$) for IPs, and to the averaged WD mass in CVs obtained by Zorotovic *et al.* (2011) ($0.83 \pm 024 M_\odot$). The 105-Month BAT Survey contains information about photon indexes Γ for all published spectra. We found a relation between Γ and the derived WD masses for all 35 considered IPs, $M/M_\odot = 1.37(\pm 0.07)/(\Gamma - 1.07(\pm 0.08))$, see Fig. 2, left panel.

Luminosity function. The distances to the IPs studied and the time averaged fluxes in the band 14-195 keV presented in 105-Month BAT Catalogue were used for deriving the luminosity function of IPs in the Galaxy. The relation between these two values is presented in Fig. 2, right panel. A line showing a distance corresponding to assumed lower detectability flux in the catalogue used (8×10^{-12} erg s^{-1} cm^{-2}; Oh *et al.* 2018) is also depicted. This line determines the maximum distance up to which IP in some luminosity range can be detectable by the BAT.

The luminosity function was found using a simplified method. We assumed that IPs are uniformly distributed in the Galactic disc with a half thickness 130 pc (Revnivtsev *et al.* 2008) and the averaged stellar density is constant over the Galactic disc and equals to value in the Sun vicinity, $\rho_{\rm St} \approx 0.04\, M_\odot\, {\rm pc}^{-3}$ (McKee *et al.* 2015). The number density of IPs and their normalized luminosity in the given luminosity range were found using the number of IPs within the luminosity range divided by the Galactic disc volume visible up to the maximum detectable distance for this luminosity range.

† https://heasarc.gsfc.nasa.gov/docs/xanadu/xspec/models/ipolar.html

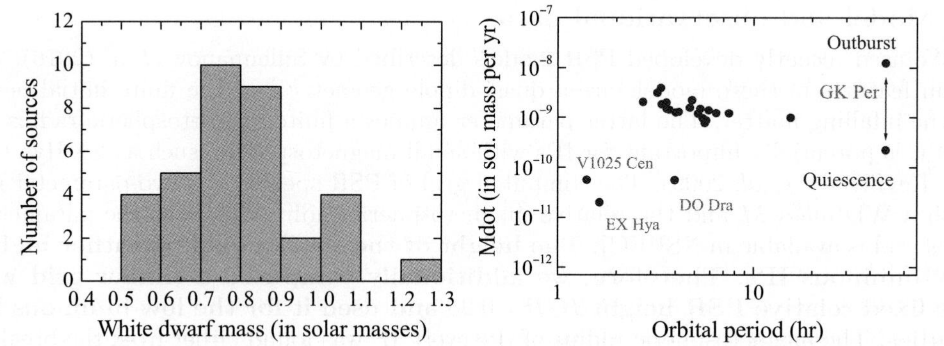

Figure 1. *Left panel:* Number distribution of WD masses. *Right panel:* Distribution of derived mass accretion rates in the studied IPs vs. their orbital periods.

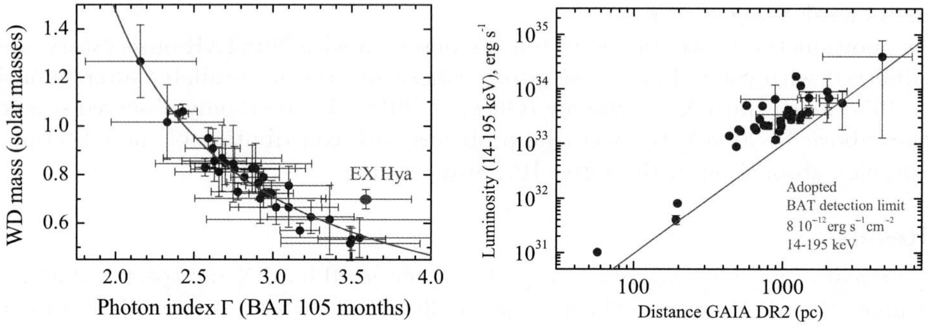

Figure 2. *Left panel:* Relation between WD masses determined using the fitting of the observed by BAT and NuSTAR spectra with the model spectra and the photon index Γ for the same IPs. The best fit of the relation is also shown. EX Hya deviates from the relation due to very small magnetospheric radius. *Right panel:* Relation between hard X-ray luminosities of the IPs studied in BAT observational band and the distances to them. The adopted lower limit of the detection for used 105-Month BAT catalogue is also shown.

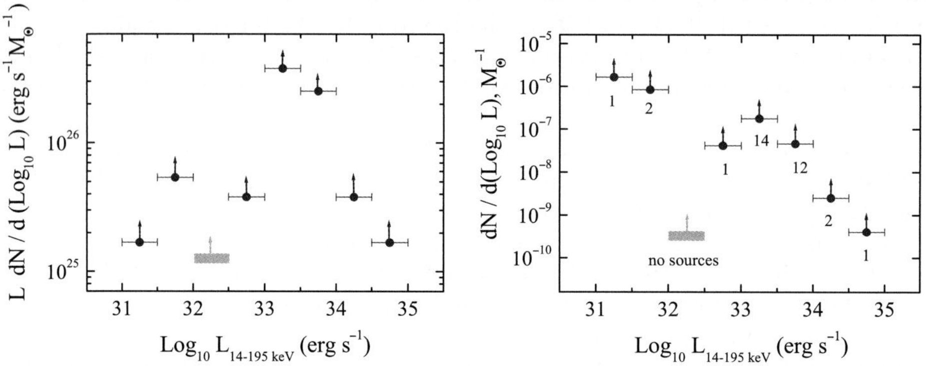

Figure 3. *Left panel:* **Differential luminosity distribution of 14-195 keV emissivity per solar mass of the investigated IPs.** *Right panel:* Luminosity function of the investigated IPs, demonstrating the number of IPs with different luminosities per solar mass at the Solar vicinity. The numbers under the points show the number of IPs in the given luminosity bin.

The results are presented in Fig. 3. The right panel shows the contribution of the given luminosity range to the total luminosity of IPs normalized to one solar mass, whereas the left panel shows the contribution of the IPs in the given luminosity range into the total number density of IPs also normalized to the solar mass. Both distributions have a two-pike structure reflecting the differences between mass accretion rates in IPs before and after the period gap (see Fig. 1, right panel). The derived luminosity functions are similar to the luminosity functions of CVs detected by INTEGRAL all-sky survey presented in Revnivtsev *et al.* (2008), see their Figs. 3 and 8.

The averaged number $\rho_{\rm N} \approx 1.14 \times 10^{-7}\,{\rm pc}^{-3}$ derived from our studies and luminosity $\rho_{\rm L} \approx 8 \times 10^{26}\,{\rm erg\,s^{-1}}\,M_{\odot}^{-1}$ densities of IPs are close to values obtained by Revnivtsev *et al.* (2008), $\rho_{\rm N} = 1.5 \pm 0.6\ \times 10^{-7}\,{\rm pc}^{-3}$ and $\rho_{\rm L} = 1.3 \pm 0.3\ \times 10^{27}\,{\rm erg\,s^{-1}}\,M_{\odot}^{-1}$ correspondingly. The value of the averaged IP number density $\rho_{\rm N} = 1^{+1}_{-0.5} \times 10^{-7}\,{\rm pc}^{-3}$ derived by Pretorius & Mukai (2014) is also close to our value. We note, however, that the luminosity functions and average densities presented here are low limits only, because the 105-Month BAT catalogue consists of at least in two times more IPs than we used here. More accurate luminosity functions will be presented in the separate paper.

Acknowlegements

The work was supported by the German Research Foundation(DFG) grant WE 1312/51-1, and the Russian Foundation for BasicResearch grant 18-42-160003 ra (VFS).

References

Aizu, K. 1973, *Progr. Theor. Phys.*, 49, 1184
Bailer-Jones, C. A. L., Rybizki, J., Fouesneau M., *et al.* 2018, *AJ*, 156, 58
Barlow, E. J., Knigge, C., Bird, A. J. J., *et al.* 2006, *MNRAS*, 372, 224
Bernardini, F., de Martino, D., Falanga M., *et al.* 2012, *Astr. and Astrophys.*, 542, A22
Brunschweiger, J., Greiner, J., Ajello, M., & Osborne, J. 2009, *Astr. and Astrophys.*, 496, 121
Cropper, M., Wu, K., Ramsay, G., & Kocabiyik, A. 1999, *MNRAS*, 306, 684
Fabian, A. C., Pringle, J. E., & Rees, M. J. 1976, *MNRAS*, 175, 43
Gaia collabotration, 2018, *Astr. and Astrophys.*, 616, A1
Gehrels, N., Chincarini, G., Giommi, P., *et al.* 2004, *ApJ*, 611, 1005
Harrison, F. A. *et al.* 2013, *ApJ*, 770, 103
McKee, C. F., Parravano, A., & Hollenbach, D. J. 2015, *ApJ*, 814, 13
Mukai, K. 2017, *PASP*, 129, 062001
Oh. K. *et al.* 2018, *ApJS*, 235, 4
Pretorius, M. L. & Mukai, K. 2014, *MNRAS*, 442, 2580
Revnivtsev, M. *et al.* 2008, *Astr. and Astrophys.*, 489, 1121
Revnivtsev, M. *et al.* 2009, *A&A*, 507, 1211
Rothschild, R. E. *et al.* 1981, *ApJ*, 250, 723
Shaw, A. W., Heinke, C.O., Mukai, K., *et al.* 2018, *MNRAS*, 476, 554
Suleimanov, V., Revnivtsev, M., & Ritter, H. 2005, *Astr. and Astrophys.*, 435, 191
Suleimanov, V., Doroshenko, V., Ducci, L., *et al.* 2016, *Astr. and Astrophys.*, 591, A35
Suleimanov, V., Doroshenko, V., & Werner, K. 2019, *MNRAS*, 482, 3622
Warner, B. 2003, *Cataclysmic Variable Stars*, Cambridge Univ. Press, Cambridge, UK
Yuasa, T., Nakazawa, K., Makishima, K., *et al.* 2010, *Astr. and Astrophys.*, 520, A25
Zorotovic, M., Schreiber, M. R., & Gänsicke, B. T. 2011, *Astr. and Astrophys.*, 536, A42

White Dwarfs as Probes of Fundamental Physics: Tracers of Planetary, Stellar and Galactic Evolution
Proceedings IAU Symposium No. 357, 2019
M. A. Barstow, S. J. Kleinman, J. L. Provencal & L. Ferrario, eds.
doi:10.1017/S1743921320000459

Realistic models of Globular Clusters with white dwarfs, neutron stars and black holes using GPU supercomputer

Bhusan Kayastha[1,2,3], Long Wang[4,5], Peter Berczik[2,3,6], Xiaoying Pang[7], Manuel Arca Sedda[3], Albrecht Kamlah[8] and Rainer Spurzem[2,3,9]

[1]University of Chinese Academy of Sciences, 19A Yuquanlu, Shijingshan District, Beijing 100049, China
email: kayastha@nao.cas.cn

[2]National Astronomical Observatories & Key Laboratory of Computational Astrophysics, Chinese Academy of Sciences, 20A Datun Road, Chaoyang District, Beijing 100101, China

[3]Astronomisches Rechen-Institut, Zentrum für Astronomie, University of Heidelberg, Mönchhofstrasse 12-14, 69120 Heidelberg, Germany

[4]Department of Astronomy, School of Science, The University of Tokyo, 7-3-1 Hongo, Bunkyo-ku, Tokyo, 113-0033, Japan

[5]RIKEN Advanced Institute for Computational Science, 7-1-26 Minatojima-minami-machi, Chuo-ku, Kobe, Hyogo 650-0047, Japan

[6]Main Astronomical Observatory, National Academy of Sciences, Kyiv, Ukraine

[7]Department of Mathematical Sciences, Xi'an Jiaotong-Liverpool University, No.111 Ren'ai Road, Dushu Lake Higher Education Town, Suzhou Industrial Park, Suzhou 215123, China

[8]Institut für Astrophysik, Georg-August-Universität Göttingen, Friedrich-Hund-Platz 1, 37077, Göttingen, Germany

[9]Kavli Institute for Astronomy & Astrophysics, Peking University, Yiheyuan Lu 5, Haidian District, Beijing 100871, China

Abstract. We present some results from the DRAGON simulations, a set of four direct N-body simulations of globular clusters (GCs) with a million stars and five percent initial (primordial) binaries. These simulations were undertaken with the NBODY6++GPU code, which allowed us to follow dynamical and stellar evolution of individual stars and binaries, formation and evolution of white dwarfs, neutron stars, and black holes, and the effect of a galactic tidal field. The simulations are the largest existing models of a realistic globular cluster over its full lifetime of 12 billion years. In particular we will show here an investigation of the population of binaries including compact objects (such as white dwarfs - cataclysmic variables and merging black hole binaries in the model as counterparts of LIGO/Virgo sources); their distribution in the cluster and evolution with time.

Keywords. globular clusters, kinematics and dynamics - stars, black holes, gravitational waves

1. Introduction

About 150 globular clusters orbit the Milky Way but more massive galaxies can have over 10,000 gravitationally bound globular clusters. As their stars have mostly formed at the same time but with different masses, globular clusters are ideal laboratories for studies of stellar dynamics and stellar evolution. Our direct N-body (DRAGON) simulations were carried out using the code NBODY6++GPU (Wang *et al.* 2015). This code is designed for an accurate computation of the motion of many bodies (e.g. in star clusters, planetary systems, galactic nuclei) under their mutual Newtonian gravity. It originates from the family of N-body codes initiated by Asrseth (1963, 1999). Even though its asymptotic complexity scales with N^2, where N is the number of particles, the code is much more efficient than any brute-force approach due to a number of important numerical and physical features (Aarseth 2003), such as a fourth order prediction-correction method using a Hermite scheme on only two points in time, hierarchical block time-steps, Kustaanheimo-Stiefel (KS) and Chain regularization of close encounters and few body sub-systems, and an Ahmad-Cohen neighbour scheme (Ahmad & Cohen 1974), which splits the gravitational force on each particle into a component originating from neighbours (irregular force) and another one from more distant particles (regular force). Each particle has two time steps, one connected with the irregular force and another one (generally much larger) for the regular force. Only the regular force computation scales with N^2 asymptotically. This algorithm has a very similar structure to smoothed particle hydrodynamics (SPH). Single and binary stellar evolution is implemented in these N-body codes using the standard recipes described by Hurley, Pols & Tout (2000) and Hurley, Tout & Pols (2002), see also Hurley (2005), with further improvements to take into account fallback at supernova explosions and stellar winds of massive stars (Belczynski *et al.* 2002) as well as neutron star kicks at formation due to Hobbs *et al.* (2005). We use the version NBODY6++GPU of the code, which has parallelization across many computer nodes (using MPI), GPU acceleration for general purpose graphical computing devices (on many nodes, using CUDA) and thread level parallelization on local multi-core CPUs (using OpenMP). Readers who are interested in more technical details are referred to (Spurzem 1999), Berczik *et al.* (2013), and Wang *et al.* (2015). All simulations were performed on the "Hydra" GPU cluster of the Max-Planck Computing and Data Facility (MPCDF), Germany. For each simulation, 8 nodes with a total of 160 Intel Xeon E5-2650 cores and 16 NVIDIA K20m GPUs were used.

2. Results

Figure 1 presents the Hertzsprung-Russell diagram of the Dragon Simulations up to 12Gyr. BSE and SSE stand for Binary Stellar Evolution and Single Stellar Evolution respectively. The zero-age main sequence is along the top-left to lower right diagonal; above the main sequence turnoff there are subgiants, red giants and supergiants seen, as a function of the age of the stellar population. In the lower left cooling white dwarfs are found. The scattered points between white dwarfs and the main sequence are the potential candidates for the cataclysmic variable stars (CVs) .The composite effective temperature for binaries is calculated as flux weighted average of the two binary members.

Figure 2 presents the mock color image (BVI) of all stars of a simulated globular cluster (central image covering about 60 pc) after 12 billion years of evolution. The surrounding panels highlight the different stellar types (from top left): main sequence stars (MS), red giants (RG) dominating the light, invisible black holes (BH), binary stars (Binary), white dwarfs (WD) and asymptotic giant branch stars (AGB). The white dwarfs (about 80,000) are unresolved in this mock image and therefore invisible. Due to the mass-segregation process, BHs form a dense subsystem in the cluster centre, as shown in Figure 2 (panel labelled with "BH", red points represent BHs in binaries). Observations of BHs can be achieved only through the effects that they have on stellar companions, via electromagnetic emission, or among themselves, via gravitational radiation, thus from an observational perspective BHs in binary systems are crucial to constrain their "detectability". In DRAGON simulations, we find afew dozens BHs in

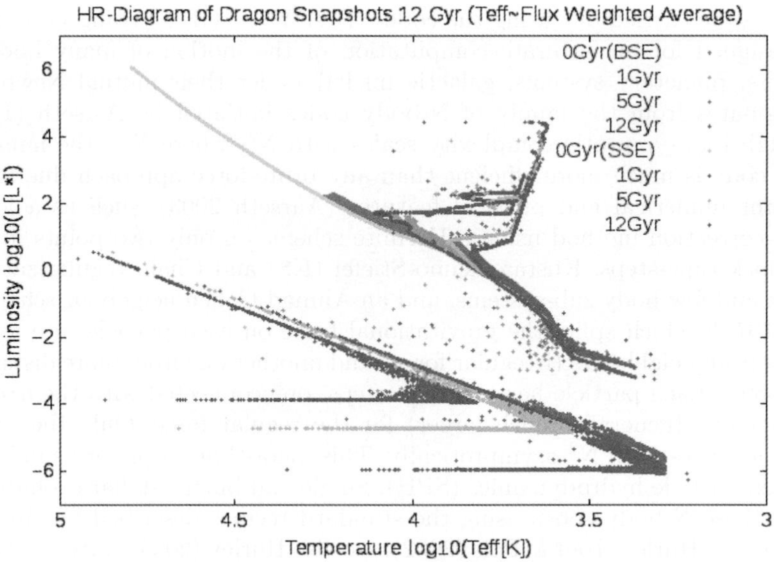

Figure 1. HR Diagram of DRAGON models up to 12Gyr.

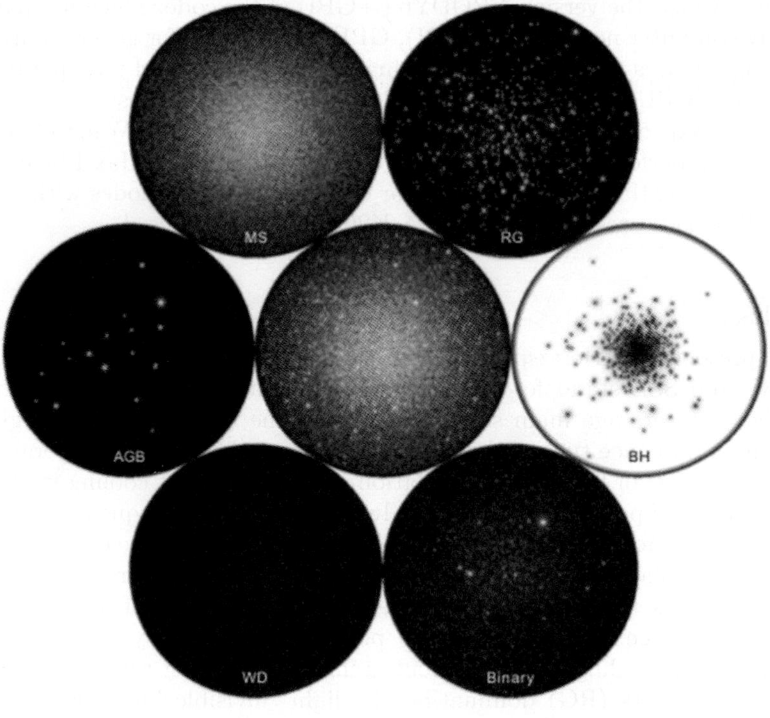

Figure 2. Snapshot of DRAGON models at 12 Gyr as observed in B, V and I bands (Wang *et al.* 2016).

binary systems. As shown in Figure 3, the BH mass spectrum in DRAGON simulations extends from 10 to 32 Msun, with BHs slowly accumulating toward the GC centre over time.

Figure 3. represents the position of the black holes in the globular cluster with respect to the density center. Black holes with 15-20 solar masses are kicked away from the

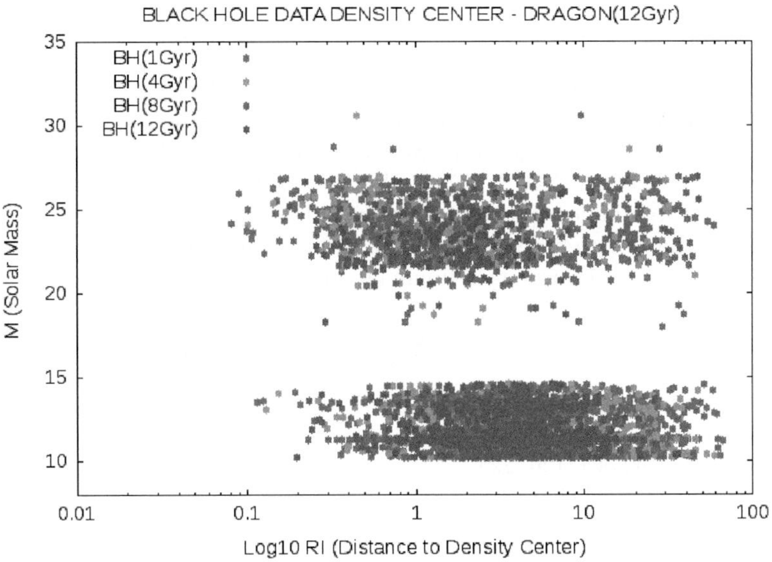

Figure 3. Position of the Black Holes from the density center in the DRAGON models up to 12Gyr.

system because of the natal kicks. Here we can see the black holes are being accumulated more and more towards the density center with time. The dynamical evolution of the central regions of the simulated clusters is dominated by hundreds of single and binary stellar mass black holes.

3. Summary & Discussions

In this paper we present the DRAGON globular cluster models which are evolved from the initial phase after gas removal up to 12 Gyr. The DRAGON clusters are four realistic direct N-body simulations of GCs with initially one million stars, including 5 % primordial binaries, both single and binary stellar evolution and tidal fields together for the first time. We have full information about the modeled cluster, regarding its population and color gradients, its projected stellar densities and kinematic properties of the luminous and compact objects. Some "gravitational wave events" observed in the Dragon simulations are in fair agreement with real LIGO/Virgo detections. Unfortunately, comparison with faster, but more approximate Monte Carlo (MOCCA) dynamical models and with a population synthesis model for galactic field binary populations shows that we cannot yet distinguish the origin of LIGO/Virgo sources (clusters or field). More simulations are required for that goal. The results will be published soon in the paper "Intrinsic Gravitational Wave Emission for Ligo/Virgo from Direct Nbody Models of Star Clusters" (Spurzem *et al.* 2020). The DRAGON simulations were done before the detection of the gravitational wave events by LIGO/Virgo so some methods and recipes used in our codes are outdated. Recently we are ready with the updated version of the NBODY6++GPU code, the main changes include the new mass fallback prescription at supernova explosions and stellar winds of massive stars (Belczynski *et al.* 2019) with low kicks at formation to electron capture supernova (Gessner & Janka 2018) to retain some neutron stars in the cluster. Results from an updated version of the NBODY6++GPU code able to interpret new astronomical discoveries on neutron stars, black holes and gravitational waves will be presented in a series of forthcoming papers.

Acknowledgements

This work has been supported by the Chinese National Science Foundation Project No. 11673032 (RS) & Sino-German Cooperation Project (No. GZ 1284). B.K. would like to thank CAS-TWAS President's PhD Fellowship Programme of the Chinese Academy of Sciences & The World Academy of Sciences, International Astronomical Union for the travel grant to attend IAUS 357. L.W. thanks the financial support from JSPS International Research Fellow (School of Science, The university of Tokyo). X.Y.P. sends her gratitude to the research development fund of Xi'an Jiaotong Liverpool University, RDF-18-02-32. M.A.S. acknowledges financial support by the Alexander von Humboldt Foundation.

References

Aarseth, S. J. 2003, *Gravitational N-Body Simulations*, Cambridge University Press
Ahmad, A. & Cohen, L. 1974, *ApJ*, 188, 469
Aarseth, S. J. 1963, *MNRAS*, 126, 223
Aarseth, S. J. 1999, *PASP*, 111, 1333
Aarseth, S. J. 2012, *MNRAS*, 422, 841
Belczynski, K., Kalogera, V., & Bulik, T. 2002, *ApJ*, 572, 407
Belczynski, K. *et al.* 2019, *ApJ*, 714, 1217
Berczik, P, *et al.* 2013, Third International Conference "High Performance Computing", HPC-UA 2013, p. 52–59
Gessner, A. & Janka, H-T, 2018, *ApJ*, 865, 61
Hobbs, G., Lorimer, D. R., Lyne, A. G., & Kramer, M. 2005, *MNRAS*, 360, 584
Huang S., Spurzem R., & Berczik P. 2015, *arXiv*, arXiv:1508.02510
Hurley, J. R. 2005, *MNRAS*, 363, 315
Hurley, J. R., Pols, O. R., & Tout, C. A. 2000, *MNRAS*, 315, 543
Hurley, J. R., Tout, C. A., & Pols, O. R. 2002, *MNRAS*, 329, 897
Hurley, J. R., Pols, O. R., Aarseth, S. J., & Tout, C. A. 2007, *MNRAS*, 363, 292
King, I. R. 1996, *AJ*, 71, 64
Kouwenhoven, M. B. N., Brown, A. G. A., Portegies, Zwart S. F., & Kaper, L. 2007, *A&A*, 474, 77
Kroupa, P. 1995, *MNRAS*, 277, 1491
Kroupa, P. 2001, *MNRAS*, 322, 231
Nitadori, K. & Aarseth, S. J. 2012, *MNRAS*, 424, 545
Spurzem, R. 1999, *JCoAM*, 109, 407
Spurzem, R. & Aarseth, S. J. 1996, *MNRAS*, 282, 19
Spurzem, R. *et al.* 2020, in press
Wang, L., Spurzem, R., Aarseth, S., Nitadori, K., Berczik, P., Kouwenhoven, M. B. N., & Naab, T. 2015, *MNRAS*, 450, 4070
Wang, L., Spurzem, R., Aarseth, S., Giersz, M., Askar, A., Berczik, P., Naab, T., Schadow, R., & Kouwenhoven, M. B. N. 2016, *MNRAS*, 458, 1450

White Dwarfs as Probes of Fundamental Physics: Tracers of Planetary, Stellar and Galactic Evolution
Proceedings IAU Symposium No. 357, 2019
M. A. Barstow, S. J. Kleinman, J. L. Provencal & L. Ferrario, eds.
doi:10.1017/S174392132000068X

Masses of white dwarfs in symbiotic binaries

Kenneth H. Hinkle[1], Francis C. Fekel[2], Richard Joyce[1], Thomas Lebzelter[3] and Oscar Straniero[4]

[1]NSF's National Optical-Infrared Astronomy Research Laboratory, P.O. Box 26732, Tucson, AZ 85726, USA
emails: hinkle@noao.edu, joyce@noao.edu

[2]Center of Excellence in Information Systems, Tennessee State University, 3500 John A. Merritt Blvd, Box 9501, Nashville, TN 37209, USA
email: fekel@evans.tsuniv.edu

[3]University of Vienna, Department of Astrophysics, Tuerkenschanzstrasse 17, 1180 Vienna, Austria
e-mail: thomas.lebzelter@univie.ac.at

[4]Istituto Nazionale di Astrofisica, Osservatorio d'Abruzzo, Via Maggini s.n.c., I-64100, Teramo, Italy
email: oscar.straniero@inaf.it

Abstract. Masses have been computed for the white dwarfs (WDs) in eclipsing, mass exchange (symbiotic), WD–red giant (RG) binaries by using single-lined spectroscopic orbits, orbital inclinations, and the RG masses. Inclinations have been measured for 13 eclipsing symbiotic binaries. Using Gaia data the mass of the RG can be found from evolutionary tracks. Since the WD evolved from the more massive star in the binary, the WD should be more massive than predicted from the mass of the current RG. Typically the WD has a lower mass than expected implying a previous mass exchange stage for these systems.

Keywords. (stars:) binaries: eclipsing, (stars:) binaries: symbiotic, stars: evolution, stars: late-type, (stars:) white dwarfs

1. Introduction

WD binary systems with a late-type stellar companion include barium stars, extrinsic S-stars, CH stars and symbiotic stars. Symbiotic binaries and extrinsic S-star binaries contain the coolest, most evolved giants. The hallmark of symbiotic binaries is a mass flow from the late-type star that is accreted onto the WD. None of the late-type symbiotic giants fill their Roche lobes. A small number of WDs in symbiotics are possible supernova progenitors with masses near the Chandrasekhar limit. The large size difference between the RG and WD plus the range of separations where mass exchange takes place results in a large number of eclipsing systems among the symbiotics. Eclipsing systems are useful in measuring a number of parameters including the orbital inclination.

2. Spectroscopic Orbits

In a series of eight papers (Fekel *et al.* 2017, 2015, 2010, 2008, 2007, 2001, 2000a,b) we measured the single-lined spectroscopic orbit of the late-type star in 30 symbiotic binary systems. Two parameters from the orbital solution, eccentricity and mass function, can be used to evaluate the group statistically. Barium star abundances show that mass transfer took place from the progenitor of the WD to the progenitor of what is now the RG barium star. Barium stars, unlike symbiotics, do not have on-going mass transfer from

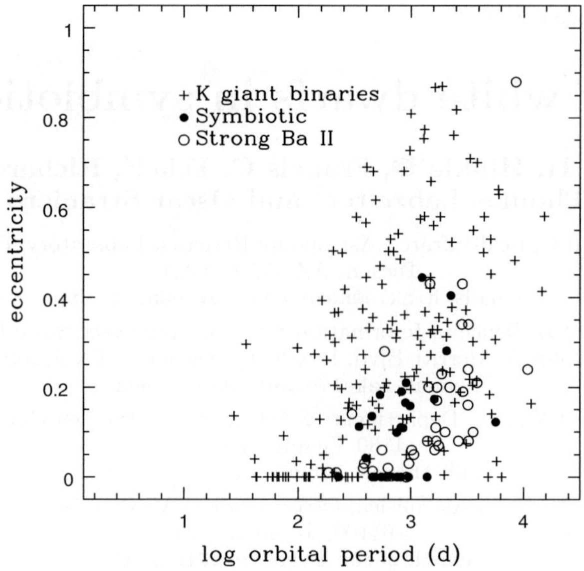

Figure 1. The eccentricity–period diagram for symbiotics taken from papers by Fekel et al. and for barium stars from van der Swaelmen *et al.* (2017). These are compared to the eccentricity–period diagram for normal K giant binaries from Mermilliod *et al.* (2007).

the RG to the WD. The eccentricities of both the barium stars and the symbiotics show the effect of the mass transfer (Fig. 1). An unanswered question is whether the symbiotic binaries underwent a previous state of mass transfer similar to that of the barium stars. Abundance analysis of symbiotic RGs (Gałan *et al.* 2017, 2016) shows some peculiarities, although much smaller than those of barium stars.

3. Mass Function

The mass function,

$$f(m) = m_{wd}^3 sin^3(i)/(m_{rg} + m_{wd})^2,$$

is a parameter that is derived from the single-lined binary orbital elements and relates the component masses and the inclination (i). For a large sample of orbits integration over the inclination produces a statistical mass ratio. The barium stars have been shown to have a nonuniform distribution of mass ratios with WD masses in the range 0.5–1 $M_\odot$ (Jorissen *et al.* 2019). The mass function of the symbiotics is clearly similar (Fig. 2). However, a larger sample of symbiotic binary orbits would be required for a detailed statistical analysis.

4. WD Masses

For eclipsing symbiotics the inclination can be solved geometrically. A few systems have precise inclinations from Raman-scattered emission lines observed with spectropolarimetry Harries & Howarth (2000). For the systems with known inclination the mass function gives the mass ratio. The giant's luminosity can be computed from the Gaia distance, K band photometry, and the bolometric correction. The effective temperature is from the spectral type. The mass of the giant was found using evolutionary tracks computed using FuNS. All of the symbiotics in this sample are on the early AGB or red giant branch, and mass loss to this point has not been large so the current and main sequence masses are similar.

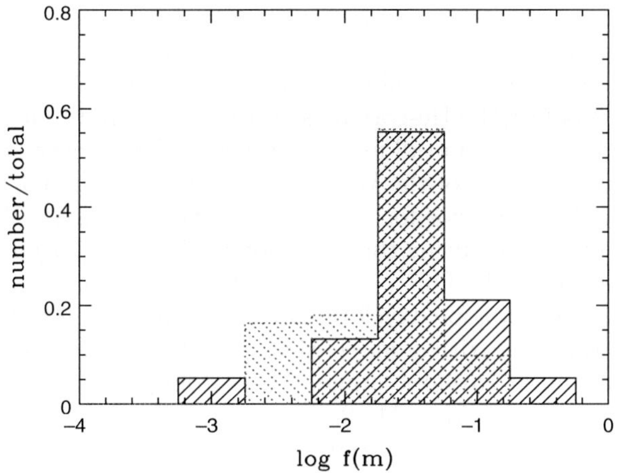

Figure 2. The solid line hatched histogram is the mass function distribution for symbiotic binaries. The dotted line hatched histogram is the strong and weak barium star mass function distribution. Symbiotic mass functions are from Fekel et al. and barium star mass functions from Jorissen *et al.* (2019).

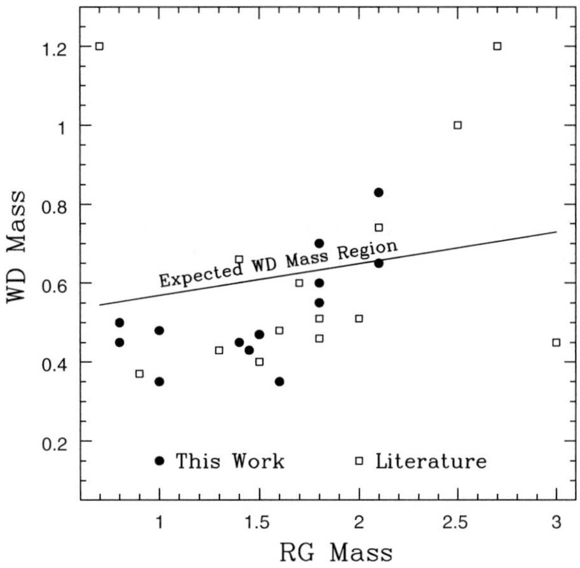

Figure 3. Masses for the WD in symbiotic binaries as a function of the RG mass. The WD evolved from the more massive member of the binary and should have a WD mass above the prediction for the WD mass from the mass of the current RG. The WD initial mass–final mass relation is the solid line (Cummings *et al.* 2018).

The resulting masses for the symbiotic binary RGs and WDs are shown in Fig. 3. The WD progenitor had to have a larger initial mass than that of the current RG so the expectation is that the WD masses sould be larger than predicted using the current RG mass. Given the uncertainties, the WD mass should scatter around the values predicted by the initial–final mass relation. However, Fig. 3 shows that systems with RG masses less than $\sim$1.6 $M_\odot$ typically contain WDs with masses considerably lower than expected. The existence of these low masses for symbiotic WDs has been previously noted in the literature. A summary of the masses determined by a variety of other techniques can be found in Mikołajewska (2003).

The initial-final mass relation shown in Fig. 3 is for single (or non-interacting) star. For single stars with low initial mass (M<1.5 $M_\odot$) a minimum CO WD mass can be produced, ~0.45 $M_\odot$. These stars develop a fully degenerate core during the first RGB and He ignition cannot occur before the He-core mass attains this limit. The limit coincides with the lower end of the theoretical initial mass-final mass relation corresponding to an initial mass of 0.6 $M_\odot$. However, most of the WD masses shown in Fig. 3 are less than 0.45 $M_\odot$.

There are two possible scenarios for the origin of these low mass WDs. Both scenarios require binary mass exchange during the evolution of the WD progenitor. The first is that the low mass WDs are He-rich. This implies that the progenitor, a star with initial mass < 1.8 $M_\odot$, lost its H-rich envelope envelope during the RGB before He ignition when the core mass was lower than the critical mass. Mass transfer could be driven by either Roche lobe overflow or a common envelope stage. The second, discussed by Prada-Moroni & Straniero (2009), concerns 1.8 $M_\odot$<M<2.5 $M_\odot$ progenitors. In this case, the He-core is only partially degenerate, so that He ignition can occur when the core mass is as small as ~0.35 $M_\odot$. If these stars lose their envelope during the He burning phase, the mass of the resulting CO WD could be as small as those shown in Fig. 3.

References

Cummings, J. D., Kalirai, J., Tremblay, P.-E., *et al.* 2018, *ApJ*, 866, 21
Fekel, F. C., Hinkle, K. H., Joyce, R. R., *et al.* 2017, *AJ*, 153, 35
Fekel, F. C., Hinkle, K. H., Joyce, R. R., *et al.* 2015, *AJ*, 150, 48
Fekel, F. C., Hinkle, K. H., Joyce, R. R., *et al.* 2010, *AJ*, 139, 1315
Fekel, F. C., Hinkle, K. H., Joyce, R. R., *et al.* 2008, *AJ*, 136, 146
Fekel, F. C., Hinkle, K. H., Joyce, R. R., *et al.* 2007, *AJ*, 133, 17
Fekel, F. C., Hinkle, K. H., Joyce, R. R., *et al.* 2001, *AJ*, 121, 2219
Fekel, F. C., Hinkle, K. H., Joyce, R. R., *et al.* 2000, *AJ*, 120, 3255
Fekel, F. C., Hinkle, K. H., Joyce, R. R., *et al.* 2000, *AJ*, 119, 1375
Gałan, C., Mikołajewska, J., Hinkle, K. H., *et al.* 2017, *MNRAS*, 466, 2194
Gałan, C., Mikołajewska, J., Hinkle, K. H., *et al.* 2016, *MNRAS*, 455, 1282
Harries, T. J. & Howarth, I.D. 2000, *A&A*, 361, 139
Jorissen, A., Boffin, H. M. J., Karinkuzhi, D., *et al.* 2019, *A&A*, 626, 127
Mermilliod, J.-C., Andersen, J., Latham, D. W., *et al.* 2007, *A&A*, 473, 829
Mikoiłajeska, J. 2003, *ASP Conf. Series*, 303, 9
Prada-Moroni, P. & Straniero, O. 2009, *A&A*, 507, 1575
van der Swaelmen, M., Boffin, H. M. J., Jorrissen, A., *et al.* 2017, *A&A*, 597, 68

White Dwarfs as Probes of Fundamental Physics: Tracers of Planetary, Stellar and Galactic Evolution
Proceedings IAU Symposium No. 357, 2019
M. A. Barstow, S. J. Kleinman, J. L. Provencal & L. Ferrario, eds.
doi:10.1017/S1743921320000915

Four new self-lensing binaries from *Kepler*: Radial velocity characterization and astrophysical implications

Kento Masuda[1], Hajime Kawahara[2,3], David W. Latham[4], Allyson Bieryla[4], Morgan MacLeod[4], Masanobu Kunitomo[5], Othman Benomar[6] and Wako Aoki[6]

[1]Institute for Advanced Study, Princeton, NJ 08540, USA
email: `kmasuda@ias.edu`

[2]Department of Earth and Planetary Science, The University of Tokyo, Tokyo 113-0033, Japan

[3]Research Center for the Early Universe, School of Science, The University of Tokyo, Tokyo 113-0033, Japan
email: `kawahara@eps.s.u-tokyo.ac.jp`

[4]Center for Astrophysics | Harvard & Smithsonian, Cambridge, MA 02138, USA

[5]Department of Physics, School of Medicine, Kurume University, Fukuoka 830-0011, Japan

[6]National Astronomical Observatory of Japan, 2-21-1 Osawa, Mitaka, Tokyo 181-8588, Japan

Abstract. In Kawahara *et al.* (2018) and Masuda *et al.* (2019), we reported the discovery of four self-lensing binaries consisting of F/G-type stars and (most likely) white dwarfs whose masses range from 0.2 to 0.6 solar masses. Here we present their updated system parameters based on new radial velocity data from the Tillinghast Reflector Echelle Spectrograph at the Fred Lawrence Whipple Observatory, and the *Gaia* parallaxes and spectroscopic parameters of the primary stars. We also briefly discuss the astrophysical implications of these findings.

Keywords. binaries: eclipsing, blue stragglers, gravitational lensing, techniques: radial velocities, white dwarfs

1. Introduction

Stars eclipsed by their compact object companions can display periodic brightening pulses in their light curves, rather than eclipse-like dips, due to gravitational microlensing. Such self-lensing binaries (SLBs) were discussed 50 years ago in the context of the search for "collapsed stars" (e.g., Trimble & Thorne 1969), but had not been known until Kruse & Agol (2014) reported the first such system, KOI-3278, from the *Kepler* photometric data. The system consists of a G-type star and a WD of 0.5 $M_\odot$ on a 88-day orbit.

The KOI-3278 system had been identified as a transiting planet candidate because the stellar light curves also showed repeating dips due to occultations of the WD, which triggered the planet-detection pipeline. On the other hand, the events with periods $\gtrsim 1\,\mathrm{yr}$ were not the focus of this pipeline, although SLBs on wider orbits generally exhibit stronger signals because of the larger Einstein radius. In Kawahara *et al.* (2018) and Masuda *et al.* (2019), we reported four additional SLBs with periods ranging from 419 to 728 days from our search (Kawahara & Masuda 2019) for long-period transiting planets that show only one or two transits in the four-year photometric data from the prime *Kepler* mission. Here we report updated parameters of these SLBs based on the

Table 1. System Parameters of the Second to Fifth Self-lensing Binaries.

	KIC 3835482	KIC 6233093	KIC 12254688	KIC 8145411
(Primary star)				
mass ($M_\odot$)	1.52 ± 0.06	1.27 ± 0.04	1.37 ± 0.07	1.11 ± 0.08
radius ($R_\odot$)	2.28 ± 0.05	2.32 ± 0.07	2.08 ± 0.05	1.27 ± 0.18
distance (kpc)	1.15 ± 0.02	1.43 ± 0.04	1.20 ± 0.03	...
(Radial velocity)				
eclipse epoch ($\mathrm{BJD_{TDB}} - 2454833$)	591.628 ± 0.005	$260.678^{+0.008}_{-0.009}$	$404.583^{+0.006}_{-0.007}$	267.97 ± 0.01
orbital period P (day)	$683.266^{+0.008}_{-0.006}$	$727.979^{+0.010}_{-0.008}$	$418.718^{+0.006}_{-0.005}$	455.83 ± 0.02
orbital eccentricity	$0.062^{+0.002}_{-0.001}$	0.123 ± 0.008	$0.180^{+0.004}_{-0.003}$	0.13 ± 0.02
argument of periastron (deg)	4 ± 3	112 ± 10	128 ± 2	-93 ± 3
RV semi-amplitude (km/s)	9.07 ± 0.02	9.5 ± 0.3	$10.44^{+0.06}_{-0.05}$	$4.59^{+0.09}_{-0.10}$
constant acceleration (m/s/day)	0.1 ± 0.1	2^{+2}_{-3}	$-0.2^{+0.4}_{-0.3}$	-0.6 ± 0.3
RV zero point (km/s)	7.45 ± 0.01	$6.61^{+0.07}_{-0.06}$	$-7.49^{+0.2}_{-0.3}$	$-0.78^{+0.06}_{-0.07}$
RV jitter (m/s)	13^{+12}_{-6}	17^{+36}_{-9}	17^{+29}_{-9}	150^{+90}_{-80}
(White dwarf)				
mass ($M_\odot$)	0.62 ± 0.01	0.61 ± 0.02	0.56 ± 0.02	0.19 ± 0.01

Note Quoted values are the median and symmetric 68.3% interval of the marginal posteriors.

radial velocity (RV) data from the Tillinghast Reflector Echelle Spectrograph (TRES; Szentgyorgyi & Furész 2007; Mink 2011) on the 1.5 m telescope at the Fred Lawrence Whipple Observatory. We also briefly discuss the astrophysical implications of these systems.

2. Updated System Parameters of the Self-lensing Binaries

We identified KIC 3835482, KIC 6233093, KIC 12254688, and KIC 8145411, as new (i.e., second to fifth) self-lensing binaries in addition to the first system KOI-3278. The periods and phases of the orbits are well determined from the self-lensing signals observed two or three times in the *Kepler* data. The light curve of KIC 8145411 has a gap in the middle of the two detected pulses that originally caused a factor-of-two uncertainty in the period, but this degeneracy has been resolved with the RV data.

For KIC 3835482, KIC 6233093, and KIC 12254688, we derived masses and radii of the primary stars by fitting stellar models to their atmospheric parameters from TRES spectra (KIC 3835482) or LAMOST DR4 (the others; Cui *et al.* 2012 and Luo *et al.* 2015), 2MASS K_s-band magnitude, and the parallax from *Gaia* DR2, assuming K_s-band extinction from a dust map. The procedure is described in Kawahara & Masuda (2019). For KIC 8145411 without available *Gaia* parallax, we obtained a high-resolution spectrum using the High-Dispersion Spectrograph (Noguchi *et al.* 2002) on the Subaru 8.2 m telescope and derived the stellar parameters by comparing the spectrum to those of touchstone stars with well-determined physical properties (Yee *et al.* 2017).

We modeled TRES RVs of the SLBs with the prior information on their orbital periods and phases from the self-lensing light curves (Figure 1). The priors were otherwise chosen to be uninformative. The resulting binary mass functions were then combined with the primary masses estimated above to derive the WD masses assuming edge-on orbits (as justified from self-lensing). The results are summarized in Table 1.

3. Discussion

Figure 2 compares the binary orbital periods and WD masses of the five known SLBs against other known WD (or sub-dwarf star) binaries with well-measured masses; see

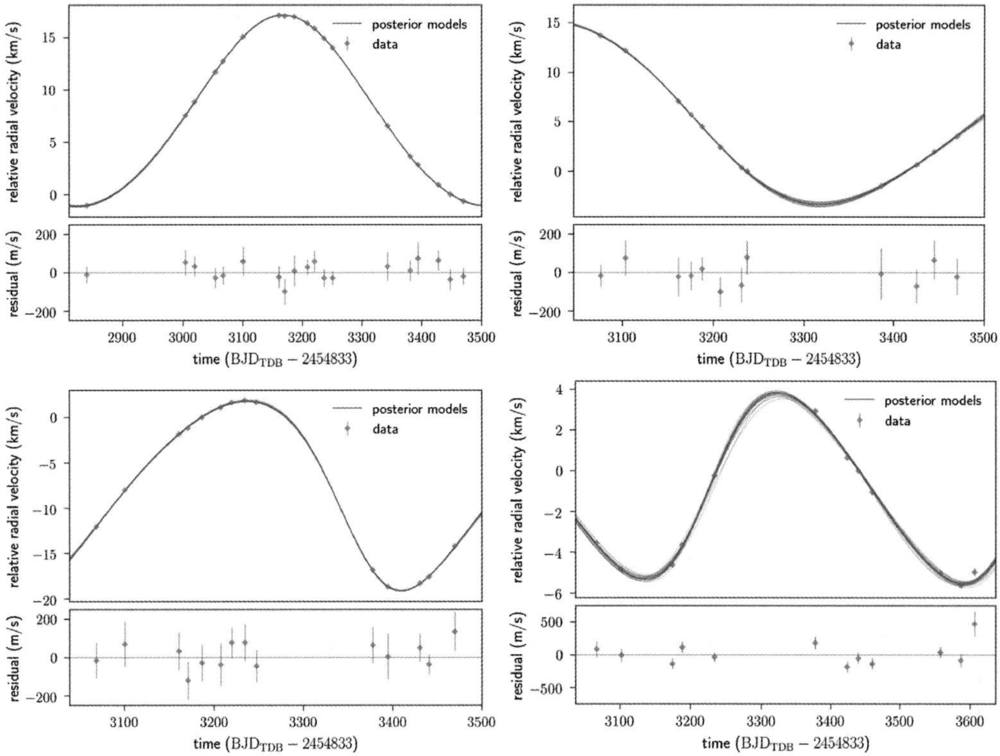

Figure 1. Radial velocity data (orange diamonds) and posterior models (blue lines) for the four SLBs in Kawahara *et al.* (2018) and Masuda *et al.* (2019). *Top-left*—KIC 3835482. *Top-right*—KIC 6233093. *Bottom-left*—KIC 12254688. *Bottom-right*—KIC 8145411.

Figure 5 of Masuda *et al.* (2019) for more details. The solid gray line shows the theoretical period–WD mass relation (Lin *et al.* 2011) for the case where the WD progenitor underwent stable mass transfer onto the companion. The good agreement between this prediction and the upper envelope of the measurements for WDs with $\lesssim 0.3\,M_{\odot}$ supports the idea that these extremely low-mass (ELM) WDs are the products of binary interactions (Marsh *et al.* 1995): such a low-mass WD cannot be formed in isolation because its progenitor would live too long to complete its evolution within the age of the Galaxy.

The ELM WD with $0.2\,M_{\odot}$ in the KIC 8145411 system challenges this view. The orbit of this system is more than 10 times wider than theoretically expected, which means that the progenitor of this WD, when its core mass was $0.2\,M_{\odot}$, should have been far too small to have filled its Roche lobe. We do not have any good explanation of how this ELM WD could have been formed — although the low orbital eccentricity of this system may still suggest that the binary interacted at some point in the past. Moreover, such systems may not be very rare among WD–stellar binaries on AU-scale orbits. The system was found because the edge-on geometry allowed us to see self-lensing, and this happens only in 1 in 200 systems for the measured orbital separation. This implies that there are some 200 such ELM WDs around 10^5 *Kepler* target stars, and this rate could be some 10% of all WDs (i.e., regardless of their masses) on AU-scale orbits around Sun-like stars. Indeed, a similar system containing an ELM "pre-WD" has recently been reported by Vos *et al.* (2018) (purple pentagon in Figure 2), and some of the field blue straggler binaries might also contain such low-mass WDs. See Section 5 of Masuda *et al.* (2019) for more detailed discussion on the occurrence rate of such systems.

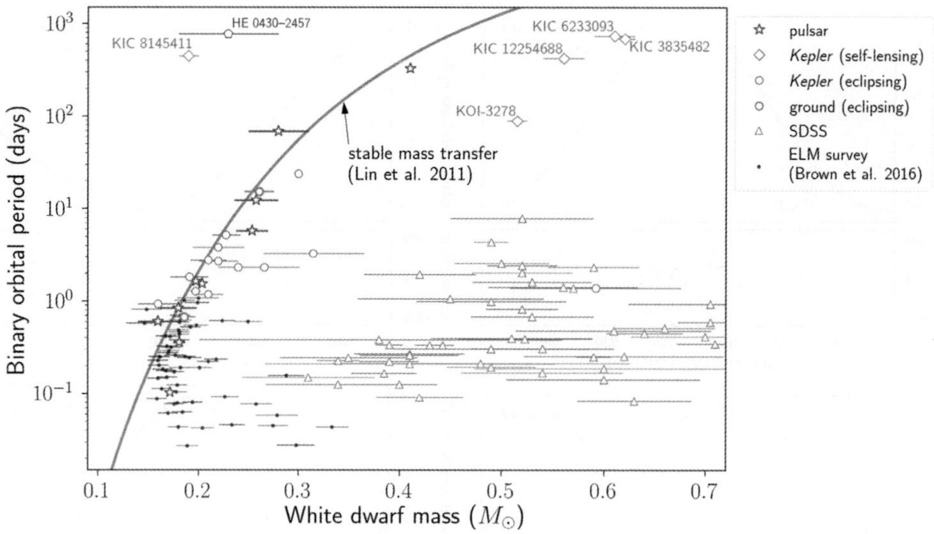

Figure 2. The masses of WDs in binaries and their orbital periods. Same as Figure 5 of Masuda *et al.* (2019), but with updated parameters for SLBs (orange diamonds) from this work and Yahalomi *et al.* (2019). Here we also show the ELM "pre-WD" in the HE 0430–2457 system (Vos *et al.* (2018)) with a purple pentagon.

The other four SLBs (KOI-3278, KIC 3835482, KIC 6233093, KIC 12254688) have WDs with rather normal masses (0.5–0.6 $M_\odot$), but their locations in Figure 2 (i.e., they are below the gray line) suggest that these binaries should have interacted when the WD progenitors were on the asymptotic giant branch. Their low orbital eccentricities support this notion, although the latter three systems do have small but non-zero eccentricities and their origin is unclear. In this regard, these systems are reminiscent of some of the blue straggler binaries in open clusters and in the field for which stable mass transfer has been proposed to explain their bluer colors than expected for their ages. In fact, the primary stars in the KIC 3835482, KIC 6233093, and KIC 12254688 systems are all more massive than typical *Kepler* targets, although their ages are not well constrained. Self-lensing enables detailed characterization of such post-interaction binaries by directly showing that the companion is a WD and by providing its precise true mass.

References

Cui, X.-Q., Zhao, Y.-H., Chu, Y.-Q., *et al.* 2012, *RAA*, 12, 1197

Kawahara, H. & Masuda, K. 2019, *AJ*, 157, 218

Kawahara, H., Masuda, K., MacLeod, M., *et al.* 2018, *AJ*, 155, 144

Kruse, E. & Agol, E. 2014, *Science*, 344, 275

Lin, J., Rappaport, S., Podsiadlowski, P., *et al.* 2011, *ApJ*, 732, 70

Luo, A.-L., Zhao, Y.-H., Zhao, G., *et al.* 2015, *RAA*, 15, 1095

Marsh, T. R., Dhillon, V. S., & Duck, S. R. 1995, *MNRAS*, 275, 828

Masuda, K., Kawahara, H., Latham, D. W., *et al.* 2019, *ApJ* (Letters), 881, L3

Mink, D. J. 2011, in Astronomical Society of the Pacific Conference Series, Vol. 442, Astronomical Data Analysis Software and Systems XX, ed. I. N. Evans, A. Accomazzi, D. J. Mink, & A. H. Rots, 305

Noguchi, K., Aoki, W., Kawanomoto, S., *et al.* 2002, *PASJ*, 54, 855

Szentgyorgyi, A. H. & Furész, G. 2007, in Revista Mexicana de Astronomia y Astrofisica, vol. 27, Vol. 28, Revista Mexicana de Astronomia y Astrofisica Conference Series, ed. S. Kurtz, 129–133

Trimble, V. L. & Thorne, K. S. 1969, *ApJ*, 156, 1013

Vos, J., Zorotovic, M., Vučković, M., Schreiber, M. R., & Østensen, R. 2018, *MNRAS*, 477, L40

Yahalomi, D. A., Shvartzvald, Y., Agol, E., *et al.* 2019, *ApJ*, 880, 33

Yee, S. W., Petigura, E. A., & von Braun, K. 2017, *ApJ*, 836, 77

White Dwarfs as Probes of Fundamental Physics: Tracers of Planetary, Stellar and Galactic Evolution
Proceedings IAU Symposium No. 357, 2019
M. A. Barstow, S. J. Kleinman, J. L. Provencal & L. Ferrario, eds.
doi:10.1017/S1743921320001374

What can ISM and non-photospheric highly ionised lines in white dwarf spectra reveal about the β CMa tunnel?

Nicolle L. Finch[1], S. P. Preval[2], M. A. Barstow[2], S. L. Casewell[2], T. Ayres[3], B. Welsh[4], M. Bainbridge[2] and N. Reindl[5]

[1]Physics and Astronomy Department, University of Leicester, LE1 7RH, Leicester, UK
email: nlf7@le.ac.uk

[2]Physics and Astronomy Department, University of Leicester, LE1 7RH, Leicester, UK

[3]Department of Astrophysical and Planetary Sciences, University of Colorado at Boulder, Boulder, CO 80309, USA

[4]Space Sciences Laboratory, University of California, 7 Gauss Way, Berkeley, CA 94720, USA

[5]Institut für Physik und Astronomie, Universitätsstandort Golm, Haus 28, Karl-Liebknecht-Str. 24/25, 14467 Potsdam, Germany

Abstract. White dwarfs are useful objects with which to study the local interstellar medium (ISM). High ionisation state absorption features that cannot be attributed to the photosphere or the ISM have been observed along the line-of-sight to a number of white dwarf stars. Suggested origins of these lines include ionisation from past supernovae, stellar winds, circumstellar disks, photoionisation from nearby hot stars or also from the white dwarf itself. In this study we consider the origin of these non-photospheric highly ionised lines in two stars towards a rarefied region of the galaxy known as the extended β CMa Tunnel. We present preliminary results from our analysis of the first of these two stars.

Keywords. Interstellar medium, spectroscopy: white dwarfs, atmospheres: white dwarfs

1. Introduction

Space is not empty. Between stars there are many clouds of cool or warm (approx 50-10000 K) gas: the interstellar medium (ISM); those clouds within 10-20 pc of our solar system are typically called the local interstellar medium (LISM). When we observe distant stars with high resolution spectroscopes (R $\sim 10^5$) we must be aware of the potential for that starlight to travel through such ISM clouds and for additional absorption features to be formed by the atoms within.

Our solar system is situated inside a Local Bubble of hot ionised gas (approx 10^6 K) which spans approx 100 pc. The Local Bubble is relatively low density (approx 5×10^{-3} atoms cm^{-3}, Dupin & Gry 1998) and is proposed to have been caused by past supernovae (e.g., Maíz-Apellániz 2001). There is a further patch of rarefied space joining the Local Bubble to another similar bubble, called the β CMa Tunnel (e.g., Lallement *et al.* 2014). It is not known how this tunnel was formed or is maintained, as it does not have the characteristic bubble-like shape caused by a supernova (e.g., Fuchs *et al.* 2006).

Many sight lines have been studied (e.g., Dupin & Gry 1998; Welsh 1991) to determine the extent of the β CMa Tunnel, as shown in Figure 1. However, we have identified two nearby white dwarfs (d = 121 pc and 137 pc respectively) that appear to have very low line-of-sight hydrogen column densities, similar to the tunnel. As they are physically

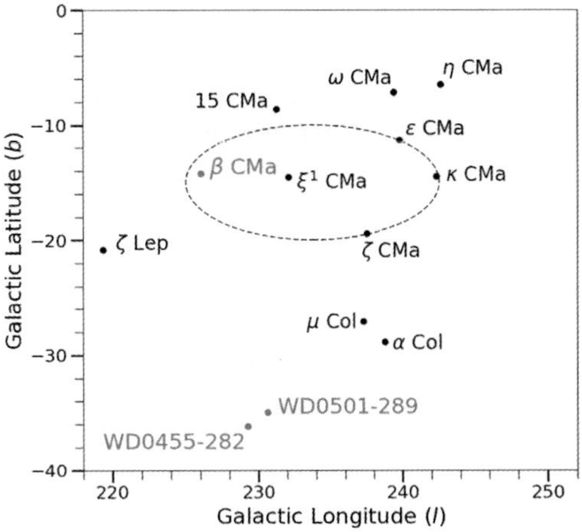

Figure 1. Updated plot of Fig. 1. of Welsh (1991) using data from SIMBAD (Wenger *et al.* 2000) showing sight lines from a study of the β CMa Tunnel. Also included are the positions of the stars in our study.

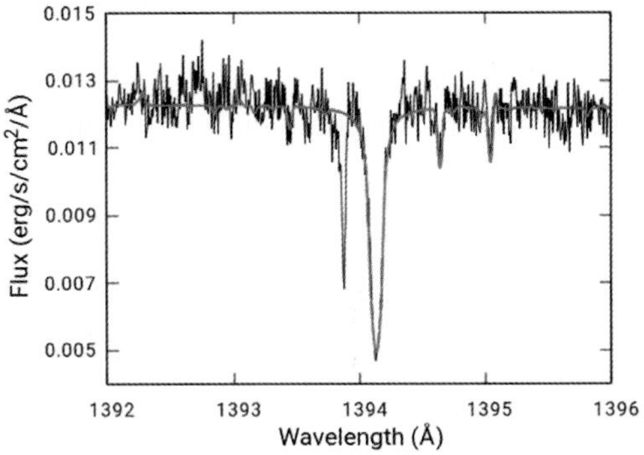

Figure 2. Example of a Si IV absorption profile featuring the photospheric component (modelled as red solid line) and an extra non photospheric but high ionisation state line at ~1393.8 Å.

close together (approx 17 pc), we have a rare opportunity to investigate the effect of their Strömgren spheres. We pose two questions. Are these stars part of an extension of the β CMa Tunnel and are they interacting with it?

The main absorption features we expect to see in the white dwarf spectra are from the photosphere itself. H-rich white dwarfs contain broad hydrogen lines, and if sufficiently hot ($>30{,}000$ K) we would also expect highly ionised photospheric metal lines in the UV spectra. If lines are formed in the photosphere then they will share a similar velocity - the photospheric velocity. However, in spectra of high enough resolution we might also see absorption lines from lower ionisation state metals; these lines originate from ISM clouds that the starlight has traversed, and are low ionisation states because the clouds are far cooler than the stellar photosphere. The ISM metal lines would have a velocity depending on the motion of the ISM cloud, thus not usually a similar velocity to that of the photosphere.

Some high ionisation lines can be present in spectra, which do not appear to originate in the stellar photosphere, as their velocities do not match the photospheric velocity. This is illustrated in Figure 2, showing a photospheric Si IV line (black), with a non-LTE stellar atmospheric model overlaid (red), and a second Si IV feature not accounted for by the photospheric model. There are a number of proposed mechanisms for creating these ionised features including ionisation from past supernovae, stellar winds, circumstellar disks, ancient planetary nebulae, photoionisation from nearby hot stars, and photoionisation from the observed star (Strömgren sphere).

2. Overview

We have identified two white dwarfs, WD 0455-282 and WD 0501-289 (see Figure 1) in the vicinity of the known β CMa Tunnel. The focus of this work is WD 0455-282. WD 0455-282 is a Teff $\approx$ 60,000 K, log g $\approx$ 7.90, H-rich white dwarf (e.g., Preval *et al.* 2019). Preval *et al.* (2019) observed WD 0455-282 with the Hubble Space Telescope (HST) using the Space Telescope Imaging Spectrograph (STIS; e.g., Riley *et al.* 2019) in the high resolution grating (R $\approx$ 114,000, see Figure 3). Due to its high temperature we detect highly ionised photospheric metal lines. We also detect lines from low ionisation states which must be due to the ISM, as well as highly ionised non-photospheric absorption lines of unknown origin. We use XSPEC (e.g., Arnaud 1996) for this analysis, specifically the built in Gaussian function GABS, to fit the absorption profiles as this accounts for the saturation of any lines.

Silicon. For the silicon ions we assess the likelihood of them arising from the same cloud. If they do, the ionisation states should have comparable Doppler width, b (km/s), values as this is a property of the cloud's temperature and turbulence. First we analyse the Si IV profiles (bottom middle and right panels of Figure 3). We can see they each have a single component and so we used a single Gaussian in each case. As these lines are a doublet, we also fit them together with the wavelength ratio fixed. Results of our models are given in rows 2-3 and 4 of Table 1, respectively. An issue with modelling the Si III profile (bottom left panel in Figure 3) is that it contains many blended components, meaning Gaussian lines need adding to reduce the chi square to an acceptable value. This in turn increases the error; so far a good fit has not been found. Lastly there are two Si II lines available (top middle and right panels of Figure 3). Si II 1260 Å presents a similar issue as the Si III profile. The weaker Si II 1304 Å has only two components in its profile, however the error found for this fit is nearly 100% (row 1 of Table 1). On inspection we found in our models that the continuum is not flat as should be expected. Following discussion with the XSPEC code author, we believe this to be a systematic error, and are taking steps to correct this before using the code further. Overall we cannot draw conclusions from these results, as the errors in some cases are too high.

Carbon. Due to the continuum issues we can only discuss the other results qualitatively. Take for example the C II 1334 Å profile (top left panel in Figure 3) which visibly appears to contain over four Gaussian components. To check this, we modelled the profile with increasing numbers of Gaussian features until a reduced chi square close to unity was obtained. First, one Gaussian absorption per side of the profile was used, which, as expected, did not fit well visibly or give a good reduced chi square value. Using two Gaussian features per side improves the fit, as has been suggested by others (e.g., Preval *et al.* 2019). However, the reduced chi square value using a fit of four Gaussian features in total is still not close to unity. Adding a fifth component to the red-shifted side yields a good reduced chi squared value.

Other elements. We can also take the velocities measured for all the ISM lines (Si II, O I, N I) and see if any of the projected velocities of LISM clouds match the values. The LISM models of Redfield & Linsky (2008) suggest that the so called Blue Cloud lies along this sight line. However, the projected velocity of 12.56 ± 1.03 km/s does not match the preliminary velocities measured in this analysis.

Table 1. Measurements of silicon ionisation states presenting the $\sim 25\,\mathrm{km\,s^{-1}}$ component.

Silicon line	**Velocity ($\mathrm{km\,s^{-1}}$)**	**Doppler width ($\mathrm{km\,s^{-1}}$)**
Si II 1304	26.05 ± 0.99	2.64 ± 2.04
Si IV 1393	25.63 ± 0.43	3.79 ± 0.67
Si IV 1402	24.82 ± 1.26	4.69 ± 2.91
Si IV doublet[1]	25.28 ± 0.49	4.99 ± 0.94
		5.02 ± 0.95

Notes: [1]Modelling was repeated fixing the wavelength ratio at its known value.

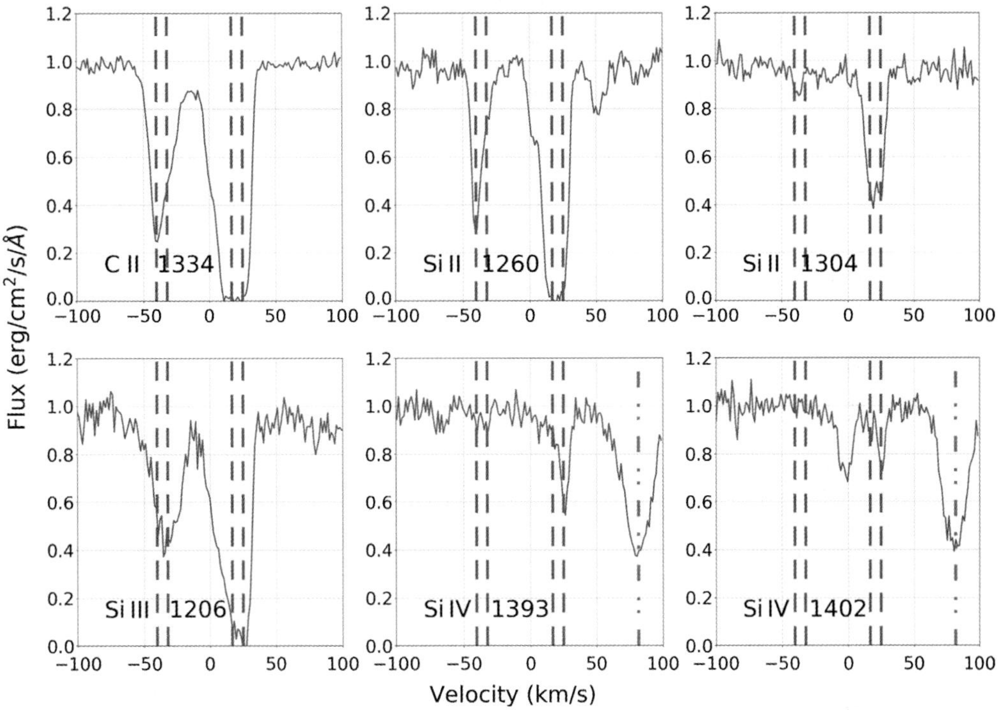

Figure 3. Absorption profiles of the carbon and silicon towards WD 0455-282, shown in velocity space. The photospheric velocity is shown by the vertical red dot-dash line, and the four component velocities reported in Preval *et al.* (2019) are shown in blue dashed lines.

3. Implications

Local space. It is important to understand how ISM along sight lines can influence our stellar spectra. Despite being a rarefied region of space, there could be as many as five non-photospheric components contributing to the absorption profiles seen in WD 0455-282, of which some or all could be due to intervening ISM clouds. As well as this, we see high ionisation states in the spectra which are not due to the photospheric. This latter point, coupled with the low neutral hydrogen column density towards this star could indicate an extension to the known β CMa Tunnel.

Stellar feedback. Analysis of ISM temperatures and composition is important for understanding how stellar material is fed back into the galaxy. The aim of this analysis is to test mechanisms for producing the components we see in this, and similar, spectra, to understand the existence of the β CMa Tunnel.

References

Arnaud, K. A. 1996, *ASP-CS*, 101, 17

Dobbie, P. D., Pinfield, D. J., Napiwotzki, R., Hambly, N. C., Burleigh, M. R., Barstow, M. A., Jameson, R. F., & Hubeny, I. 2004, *MNRAS* (Letters), 355, L39

Dupin, O. & Gry, C. 1998, *A&A* 335, 661

Fuchs, B., Breitschwerdt, D., de Avillez, M. A., Dettbarn, C., & Flynn, C. 2006, *MNRAS* 373, 993

Lallement, R., Vergely, J.-L., Valette, B., Puspitarini, L., Eyer, L., & Casagrande, L. 2014, *A&A*, 561, A91

Maíz-Apellániz, J. 2001, *ApJ* (Letters), 560, L83

Preval, S. P., Barstow, M. A., Bainbridge, M., Reindl, N., Ayres, T., Holberg, J. B., Barrow, J. D., Lee, C.-C., Webb, J. K., & Hu, J. 2019 *MNRAS*, 487, 3470

Redfield, S. & Linsky, J. L. 2008, *ApJ*, 673, 283

Riley, A. *et al.* 2019, "STIS Instrument Handbook" Version 18.0, (Baltimore: STScI)

Welsh, B. Y. 1991, *ApJ*, 373, 556

Wenger, M. *et al.* 2000, *A&AS*, 143, 9

White Dwarfs as Probes of Fundamental Physics: Tracers of Planetary, Stellar and Galactic Evolution
Proceedings IAU Symposium No. 357, 2019
M. A. Barstow, S. J. Kleinman, J. L. Provencal & L. Ferrario, eds.
doi:10.1017/S1743921320000216

Geometry of nova ejecta

M. Pavana[1,2], G. C. Anupama[1], Ramya M. Anche[1], U. S. Kamath[1] and G. Selva Kumar[1]

[1]Indian Institute of Astrophysics, Koramangala block II, 560034 Bangalore, India
[2]Pondicherry University, R. V. Nagar, Kalapet, 605014 Puducherry, India
email: pavana@iiap.res.in, murali.pavana@gmail.com

Abstract. We present photo-ionization and morpho-kinematic analyses of the ejecta of novae. The sample consists of ten novae belonging to the Fe II, He/N and hybrid classes. The Fe II class of novae in the sample have bipolar cone-like structures, with or without equatorial rings with inclination angle in the range of 40°–60°. The He/N novae have bullet-nose curve along with bipolar cone-like structures and equatorial rings with an inclination angle of ∼80°. The hybrid nova in the sample is a bipolar frustum of prolate spheroid along with bipolar cone-like structures and equatorial rings with an inclination angle of 63°.

Keywords. novae, cataclysmic variables - stars : individual - techniques : spectroscopic

1. Introduction

Geometry of nova ejecta provides useful information on the nature of the binary system. The complex structures in the ejecta are not well understood. These may be formed as a consequence of several mechanisms such as interaction of the outflow from white dwarf (WD) with the secondary, rotation of the WD, prominent magnetic field of the WD, and interaction of the ejecta with a pre-existing circumbinary material. A sample of ten novae is used here to understand the geometry of nova ejecta. Seven novae among these belong to Fe II class and two to He/N and one to the hybrid class of nova. The details of the sample are given in Table 1. The novae in the sample are also those with distinct features such as γ-ray emission, ^{7}Be II detection, secondary maxima or few rebrightenings in the optical light curve, complex velocity profiles, dust formation and observational evidence of pre-existing circumbinary material.

The optical spectra used in this study are obtained using the Himalayan Faint Object Spectrograph Camera (HFOSC) with R = 1300 and 2200, and the Hanle Echelle Spectrograph (HESP) with R = 30000 and 60000 at the 2 m Himalayan Chandra Telescope (HCT), and the high resolution echelle spectrograph with R = 27000 at the 2.3 m Vainu Bappu Telescope (VBT).

2. Analyses

The photo-ionization code CLOUDY, C17.01 (Ferland *et al.* 2017) was used to model the low resolution optical spectra. The 3D ionization structure of the ejecta was obtained using pyCloudy (Morisset 2013), as described in Pavana *et al.* (2019). The morpho-kinematic application, SHAPE (Steffen *et al.* 2011) was used to analyse and understand the morphological geometry of the nova ejecta using Hα velocity profiles. The morphologies obtained for all the novae are as shown in Figs. 1 and 2.

Table 1. Details of individual objects in the sample.

Object	Date	Days since discovery	Type	Resolution	Lines used	Analysis
V339 Del	23 Sept 2013	39	Fe II	27000	Hα	Morpho-kinematic
V1369 Cen	22 Dec 2013	20	Fe II	27000	Hα	Morpho-kinematic
V5857 Sgr	12 Apr 2018	04	Fe II	30000	Hα	Morpho-kinematic
V3665 Oph	10 Mar 2018	05	Fe II	60000	Hα	Morpho-kinematic
V5587 Sgr	06 Jun 2011	132	Fe II	1300	Hα	Morpho-kinematic
V2362 Cyg	25 Jul 2007	480	Fe II	1300	He I, He II, [O I], [O III], [N II], Hα, Hβ	Photo-ionization
V2944 Oph	24 Feb 2017	699	Fe II	1300	He I, He II, [O III], [N II], Hα, Hβ	Photo-ionization
V477 Sct	23 Nov 2005	43	He/N	1300, 2200	He I, He II, [O III], [N II], Hα, Hβ	Photo-ionization
M31N 2008-12a	08 Nov 2018	1.8	He/N	1300	He I, He II, [N II], Hα, Hβ	Photo-ionization
V5588 Sgr	09 May 2011	43	Hybrid	1300, 2200	Fe II, He I, He II, [O I], [N II], Hα, Hβ	Photo-ionization

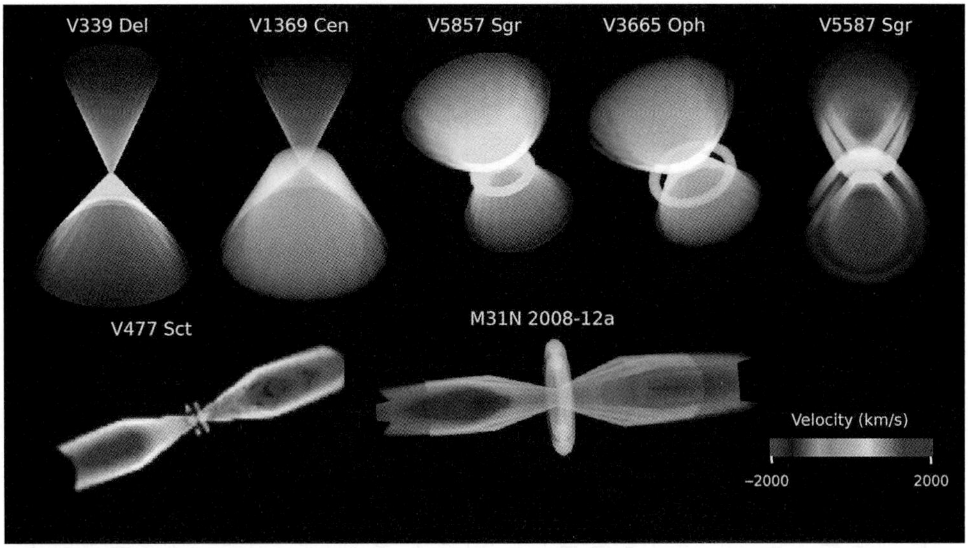

Figure 1. Morphology of the nova ejecta generated using Hα velocity profiles as seen in SHAPE corresponding to their inclination angles.

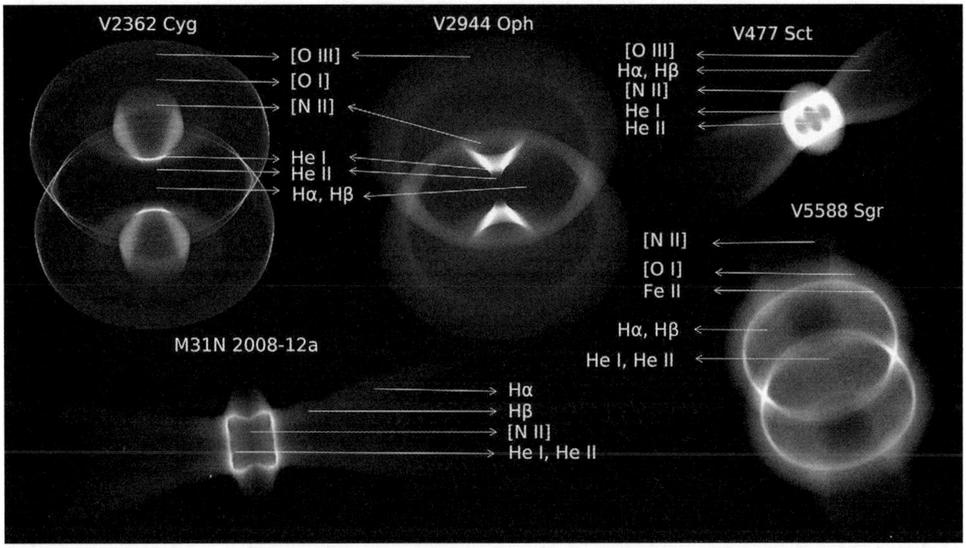

Figure 2. Ionized geometrical structure of nova ejecta generated using the optical spectrum corresponding to their inclination angles.

3. Discussion

In order to understand the mechanisms resulting in the formation of complex structures of nova ejecta, a sample of ten novae was considered. A robust model is required for novae to understand the complex velocity profiles. This can be achieved by choosing a sample with different characteristic features seen in the system as it evolves.

It was suggested by Ribeiro *et al.* (2013a) that morphology of He/N type, Nova Mon 2012 is bipolar with a high inclination angle of 82° ± 6°. The best-fitting remnant morphology of He/N type, Nova Cyg 2008 #2, was found to be polar blobs and an equatorial ring with an high inclination angle of 80° by Ribeiro *et al.* (2011). The best-fit

Table 2. Geometry of nova ejecta.

Object	Type	i (°)	Morphology	Comments
V339 Del	Fe II	58	Asymmetric bipolar cones	Asymmetry is attributed to the optical depth effects (Ribeiro *et al.* 2011) as the system is still in its early decline phase.
V1369 Cen	Fe II	50	Asymmetric bipolar cones	
V5857 Sgr	Fe II	41	Asymmetric bipolar cones with an equatorial ring	Asymmetry attributed to the optical depth effects as the system is in the optical maximum phase where the optical depth is higher, fast moving material at this angle could give rise to equatorial ring (Ribeiro *et al.* 2011).
V3665 Oph	Fe II	45	Asymmetric bipolar cones with an equatorial ring	
V5587 Sgr	Fe II	54	Bipolar cones	Almost symmetric structure and cones might be due to the interaction of wind with the secondary (Gill *et al.* (1999) and O'Brien *et al.* (1995)).
V2944 Oph	Fe II	51	Bipolar cones with equatorial rings	Equatorial ring due to the interaction of the outflow with pre-existing circumbinary material (Munari & Walter 2016).
V2362 Cyg	Fe II	48	Bipolar frustum of prolate spheroid, bipolar cones with equatorial rings	Origin of forbidden lines like [O III] and [N II] is found to be from the outer regions implying that it is coming from the optically thin expanding shell as rightly known that they do not come from primary, secondary or the accretion disc.
V477 Sct	He/N	83	Bullet-nose curve, bipolar cones with equatorial rings	Helium lines are coming from the accretion disc.
M31N 2008-12a	He/N	81	Bullet-nose curve, bipolar cones with an equatorial ring	Helium lines are coming from the inner regions of common envelope.
V5588 Sgr	Hybrid	63	Bipolar frustum of prolate spheroid, bipolar cone-like with equatorial rings	[N II] structure may be generated as a consequence of frictional interaction of wind with the secondary. Presence of equatorial ring due to low expansion ejecta velocity of Fe II lines (Lloyd *et al.* 1997).

i – inclination angle, angle between the ejecta axis and line of sight.

Hα profile corresponding to the observed profile of KT Eri obtained by Ribeiro *et al.* (2013b) resulted in a dumbbell shaped structure with an inclination angle of 58°. The ionized structure of the ejecta of the well-known RN of hybrid class, T Pyx, was found to be a bipolar conical structure with equatorial rings with a low inclination angle of ~15° (Pavana *et al.* 2019). The best-fit morphology of the hybrid nova, V906 Car was found to be asymmetric bipolar blob-like structure with an inclination angle of about 53° (Pavana *et al.* [in preparation]).

There is a clear distinction between the morphologies obtained for Fe II, He/N and hybrid class of novae (Table 2). This indicates that it evidently depends on the ejecta mechanism. However, confirmation on this can be obtained only after increasing the sample. One interesting feature seen in all novae in the sample is the presence of cone-like structures. From the numerical calculations and suggestions in the literature by Gill *et al.* (1999) and O'Brien *et al.* (1995), cones are generated as a consequence of frictional interaction of wind with the secondary, and rings as a consequence of sweeping up conical regions of enhanced density in the ejected mass from the primary. In the case of HR Del, Hutchings (1972) suggested the sharp and low velocity of the central peaks in the line profile are an indication of an inner cone of low velocity, either with a small cone angle, or viewed pole-on, while equatorial ring gives rise to less-sharp peaks and a more filled-in profile, even when it is thin. It is noticed from our sample that [N II] emission is from the equatorial ring components in the case of V477 Sct and also T Pyx (Pavana *et al.* 2019), rather than from the bipolar regions as in the case of V5588 Sgr. This might be due to variation in ionization rather than the density as also seen and suggested in the case of V705 Cas by Gill *et al.* (1999).

4. Summary

The important results are summarised as follows.

(*a*) The Fe II class of novae in our sample have bipolar cone-like structures with or without equatorial rings with inclination angle in the range of 40° – 60°.

(*b*) The He/N novae have bullet-nose curve along with bipolar cone-like structures and equatorial rings with an inclination angle of ~80°.

(*c*) The hybrid nova in the sample (V5588 Sgr) is a bipolar frustum of prolate spheroid along with bipolar cone-like structures and equatorial rings with an inclination angle of 63°.

(*d*) This study suggests there could be some differences in the ejecta mechanism for the different types of novae.

References

Ferland, G. J. *et al.* 2017, *ArXiv preprint*, (arXiv:1705.10877)
Gill, C. D. & O'Brien, T. J. 1999, *MNRAS*, 307, 677
Hutchings, J. B. 1972, *MNRAS*, 158, 177
Morisset, C. 2013, *Astrophysics Source Code Library* (ascl:1304.020)
Munari, U. & Walter, F. M. 2016, *MNRAS*, 455, L57
Lloyd, H. M., O'Brien, T. J., & Bode, M. F. 1997, *MNRAS*, 284, 137
O'Brien, T. J., Lloyd, H. M., Slavin A. J., *et al.* 1995, in Harpaz A., Soker N., eds, *Annals of the Israel Physical Society* Vol. 11, Asymmetrical Planetary Nebulae. p. 258
Pavana, M., Anche, R. M., Anupama, G. C., Ramaprakash, A. N., Selvakumar, G., *et al.* 2019, *A&A*, 622, A126
Ribeiro, V. A. R. M. *et al.* 2011, *MNRAS*, 412, 1701
Ribeiro, V. A. R. M., Munari U., Valisa P. *et al.* 2013a, *ApJ*, 768, 49
Ribeiro, V. A. R. M. *et al.* 2013b, *MNRAS*, 433, 1991
Steffen, W., Koning, N., Wenger, S., Morisset, C., Magnor, M., *et al.* 2011, *IEEE Transactions on Visualization and Computer Graphics*, Volume 17, Issue 4, p.454–465, 17, 454

White Dwarfs as Probes of Fundamental Physics: Tracers of Planetary, Stellar and Galactic Evolution
Proceedings IAU Symposium No. 357, 2019
M. A. Barstow, S. J. Kleinman, J. L. Provencal & L. Ferrario, eds.
doi:10.1017/S1743921320001477

Main conclusions from Symposium discussions

M. A. Barstow
School of Physics and Astronomy, University of Leicester, Leicester LE1 7RH, UK
email: mab@le.ac.uk

Abstract. The programme of this IAU Symposium, number 357, consisted of sessions organized around a number of key themes, as detailed in the preceding text. Each session included one or two invited keynote talks plus a number of contributed papers. Time was set aside for extensive discussion following the sessions associated with each of them. These were moderated by members of the Science Organising Committee, posing a number of questions to the audience to stimulate the discussion. The nature of such discussions makes them hard to record in detail, but a number of key points have been extracted and incorporated into this short concluding paper.

Keywords. white dwarfs

1. Introduction

White dwarf stars are interesting in their own right, as the end stage of the evolution of most stars ($M >\approx 8M_\odot$). However, they are also very important players in other branches of astrophysics. White dwarfs are clearly implicated in the mechanism(s) of Type Ia Supernovae, although proven progenitors have yet to be identified. Measurement of the ages of the oldest white dwarfs provides a limit on the age of the galaxy and its components, while the white dwarf luminosity function can be used to trace episodes of historical star formation. Study of their cooling rates and pulsations can shed light on the behaviour of matter under the extreme physical conditions found in their envelopes and cores. Several decades of studies of their photospheric composition have concluded that many are accreting material from the debris of past planetary systems. Therefore, currently, they provide the **only** way of exploring the range of bulk compositions of extra-solar planetary material. The presence of large numbers of absorption lines of highly ionised elements such as iron and nickel provides an opportunity to search for tiny differences between laboratory and observed line wavelengths that might indicate variation in the fine structure constant α. Similar anomalies in the wavelengths of molecular H_2 lines could, if present, point to a gravitational dependence of the fundamental proton-electron mass ratio (μ). White dwarfs can also act as background sources for observation of absorption lines which can be used to probe the structure and state of interstellar and circumstellar gas across a wide range of temperatures and densities.

The aim of this symposium was to bring together scientists covering the wide range of topics outlined above, allow them to present their work and stimulate discussion across the disciplinary barriers. As conferences are generally narrower in scope than this particular meeting, thought was also given to how this dialogue could be continued in the future.

2. Scientific conclusions

It is clear that the "exotic" and interesting white dwarfs that might provide the strongest evidence for, or constraints on, a particular physical phenomenon are often small in number, even in the current samples ($\approx$10,000 stars) that have been expanded by recent surveys such as SDSS, Pan-STARSS and LAMOST. For example, the potential channels for SN Ia may be more diverse and complicated than we thought. Although there is a regular stream of candidates, it is still the case that no clear progenitor system has so far been identified. We do need to find the progenitor systems. Similarly, in studying accretion of exotic-planetary material, we need to observe active systems with transiting debris to test the models and mechanisms proposed. However, we currently have only a single example where transiting orbiting debris is detected.

Over a period of more than 20 years, spectroscopic observations of several hundred white dwarfs have been carried out with a variety of ground- and space-based telescopes to understand their composition. However, these observations are trying to address many variables across the white dwarf temperature and mass range, besides the many elements detected. Therefore, the observational statistics often limit any strong conclusions from being drawn. The Gaia DR2 catalogue contains an estimated 200,000 white dwarfs, more than a factor 10 increase on current numbers. It is gratifying to see, as a member of the Gaia Data Processing and Analysis Consortium (DPAC), that Gaia data appears in almost every paper in these proceedings, from large scale surveys to studies of small groups and individual objects. However, to make use of this increase, a significant dedicated follow-up effort from both ground- and space-based telescopes is required. While this might be feasible on the ground with multi-object spectroscopy, the limited time available on space telescopes (currently only Hubble) is an insurmountable problem.

Photometry with space telescopes such as Kepler and Tess has dramatically increased the number of pulsating white dwarfs known. However, several presenters demonstrated that understanding the complexities of individual objects is very time-consuming. Therefore, there is a need to focus on the pulsators that are already known and well-studied to provide follow-up data and detailed analysis for these.

3. Organisational and programmatic conclusions

As indicated in the previous section, one of the greatest challenges for future studies of white dwarfs is access to telescope time for follow-up work. Astronomy is an ever expanding subject: 25 years ago, few telescopes were observing extra-solar planets; surveys increase in breadth and depth and produce ever larger catalogues of objects. Against this, as larger facilities are developed, smaller one come under increased financial pressure and may close or be repurposed into narrower scope, single-instrument activities. An important message in the discussion was to limit competition between members of the "white dwarf"community through more coordination and cooperation. The community needs to generate coordinated cases for future observing capabilities to observatories (or their funders), facility directors and oversight committees.

Attendees were encouraged to share the existing data that we have through the Montreal White Dwarf Database (MWWD, http://www.montrealwhitedwarf database.org). More sharing of relevant astrophysical techniques was also encouraged, as some are already doing. However, at the moment there is not a clear forum that would simplify this. An important related topic is access to improved models across the board, including atmospheres, pulsations and structure/evolution. Some of these are already available, with a shining example being the Montreal evolutionary calculations made available through the MWWD, but more coordination and on-line publication is needed. The generation and use of improved models is placing increased demand on computational resources. A good example is the need to run many trials of pulsation models to find a match to the complex modes in real stars and understand the uniqueness of such a match. There is a need to discuss new approaches to computation and data storage to address these problems. Extending our discussions to include Informatics and AI experts may offer some solutions.

4. The community

While most of the discussion focussed on the scientific questions addressed by the symposium, some of the debate was drawn to the issues for the scientists working in the white dwarf community. These are not unique to white dwarf researchers, but, as a relatively compact group of scientists who are mostly well known to each other, it should be possible to identify common issues and work on them together. While the conference attendees work in many different academic systems, each one has a shortage of funding in varying degrees. Common to everyone is that the situation is driven by the political factors of government policy and economics and the programmatic tensions between project funding (building large new facilities) and exploitation (money for people to do research). Changing these things is very hard, but coordinated advocacy for our subject can help. We should also collaborate outside our discipline, seek new applications for our work and then try to access funding sources beyond those we usually approach. A good example is how the University of Texas, Austin group has branched out into simulating the physical conditions of white dwarf atmospheres using the Z machine at Sandia National Laboratory.

Cross-disciplinary working is seen to be increasingly important, but interdisciplinary meetings like this IAU Symposium are rare. The bi-annual European White Dwarf Workshop already has a very packed programme and bringing in new topics will inevitably force out others. More focused meetings on a single WD-related topic could be envisaged or WD researchers could attend the meetings of other subjects, but a proliferation of meetings is undesirable due to cost, time and the need to all lower our travel carbon footprint. It is essential that we develop solutions to holding large remote meetings that are able to recreate some of the positive advantages of a face-to-face conference.

Acknowledgments

I would like to thank the participants of IAU Symposium 357 for the useful and highly stimulating discussions during the meeting, which provided the main input for this summary.

Author index